I0037381

By-Products of Palm Trees and Their Applications

1st World Conference on By-Products of Palm Trees and Their Applications (ByPalma), Aswan, Egypt, 15-17 December, 2018.

Hamed El-Mously[1], Mohamad Midani[2] and Mohamed Wagih[1]

[1] Faculty of Engineering, Ain Shams University, Egypt
[2] Faculty of Engineering and Materials Science, German University in Cairo, Egypt

Peer review statement

All papers published in this volume of "Materials Research Proceedings" have been peer reviewed. The process of peer review was initiated and overseen by the above proceedings editors. All reviews were conducted by expert referees in accordance to Materials Research Forum LLC high standards.

Published under License by **Materials Research Forum LLC**
Millersville, PA 17551, USA

Published as part of the proceedings series
Materials Research Proceedings
Volume 11 (2019)

ISSN 2474-3941 (Print)
ISSN 2474-395X (Online)

ISBN 978-1-64490-016-1 (Print)
ISBN 978-1-64490-017-8 (eBook)

This book contains information obtained from authentic and highly regarded sources. Reasonable efforts have been made to publish reliable data and information, but the author and publisher cannot assume responsibility for the validity of all materials or the consequences of their use. The authors and publishers have attempted to trace the copyright holders of all material reproduced in this publication and apologize to copyright holders if permission to publish in this form has not been obtained. If any copyright material has not been acknowledged please write and let us know so we may rectify in any future reprint.

Distributed worldwide by

Materials Research Forum LLC
105 Springdale Lane
Millersville, PA 17551
USA
http://www.mrforum.com

Manufactured in the United State of America
10 9 8 7 6 5 4 3 2 1

Table of Contents

Bio-Composites

Biomedicine and Biotechnology

Fiber, Paper, and Textile

Food Applications

Design and Architecture

Preface

The 1st World Conference on By-Products of Palm Trees and their Applications (ByPalma), Aswan, Egypt, 2018 December 15 – 17.

Over the three days of the conference, 15-17 December 2018, ByPalma brought together scientists, experts, researchers, palm growers, designers, artisans, and policy makers. ByPalma combined expert short presentations by specialists with structured discussion panels on palm by-products utilization.

The objective of the conference was threefold:

1. To bring together leading academic scientists, researchers, artisans, entrepreneurs and industry professionals active in the area of palm by-products R&D, manufacturing, and crafts from all around the globe to exchange recent developments, technologies, innovations, trends, concerns, challenges and opportunities.
2. To rediscover palm by-products and maximize their added-value and create an economical resource that can help in the sustainable development of vast rural areas in different countries in the world.
3. To establish an international network of scientists, artisans, and industry professionals active in the area of palm by-products R&D, manufacturing, and crafts.

The main organizing institution was the Faculty of Engineering – Ain Shams University, a world-renowned engineering school, and a leader in date palm by-products research and development.

Conference Chairman, Professor Dr. Hamed El-Mously, Ain Shams University

Conference Co-chair, Dr. Mohamad Midani, German University in Cairo

Topics Include the utilization of palm by-products in the following applications:

- Wood Alternatives and Panels
- Sustainable Energy and Fertilizers
- Bio-Composites
- Biomedicine and Biotechnology
- Fiber, Paper, and Textile
- Food Applications
- Design and Architecture

Committees

Organizing Committee

Dr. Hamed El-Mously (Chair), Ain Shams University - Egypt
Dr. Mohamad Midani (Co-Chair), German University in Cairo - Egypt
Dr. Othman Alothman, King Saud University - KSA
Dr. Amgad El-Kady Food and Agro Industries Technology Center - Egypt
Dr. Sherif El-Sharabasy, Central Laboratory for Date Palm Research and Development, Egypt
Dr. Mahmoud Farag, American University in Cairo - Egypt
Dr. Mohammad Jawaid, Universiti Putra Malaysia - Malaysia
Dr. Ramzi Khiari, Institut Supérieur des études Technologiques de Ksar-Hellal - Tunisia
Khaled Mostafa, Ain Shams University - Egypt
Dr. Julia Baruque Ramos, University of Sao Paulo - Brazil
Dr. Navin K. Rastogi, Central Food Technological Research Institute - India
Mohamed Wagih, Ain Shams University - Egypt
Dr. Irene Xiarchos, Food and Agriculture Organization - NENA Region
Dr. Mohamed Yacoub, Food and agriculture Organization – Egypt

Scientific Committee

Dr. S.M. Sapuan, Univeristi Putra Malaysia - Malaysia
Dr. Paridah Md Tahir, Univeristi Putra Malaysia - Malaysia
Dr. Hazizan Md Akil, Universiti Sains - Malaysia
Dr. HPS Abdul Khalil, Universiti Sains, Malaysia
Dr. Mohamed Hemida Abd-Alla, Assiut University - Egypt
Dr. Mohamed Al-Farsi, Plant Research Centre - Oman
Dr. Khalid El-Shoaily, Agricultural Research Center - Oman
Dr. Abdelouahed Kriker, University of Ouargla - Algeria
Dr. Ahmed Hassanin, Alexandria University - Egypt
Dr. Carlos A.Cardona, Universidad Nacional de Colombia Sede Manizales - Colombia
Dr. Tamer Hamouda, National Research Center - Egypt
Dr. Sasa Sofyan Munawar, Indonesian Institute of Sciences - Indonesia
Dr. Manuel Abad, Universidad Politécnica de Valencia - Spain
Dr. A. Olorunnisola, University of Ibadan - Nigeria
Dr. Luiz Augusto H Nogueira, Federal University of Itajuba - Brazil
Dr. Maiada Mohamed El-Dawayati, Agricultural Research Center - Egypt
Dr. Edi S. Bakar, Universiti Putra Malaysia – Malaysia

Conference Logo

Organizer Logo

Sponsors Logos

With the technical Cooperation of the F.A.O

Partners Logos

PalmwoodNet

UPM
UNIVERSITI PUTRA MALAYSIA

Sci
Dev
Net

أكاديميـة البحـث
العلمي والتكنولوجيا
Academy of Scientific
Research & Technology

IDDC
INTERNATIONAL DRYLAND
DEVELOPMENT COMMISSION

غرفة صناعة الحرف اليدوية
CHAMBER OF HANDICRAFTS

EGYPTIAN EXPORT COUNCIL
FOR HANDICRAFTS

By-Products of Palm Trees and Their Applications
Materials Research Proceedings **11** (2019)

Materials Research Forum LLC
doi: https://doi.org/10.21741/9781644900178

Keynotes

By-Products of Palm Trees and Their Applications
Materials Research Proceedings 11 (2019) 3-61

Materials Research Forum LLC
doi: https://doi.org/10.21741/9781644900178-1

Rediscovering Date Palm by-products: an Opportunity for Sustainable Development

H. El-Mously

Dept. of Design and Production, Ain Shams University, Egypt

hamed.elmously@gmail.com

Keywords: date palm, palm midrib, endogenous development, rediscovery of local resources, palm by-products, bioeconomy, date waste, lignocellulosic by-products

Abstract. The date palm was the pivot of cultural, social and economic life for long centuries in rural areas in the Arab region. The basic needs of millions of people in rural areas were being satisfied relying on the by-products of date palms. With the drastic change of the style of life most of these byproducts became redundant leading to the neglect of pruning of date palms, and thus becoming a direct cause of fire accidents and infestation by dangerous insects. This situation represents a real challenge to those concerned with development. How to compose a new vision to palm by-products transcending the traditional forms of utilization of these by-products being treated as waste? The path of rediscovery of these by-products is paramount. How to develop new forms of utilization of palm by-products to satisfy modern demands on the local, national and international levels? An approach has been suggested for the industrial utilization of date palm by-products.

The research conducted at the premises of the Faculty of Engineering, Ain Shams University has proven that the date palm midribs enjoy mechanical properties similar to those for imported wood species. It was also proven that the date palm midrib can be used as a core layer for the manufacture of blockboards competing with those manufactured from wood. Lumber-like blocks have been successfully made from palm midribs. The palm midribs were successfully used for the production of Mashrabiah (Arabesque) products as a substitute for beech wood. Particleboards and MDF boards satisfying the international standards have been also manufactured from palm midribs. Poultry and livestock feed, as well as compost have been produced using the date palm midribs. Space trusses and claddings have been successfully made from palm midribs. New machines have been successfully designed and manufactured for the conversion of palm midribs into strips of regular cross-section. There are wide future prospects for the use of date palm by-products us a substitute for wood, for paper manufacture and for the reinforcement of polymers. Within the framework of bioeconomy there are high potentialities for the use of the date waste, as well as the ligne-cellulosic by-products in a wide spectrum of bio-industries. To guarantee the continuation of endeavors to support the use of palm by-products on the international level it is necessary to establish The International Association For Palm By-Products as a forum for all parties interested and involved in the use of palm by-products.

Introduction

The palm plantations were the pivot of cultural, social, and economic life for long centuries in rural areas in the South: in Latin America, Africa, and Asia. The basic needs of millions of people in rural areas were being satisfied relying on the by-products of palms (PBP). The way of life in these vast areas was woven using these available indigenous sustainable secondary products of palms. Via this process very rich technical heritage blossomed, being mainly the

By-Products of Palm Trees and Their Applications   Materials Research Forum LLC
Materials Research Proceedings 11 (2019) 3-61   doi: https://doi.org/10.21741/9781644900178-1

property of the poor in the local communities in rural areas. The shift from the prevailing subsistence economy to the capitalist mode of production and the dominance of the cash crop ideology, together with the propagation of the Western life style has led to the negligence of PBP and the freezing of the associated technical heritage. Thus, the rural populations in many countries in the South turned from producers and active participants in development of their local communities to mere consumers of whatever could be purchased from cities or abroad.

This situation represents a real challenge to those concerned with development. How to compose a new vision to PBP transcending the traditional forms of utilization of PBP and imagining new modern avenues of utilization of these renewable almost priceless and voluminous materials being, presently regarded as waste?

The path of rediscovery of PBP is paramount in this context. How to develop new forms of utilization of PBP to satisfy modern demands on the local, national and international levels? Adopting a participatory approach this trend of thought may provoke waves of innovation: beginning from the rural areas and reaching urban areas. The economic utilization of PBP will provide labor opportunities on the local level, attract the youth back to village as innovators and entrepreneurs and transform the village from pure reliance on the agricultural activity, subject to the fluctuations of prices of the agricultural products, to a wide sphere of economic activity including beside agriculture, industry and trade. Thus, the economic utilization of PBP will return vitality to local communities, where the palm plantations exit and provide appropriate conditions for sustainable development.

The date palm in the cultural history of the Arab region
The local materials are nothing but the material milieu, through which cultures were able to express themselves. Proceeding from the historical perspective, the different cultures of the world were born and developed in company with different materials. Who could deny the relation between the ancient Egyptian culture and papyrus, lotus, lime stone and granite, nor between the Asian cultures and bamboo, rattan and rice? It is extremely important to capture the relation between culture and local materials as an important asset for development. The linking between development and local materials means that you are building on the existing culture of interaction with these materials, i.e., you do not begin development from a zero datum, but with what people – members of each local community – have at hands (the local materials), as well as in minds (psychological familiarity with these materials, as well as technical heritage, associated with their production, manufacture and use in the different walks of life). In this concern the date palm (Phoenix dactylifera L.), represents an eloquent example. It is an authentic element of the region's flora, which accompanied our historical march for thousands of years. It is our duty now as researchers and intelligentsia to rediscover the date palm as a pivot for our life at present, as well as for the future generations.

It may be difficult to record the first emergence in history of the date palm, but it was well known 4000 years BC, where it was used to build the moon temple near to Ore, south of Iraq [50]. The second proof of the ancient presence of the date palm comes from the Nile valley, where the date palm was taken as the symbol of the year and the palm midrib as a symbol of the month in the hieroglyphic Egyptian language. But the cultivation of the date palm in Egypt was 2000 – 3000 years later than in Iraq [50]. The date palm was one of the pivots of economic and, hence, social and cultural life in this region from ancient times. In ancient Egypt the heads of pillars in temples were made resembling the growing top of the date palm. The date palm appeared frequently on walls of temples in different contexts revealing its significance in life in Egypt. According to Nubian (south Egypt) traditions, when a child is born, they plant a date palm for him. When he has matured, the date palm will have grown to a number of palms, providing a basis for his future economic life. In Upper Egypt each village has evolved beside its life supporting palm plantations. The date palm is well adapted to our environment. It is grown well in the Nile Valley, where it gives gentle shade against the sun and protection from the wind to crops growing below it. It tolerates the harsh climate of the Sahara, making possible the life of Bedouins; it even tolerates high levels of salinity, growing along the seashore in Egypt. It needs much less water and service and is less subject to diseases and parasites than other trees.

Date, the primary product of the palm is rich in protein, vitamins, and mineral salts. That is why it represents an essential element of diet for the cultivator himself and his animals (the low-grade date with kernel). All secondary products of the palm result from annual pruning (Fig. 1) and have essential uses for the cultivator. Thus, no waste results from the growing of palms. The date palm's midribs of grown palms after being woven in a mat using coir ropes are used in roofing[1]. Crates for the transportation of vegetables and fruits are also made from the palm midrib, as well as furniture items, manual fans, doors of gardens and coops for chickens and rabbits.

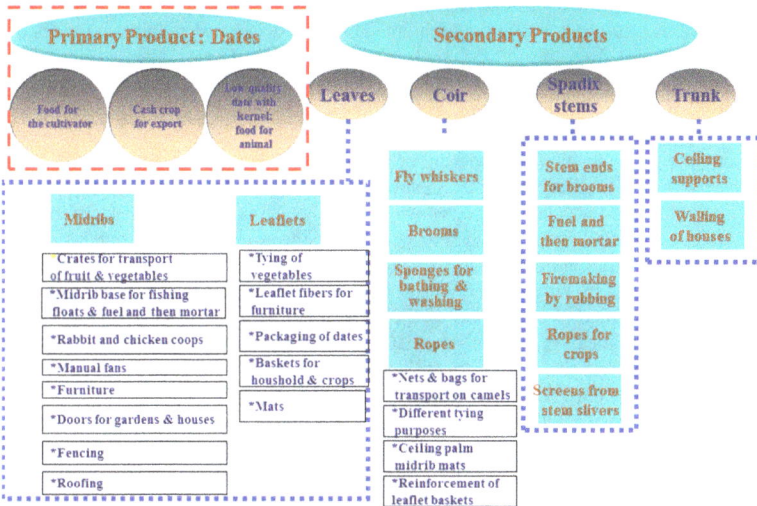

Fig. 1: The date palm: a pivot for the ways of life in our traditional communities [27].

Midribs of young palms are used in fencing gardens. The midrib base is used as floats for fishing nets or for fuel in rural ovens: the ashes being used afterwards in mortar. The leaflets are used after being woven, in mat making, as well as in a very wide variety of baskets for use in the cultivator's household, as well as for transportation of various agricultural crops and packing of

---

1 - This was my first acquaintance with palm midribs. Roofs in El-Arishi traditional houses, made of midribs, lasting for centuries, were my first natural proof of their durability and good mechanical properties.

dates. They are also used for the manufacture of screens for households and as ropes for typing up vegetables. The leaflet fibers are used in the manufacture of carina used for stuffing of upholstered furniture. The coir is being used for making washing and bathing sponges, as well as for the manufacture of ropes for different uses. From coir, rope nets, and bags for the transportation of agricultural crops on camels are being made. Household brooms and fly whiskers are also made from coir. The spadix stem is crushed to obtain very strong fibers for tying up agricultural crops. The spadix stem ends with fruit stalks are used as brooms in streets. Spadix stems of certain palm species were even used for fire making by rubbing. They were also used as coat hangers, and after being sliced into strips, were used for making screens for household use. The palm trunk is being used, after cutting it into halves or quarters, as beams for ceilings or walling in rural and desert regions. *Thus, the date palm in our traditions (Fig. 1) represented an eloquent example of integrated sustainable use of renewable material resources.*

How we got in touch with the palm midrib and why we focused our research on the palm midrib

We – as city dwellers – are accustomed to look with scorn to the agricultural materials coming from the village. During the years of engineering education we learned to respect the industrial materials, such as steel, cast iron and aluminum. Thus, we developed negative attitudes towards the materials coming from the village, such as palm midribs. During a scientific journey we made to North Sinai in November, 1979 I [1] saw – for my first time – the palm midribs in the roofs (Fig. 2) of the traditional houses in El-Arish. I was astonished to learn that such roofs remain in a good state for hundreds of years. This meant for me a natural experiment emphasizing the good mechanical properties of the palm midrib.

Our extensive field studies, conducted in several governorates in Egypt, showed that there are very rich potentialities for development, which need to be activated. A wide spectrum of natural material resources is available everywhere depending on the geology, climate and history of each locality. These resources are not by themselves of value, isolated from the community. They acquire their real value by virtue of the wealth of experience and spirit of endogenous creativity of each local community. We perceive of the local community as a socio-cultural-ecological subsystem, i.e., the local group in its relation with the surrounding environment. Thus, endogenous development could begin in each local community with what people have more of and know better about[50]. Table (1) illustrates that the date palm exists almost in all Egypt governorates. Secondly we got a very strong evidence that the drastic change in the way of life in Egypt has made palm midribs redundant: people stopped using it in ceiling, fencing, doors and even to a large extent in crates. Thus, palm midribs in many governorates turned to be "waste": a burden and source of pollution to the environment and sometimes a cause of fire in the palm gardens. Since the price of date in the palm garden is very low, we found that this situation may threaten the economic feasibility of cultivation of the date palm as a whole, which may deprive people, particularly the poor from an essential element of their diet. Besides, I was so strongly convinced with the idea of endogenous development and felt that the palm midribs – extensively available in Egypt and the Arab world – may serve as the material milieu for the application of the idea of endogenous development in local communities possessing date palms.

Fig. 2: The ceiling of a traditional Arishi house, made from palm midribs [2]

Table 1: Estimation of the production of date in Egypt, 2015 [5].

Governates	Area	fruitful palm (palm)	Productivity (Kg\ palm)	Production (ton)
Alexandria	436	82563	83.706	6911
Beheira	14327	1371794	168.436	231060
Garbiya	315	48368	103.560	5009
Kafr El Sheikh	5159	343427	128.062	43980
Dekhalia	674	216716	110.010	32841
Damiatta	15	866216	99.166	85899
Sharkia	260	1211196	171.536	207764
Ismailia	1327	670532	131.809	88382
port said	-	11195	84.413	945
El suez	456	93879	94.590	8880
Monofia	75	163339	100.656	16441
Qalyubia	547	203469	125.729	25582
Cairo	810	37586	53.424	2008
Lower Egypt (Total)	24392	5320280	140.350	746702
Giza	21089	1813322	130.798	237178
Beni Suef	61	320783	92.324	29616
Faiyum	1158	643074	133.832	86046
Minya	586	337608	110.000	37137
Middle Egypt (Total)	22894	3114787	125.208	389995
Asyut	400	462501	95.431	44137
Sohag	799	414071	93.291	38713
Qena	1039	361346	61.368	22175
Luxor	552	192360	70.254	13514
Aswan	24840	2477458	90.840	225054
Upper Egypt (Total)	27630	3908663	87.906	343593
Inside Valley (Total)	74916	12343730	119.922	1480290
New Valley	18482	1262475	81.681	103120
Matruh	7207	330674	90.001	29761
Red Sea	134	39528	42.856	1694

North Sinai	9076	320650	53.429	17132
South Sinai	-	91304	39.998	3652
Nubariya	5795	567970	86.744	49268
Outside Valley (Total)	40694	2612601	78.323	204627
Total	115610	14956331	112.656	1684917

Rediscovery of renewable materials: the palm midrib as an example

I think that we researchers should not take the path of imitation of Western scholars in selection of our subjects of scientific interest. Our natural materials are the material milieu, via which our scientific and technological capabilities could mature. Many of these materials, such as palm midrib, found so many genuine forms of utilization in our traditional ways of life [27]. The study of these material's traditional uses could be very useful in order to learn more about their physical and mechanical properties. However, these materials are often produced and/or consumed only in certain social strata in villages. For example, the palm by-products in Southern countries are usually associated with the poor in rural areas, whether as producers, manufacturers or consumers. With the change of the way of life the demand on products from these materials decreases. This leads to the neglect of these materials and the collapse of the primary activity producing them, leading in turn to the marginalization of rural areas, poverty and the collapse of the local community. What is needed is a new perception of local materials in a context much wider than the traditional context. This means:-

(a)   designing products or services for the satisfaction of new contemporary needs on the level of the local community in villages;

(b)   or widening the circle of consumption to include higher (medium and upper) classes in cities locally or abroad. This includes the design of new products to satisfy the needs of new markets;

(c)   introducing more advanced modern technologies with the objective of improving product quality, increasing productivity, introducing new product characteristics or functions, and developing new technologies for totally new lines of products.

Thus, what is called the rediscovery of renewable materials means: imagining, thinking, designing, carrying out research and producing new product prototypes, conducting market research, etc. in a very interactive fashion as shown in Fig. 3, with the objective of developing new uses of the local materials. Returning to the local community level this means the creation of a new pattern of sustainability and the transition from subsistence to market economy and from the traditional to an authentic form of modernity for the local community.

Fig. 4 illustrates the different alternatives of rediscovering the palm midrib and the associated product versions.

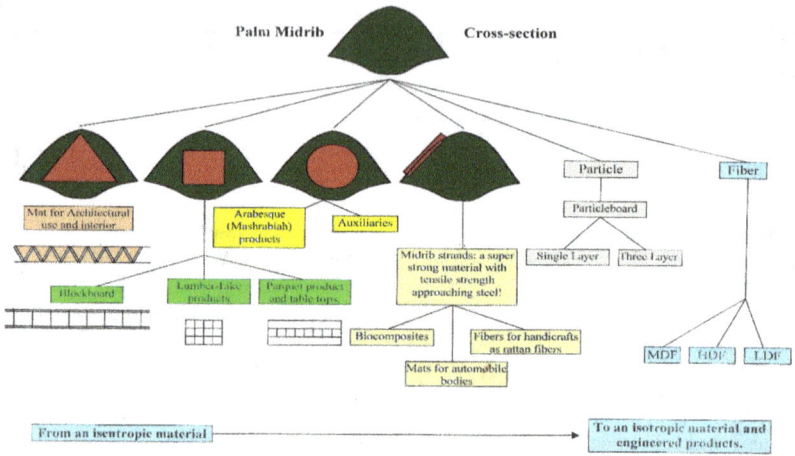

Fig. 3: A synergetic dialogue has been lunched between the designers and EGYCOM technical team leading to the innovation of new products from palm midribs and leaflets.

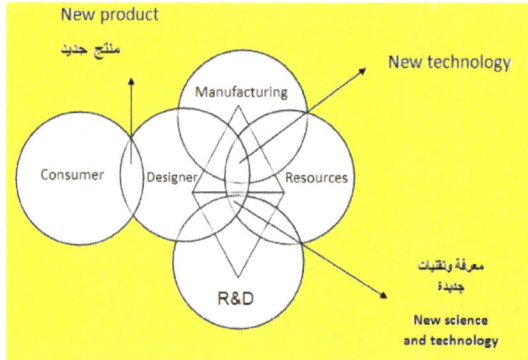

Fig. 4: Different alternatives of rediscovering the palm midrib and the associated products vision.

Interfaces of innovation

We could discern two interfaces of innovation.

1. The first interface lies between our perception of the physical and mechanical properties of the date palm by-products and our perception of the performance criteria of different products or services. *The outcome of this interface is an idea of a new product or service.*

2. The second interface of innovation lies between the idea of a certain product or service and the socio-economic and cultural aspects of the local society, where the product or service is to be produced. *The outcome of this interface is an appropriate technology that the local*

community could assimilate, digest, and thus becomes stronger, so that it could afterwards develop it independently.

The approach we followed for the industrial utilization of the date palm by-products
The suggested approach proceeds from the principle of rediscovery of our local materials. This means that we should direct our forces of imagination and thinking to find new avenues for the utilization of these materials within a vision for development and modernization we set by ourselves. Therefore, we do not resort to replicate the ways of life and patterns of production our forefathers have had, nor to the blind imitation of the Western pattern of development and modernization. This approach leads us to an intermediate situation, whereby we could make maximum use of our development potentialities and comparative and competitive advantages and utilize the Western scientific and technological achievements, while giving due attention to the specificity of our situation and cultural vision. In addition, the principle of rediscovery of our local materials means that we put each generation before the challenge of innovation to develop its own edition of use of local materials, which will unleash its potential energies and creative power and push it forward to participate actively in the development of society. *The suggested approach includes the following steps.*

1. The conduction of tests to determine the physical and mechanical properties of the palm by-products according to the latest international standards. This is thought to give a new identity to these materials and put it in comparison with the similar materials in the international market.

2. Development of new products from the palm materials proceeding from the knowledge about their physical and mechanical properties, and chemical composition, as well as from the specificity of local, national and international needs.

3. Design of the manufacturing processes and machines for the aforementioned products taking into consideration the features of the socio-cultural context, where the industrial activity will be located. The image of industrial establishment may not take the image of the orthodox factory. The village with its palm gardens and houses with inner courts may be an appropriate context for industrialization. It is a new mode of engineering trying to initiate a dialogue with that culturally specific context, so that the new scientific and technological inputs become harmonious with that context, which could be thus able to digest and assimilate them and get stronger and more developed.

4. Determination of the appropriate manufacturing process parameters. This necessitates the conduction of applied research to attain definite required results: the determination of the values of these parameters. This we call development and research (D & R and not R & D). Firstly, efforts are devoted to find a new avenue or use of the resource (e.g. product prototypes). When we succeed in these exploratory experiments in manufacturing a product satisfying the needed consumption criteria and get clear economic indices of success of marketing of the product via market research, we go back to the manufacturing process and conduct research to find the optimum conditions of manufacturing the product: both technically and economically. This approach is much more appropriate in Southern countries having limited funds, allocated to research.

5. Design, manufacturing and testing of the prototypes of equipment needed for manufacture and their modification if necessary, to cope with the needed performance requirements.

Materials Research Forum LLC
doi: https://doi.org/10.21741/9781644900178-1

6. The establishment of an experimental pilot unit for the new products putting the new industrial potentiality under test: technologically and economically. This is the first step to convince the entrepreneurs to invest in this new field of industrial activity, the market specialists to open new markets for these products, as well as the consumers to accept the new products, made from the palm by-products.

7. The conduction of technical and economic feasibility studies to establish industrial projects relying on the palm by-products. These studies should give the detailed technical information (specifications of equipment and operations sequence, etc.), as well as the indices of economic feasibility (profit, return on the capital, etc.) necessary to guarantee the profitability of the industrial utilization of these materials.

Research findings

1. Characterization of date palm by-products

1.1.   Date palm midrib

1.1.1. Date palm midrib geometry

The midrib represents the central part of the palm leaf. Fig.5 illustrates a general view of the palm midrib, divided into three parts: the basal, the middle and the top, as well as the form of the cross-section of its parts. Fig. 6 illustrates the variation of the area of midrib cross section, the diameter of the largest circle drawn within the cross section and its area: from the base of the midrib to the top [24]. Table 2 represents a summary of the weights, lengths and dimensions of the cross sections of midribs of palms of different species (Siwi, Amhat and Balady) in El-Fayoum governorate [24]. It is clear that there are significant differences among the species, as well as among different localities for the same species.

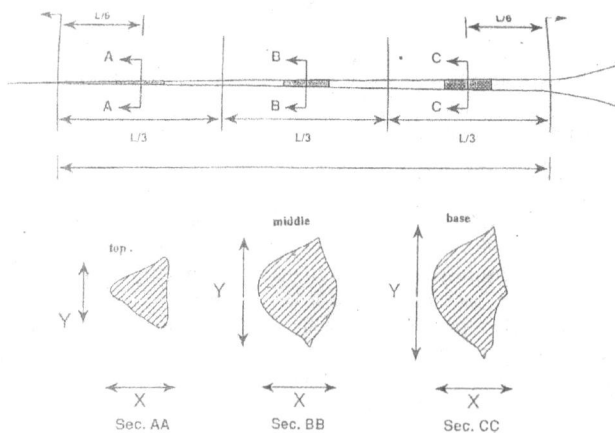

Fig. 5: A diagrammatic sketch showing the palm midrib, as divided to three distinguished parts: base, middle and top [24].

1.1.2. Specific gravity of the palm midrib

The value of specific gravity was determined for samples taken from the central parts of palm midribs, obtained from the New Valley and Asiut governorates [2]. It was found that these values fluctuate between 0.51 to 0.79 with an average value of 0.66 at a moisture content ~ 8%. The corresponding value for beech was found to be 0.65 and for spruce 0.35 [2].

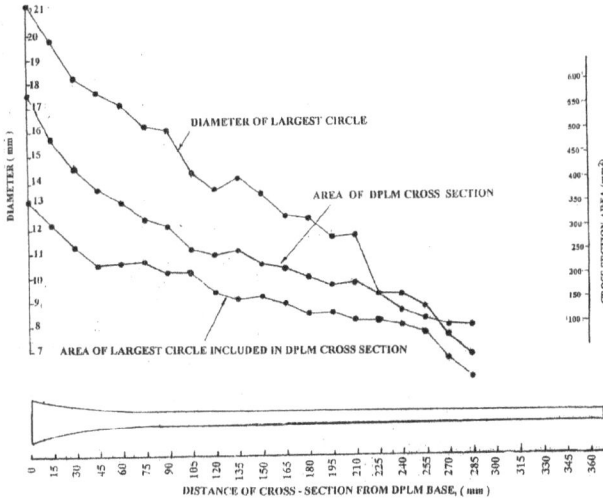

Fig. 6: The variation of cross section, value of the maximum diameter and the area of the largest circle inside the cross section for Siwi palm midribs from the base to the top [23].

Table 2: Summary of weights, lengths and dimensions of cross sections for palm midribs (Siwi, Amhat and Baladi species from El Fayoum[24].

Species	Locality	n	Weight, (gm)	Length, (cm)		Cross section, dimensions (mm)		
						Base	Middle	Top
Siwi	Garfis	22	362	1.62	X	17.90	15.23	12.19
					Y	14.62	13.20	11.98
Amhat		14	528	2.05	X	20.96	16.70	13.86
					Y	17.20	16.25	14.40
Balady		10	339	1.59	X	19.36	16.75	13.23
					Y	16.38	14.19	13.00
Siwi	El-Kaabi	33	381	2.35	X	31.87	21.48	16.78
					Y	26.15	18.84	17.10
Amhat		22	629	2.59	X	25.77	19.97	15.10
					Y	21.91	17.37	15.72
Balady		19	493	2.56	X	24.51	18.83	14.81
					Y	20.06	16.18	15.35
Siwi	Abou-Ksah	19	657	2.73	X	39.78	23.27	17.07
					Y	25.42	18.84	16.26
Amhat		21	666	2.75	X	32.90	21.90	17.83
					Y	23.02	18.15	15.54
Balady		20	451	2.52	X	27.71	19.41	15.64
					Y	19.26	16.08	14.72

1.1.3. Date palm midrib dimensional stability

The research findings have shown the high tendency of palm midribs to swell when immersed in water as compared with wood. The percentages of volumetric swelling of midribs of palm of different species (Siwi, Mantour, Tamr and Males) after immersion in water for 38 hours ranged between 980% for Males and 176% for Siwi, whereas the corresponding values for wood of different species (beech, spruce, pitch pine and mahogany) ranged between 13% and 29% [2].

1.1.4. Mechanical Properties of Palm Midribs

The results of tests, conducted on samples taken from the central parts of midribs obtained from different palm species and geographic locations, indicate that the values of mechanical properties of palm midribs vary with the palm species and geographic locations. Table 3 illustrates the results of tests conducted on midribs of palms of Siwi, Amhat and Baladi species in different localities in El Fayoum governorate. The modulus of rupture (MOR) in bending varied from ~ 70 for Amhat, ~ 76 for Baladi to 82 N/mm^2 for Siwi palm species. The compressive strength varied from ~ 34 for Amhat, 36 for Baladi to ~ 40 N/mm^2 for Siwi. The tensile strength varied from ~ 66 for Amhat, ~ 71 for Baladi to ~ 75 N/mm^2 for Siwi palm species [24].

Table 3: Mean values of the main mechanical properties of palm midribs (Siwi, Amhat and Baladi species from El Fayoum Governorate) [24].

Species	Locality	Static bending, N/mm^2			Comp.//to grain, N/mm^2			Tension//to grain, N/mm^2		
		n	MOR	MOE	n	CS$_{max}$	MOE	N	UTS	MOE
Siwi	Garfis	13	76.17 (13.40)	5266 (1247)	28	39.76 (5.30)	20.35 (4.92)	22	69.23 (8.47)	40.61 (5.74)
	El-Kaabi	12	76.60 (12.24)	4349 (872)	17	37.66 (6.12)	17.18 (3.57)	18	79.69 (12.54)	-
	Abou-Ksah	9	96.10 (12.66)	5731 (718)	10	44.02 (9.89)	19.47 (4.55)	6	83.00 (13.98)	49.91 (11.06)
	Species average	34	81.6			39.89		46	75.12	
Amhat	Garfis	35	55.68 (9.04)	3648 (831)	51	30.16 (11.15)	13.53 (4.44)	40	52.61 (9.45)	-
	El-Kaabi	50	77.06 (11.93)	5128 (1085)	64	37.35 (7.24)	18.66 (4.56)	39	76.67 (10.19)	-
	Abou-Ksah	21	73.09 (8.50)	5042 (889)	56	34.08 (5.37)	18.07 (4.37)	30	71.10 (9.83)	-
	Species average	119	69.64		171	34.13		109	66.31	
Balady	Garfis	26	68.26 (12.28(	4355 (744)	43	34.01 (5.82)	16.75 (3.39)	38	64.48 (9.22)	--
	El-Kaabi	21	72.43 (8.85)	4946 (984)	32	36.30 (5.21)	19.25 (4.60)	45	73.25 (13.39)	-
	Abou-Ksah	22	88.98 (11.42)	6096 (955)	27	38.98 (4.67)	39 (4.53)	39	76.20 (14.65)	
	Species average	69	76.14		102	36.04		122	71.46	

- Means with the same letters are not significantly different according to Duncan's multiple range test ($p = 0.05$), values in parenthesis represent standard deviations, and n is the number of specimens.

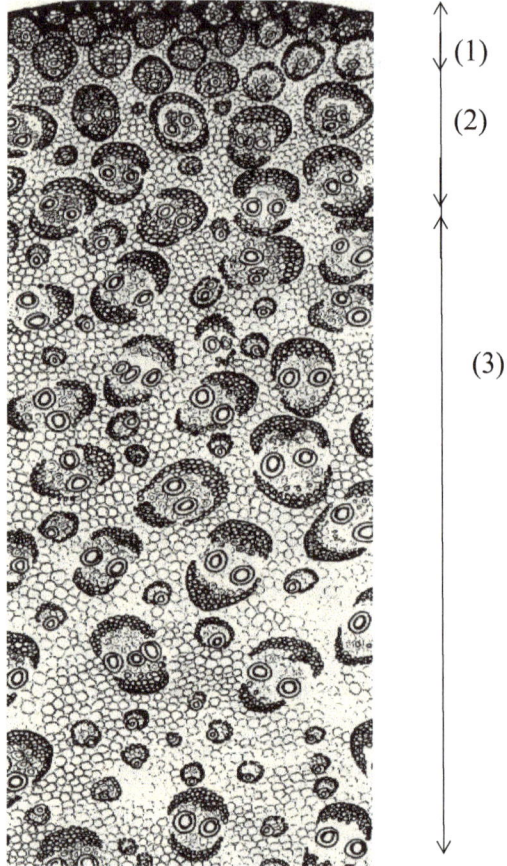

Fig. 7: A diagrammatic illustration of a transverse section of the palm midrib showing the peripheral (1), transitional (2) and core zone (3) (25X) [38].

By-Products of Palm Trees and Their Applications   Materials Research Forum LLC
Materials Research Proceedings 11 (2019) 3-61   doi: https://doi.org/10.21741/9781644900178-1

1.1.5. The Anatomical[(2)] Study of the Palm Midrib

A study has been conducted on the anatomy of the midrib, (Fig. 7), [24]. This study has shown that the palm midrib belongs to the class of monocotyledons, where cross linking is absent in the cross section. The fibro-vascular bundle is the structural unit of the cross section. The cross sectional area of the fibro-vascular bundles, as well as the density of their distribution differ greatly from the epidermal layer to the core of the midrib. In was possible in this study to classify the midrib cross section into three regions: peripheral, transitional and core layers.

It was possible to evaluate the thickness of both the peripheral and transitional layers, where the ratio of fibers increases with adherence to the periphery, by ~ 1.25 mm. It was also possible to evaluate the average length of fibers by 1.366 mm for Baladi midribs and 1.288 mm for Siwi midribs.

1.2.   Date Palm Leaflets

The leaflets emerge on both sides of the midrib of the palm leaf. Fig. 8 illustrates the external view of the leaflet. The leaflet represents a green blade resembling the sward bending longitudinally in its middle and having a swell at its base, where the leaflet is connected with the midrib.

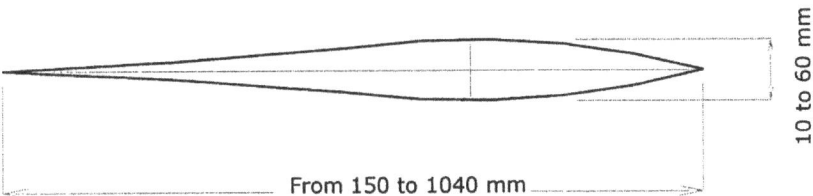

Fig. 8: A diagrammatic sketch illustrating an overall view of the leaflet [7].

1.2.1. Dimensions, Density and Mass of Leaflets

A study has been conducted on the leaflets of Siwi palms from Giza governorate [7].

The following results were obtained:

• Number of leaflets in one midrib varies: from 166 to 197 with an average of 176.

• The width of the leaflet changes from zero at the top to its highest value near the base ranging from 10 to 60 mm as shown in Fig. 8.

• The thickness of the leaflet varies from 0.305 to 0.508 mm with an average of ~ 0.4 mm.

• The density of the leaflets at moisture content 12% is equal to 0.56 gm/cm^3.

---

[2] -   This study has been conducted in the department of Forestry, the Faculty of Agriculture, Alexandria University by Prof. Dr. M.M. Megahed.

- The mass of the leaflets for each midrib is equal to ~ 0.5 kg (at a moisture content of 12%).

1.2.2. Mechanical Properties of Leaflets

Fig. 9 illustrates a comparison between the values of the tensile strength for leaflets taken from different palm species: Males, Mantour, Saeidy, Sohagi (dry and green). It is evident from this figure that these values range from 50 to 120 N/mm^2 and thus fall within the same range of the corresponding values of the palm midrib.

1.2.3. Anatomical Study of the Leaflets

An anatomical study has been conducted on the leaflets of Balady palms, brought from Giza governorate [24]. Fig. 10 illustrates a section in a leaflet showing the pattern of distribution of the fibro-vascular bundles. This study has shown that the average fiber length in the leaflets is 1.181 mm, which falls within the same range of averages for the clade of dicotyledons and hardwoods and softwoods and less than that of the palm midrib (1.325 mm).

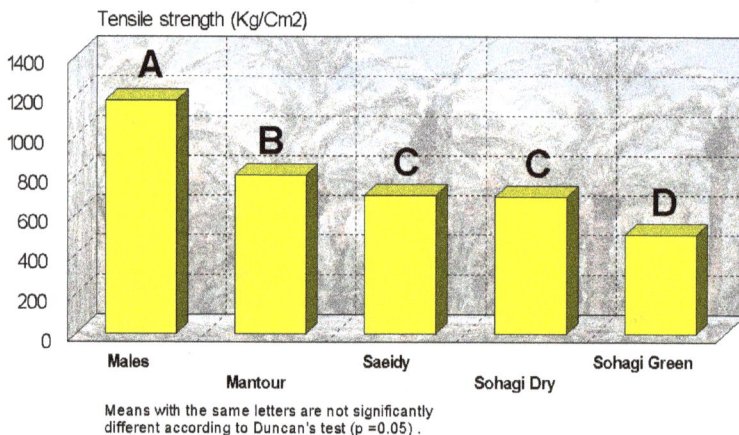

Means with the same letters are not significantly different according to Duncan's test (p =0.05) .

Fig.9: Comparison between the values of tensile strength for leaflets taken from palms of different species: males, Mantour, Saeidy (Siwi) and Sohagi dry and green [7].

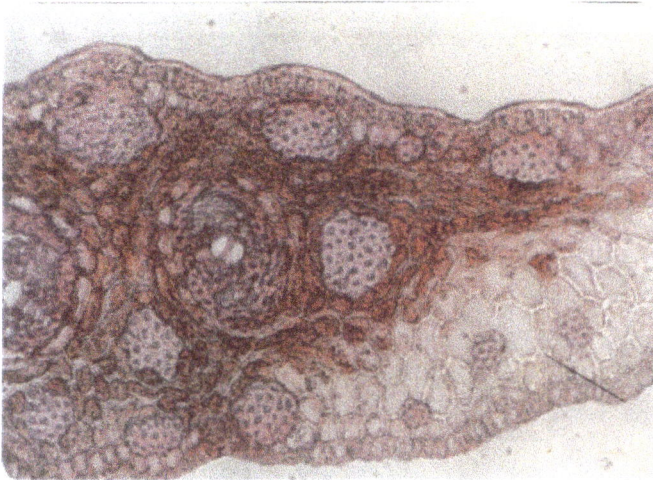

Fig.10: A cross section in the leaflet illustrating the distribution of fibro-vascular bundles (26X) [24].

1.3. The Spadix Stem
1.3.1. Spadix Stem Geometry

The spadix stem carries the date amounting to more than 40 kg. The length of the spadix stem ranges from 1-2 meters. Its cross section has the form of an ellipse with a total area of about 500 mm^2.

1.3.2. The Anatomical Study of the Spadix Stem

An anatomical study has been conducted on a spadix stem from Balady palm, taken from Giza governorate [24]. Fig. 11 illustrates a cross section in the spadix stem. It is clear that the intensity of the fibro-vascular bundles increases and their diameter decreases and the ratio of fibers increases as we move from the core to the periphery. The study has shown that the average value of fibers length in the spadix stem is about 1.114 mm, which falls within the average values of the clade of dicotyledons and the hardwoods and softwoods and less than that of palm midrib fibers (1.325 mm).

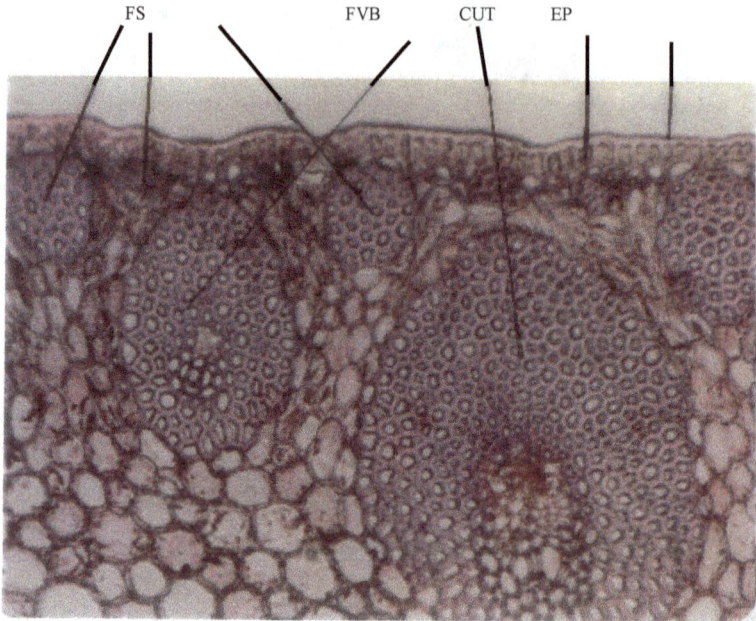

Fig. 11: Photomicrograph of transverse section of date palm spadix stem in the outer zone showing cuticle layer (CUT), epidermis (EP), cortex, fiber strands (FS) and outer fibro-vascular bundles (FVB) (265X) [24].

1.3.3. Coir

The base of the midrib is linked to the trunk by a sheath of coir surrounding the whole trunk. The coir sheath consists of white tissues interspersed by fibro-vascular bundles [24]. With the growth of the palm leaf most of the white tissues disappear leaving the dry dark fibro vascular bundles forming a coarse rough sheath surrounding the trunk. This sheath should be seasonally pruned, otherwise it may increase the danger of fire in the palm garden due to its high susceptibility to fire, especially in the dry state.

1.3.4. Mechanical Properties of Coir

The tensile strength was measured for a coir cord [3] and was found to be ~ 60 N/mm². This value falls within the range of the corresponding values for the palm midrib.

1.3.5. The Anatomical Study of Coir

An anatomical study has been conducted on the coir of a Balady palm. The results of this study [24], have shown that the average length of fibers of coir is ~ 1.143 mm, which falls within the same range of dicotyledons and hardwoods and softwoods and shorter than that of the palm midrib (1.325 mm).

1.4. Date Palm Trunk Mechanical Properties

Three points bending test has been performed on air dried specimens of size 50 x 20 x 450 mm from palm trunk, red European pine and beech (3 similar specimens from each material). The span length was 400 mm. The test was performed on a universal testing machine at a crosshead speed of 5 mm/min in compliance with ASTM D1037-72.

The results of tests are shown in Table (4). It is clear from this table that the palm trunk has an overage bending strength amounting to 82.4 MPa, which is 15% lower than red European pine and 36% lower them the beech wood. The value of the Young's modulus in bending of the palm trunk is 6311.6 MPa, which is about 19% lower than red European pine and 31% lower than beech wood. *This opens the potentiality of using date palm trunks as a substitute for imported wood.* This may help in satisfying the basic needs of millions of people in Southern countries of furniture doors, windows, etc. at a much lower cost and without cutting trees from forests leading to the green house phenomenon and global warming.

1.5. Comparison Between Some Physical and Mechanical Properties of Date Palm by-products

Table (5) illustrates a comparison between some physical and mechanical properties of date palm by-products, as well as with known wood species (e.g. beech, pine and spruce).

Table (4): The average bending strength, and Young's modulus in bending for the date palm trunk and other wood species.

No	Sample Name	σ_b (MPa)	E_b (MPa)
1	Trunk	82.412 (13.59)	6311.638 (974.34)
2	RED European pine	97.144 (16.32)	7807.751 (1524.57)
3	Beech	128.52 (29.27)	9176.618 (1212.68)

Table 5: Comparison between physical and mechanical properties of date palm by-products [2].

No.	Type	Fiber length, mm	Specific wt.	Static bending		Tension test parallel to fibers		Compression test parallel to fibers	
				Mod. Of elasticity, N/mm²	Mod. Of Rupture, N/mm²	Mod. Of elasticity N/mm²	Tensile strength, N/mm²	Mod. Of elasticity N/mm²	Max. compressive strength N/mm²
1	Midribs (different species)	1.29-1.37	0.51-0.79 (average 0.66)	69.6-81.6	3648-6096	66.3-75.1	-	34.1-39.9	-
2	Leaflets	1.18	0.56	-	-	50-120	-	-	-
3	Palm trunks	-	0.48	38	-	-	-	-	-
4	Spadix stem	1.1	-	-	-	-	-	-	-
5	Coir	1.14	-	-	-	-	-	-	-
6	Beech wood [2]	-	0.65	137	11700	97	-	74	-
7	Scots Pine wood [47]	-	0.53	98	11760	102	-	54	-
8	Spruce wood [2]	-	0.35	93	9300	47	-	42.5	-

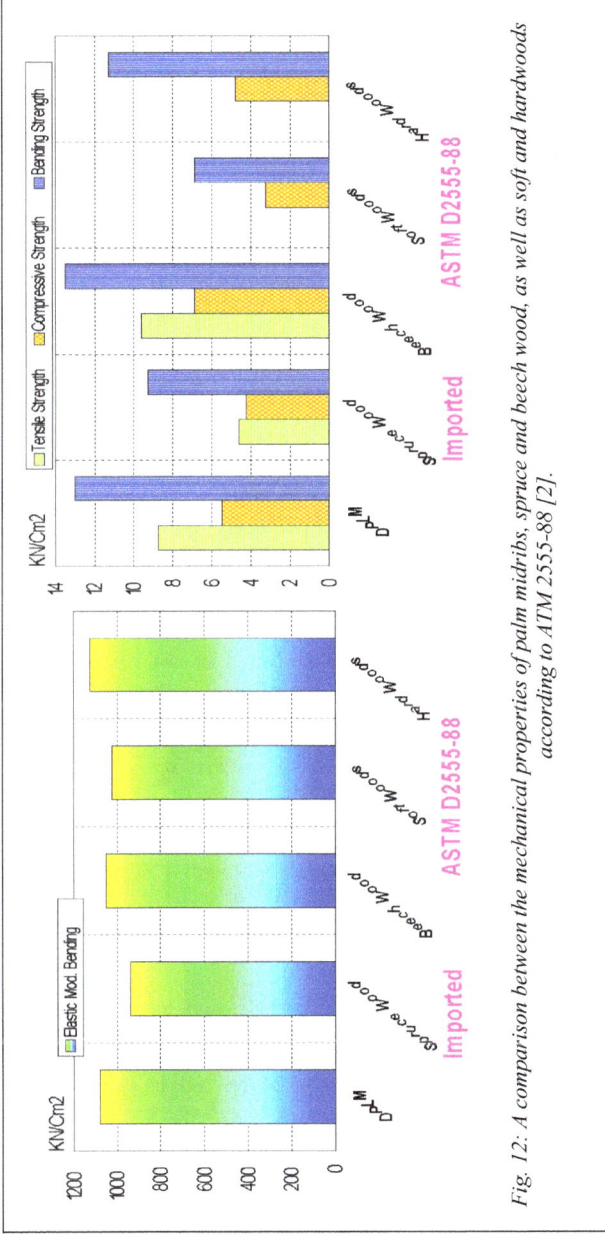

Fig. 12: A comparison between the mechanical properties of palm midribs, spruce and beech wood, as well as soft and hardwoods according to ATM 2555-88 [2].

1.6.   Study of Insects Attacking Palm Midribs

A study has been conducted to diagnose the most prominent insects attacking the palm midrib (midrib borers) and seasons of severe infestation, as well as to find out the most appropriate means for protection from – and struggle against – infestation [8]. The most important of these midrib borers were found to be:

- *Phonapate Frontalis*

  It is an insect of large size of length ~ 15 mm having, two projecting toothed anvils protruding forward, beside many short sharp ends in its frontal part. It is of black color and the infestation takes the form of big corridors leading to the fracture of the midrib.

- *Enneadesmus Trispinosus*

  It is an insect of length 3 millimeters and is dark brown in color. The appearance of infestation takes the form of round holes 1 mm in diameter, beside the existence of corridors inside the midrib as well as the appearance of "flour" near to the corridors, made by the insect.

2. Industrial Applications and Products From Date Palm by-products
2.1.1. Use of Palm Midrib as a Core Material in Blockboards

Why we chose this topic of research? The blockboard is highly needed in Egypt for the manufacture of high quality furniture, as well as wall and ceiling paneling, interior fittings, etc. The local blockboard industry could satisfy only 15% of the local demand, the rest being imported. We found that this industry is in a shaky position, because it totally relies on imported wood, the price of which is continuously increasing due to the success of environmental movements world-wide putting increasing pressure on the cutting of wood from forests. The research, conducted by the faculty of engineering, Ain Shams Univ. collaboration with the Academy of Scientific Research and Technology has proven that the date palm midrib enjoys mechanical properties comparable with those for imported wood (e.g. spruce and beech) (Fig. 12) If we take a blockboard of thickness 16 mm; the wood core layer will be ~ 13 mm. Here comes the spark of rediscovery of the palm midrib: could we replace the inner core of the blockboard with palm midribs and thus save ~ 80% of the wood being imported?. Thus, a master thesis has been registered with the headline: "A Study on the Appropriate Conditions of the Press-cycle for the Manufacture of Palm-midrib Core Blockboards [37]. The appropriate conditions of the press cycle were determined and found as follows:

Pressure: 0.8 N/mm^2, temperature 120°C

Press time: 5 minutes [37].

Side by side with conducting research the Centre for Development of Small-Scale Industries, Ain-Shams University launched a project for the establishment of a blockboard pilot unit in El-Kharga, the New Valley Governorate by a grant from GTZ. The date palm midrib core blockboard specimens were sent to Munich Institute for wood Research in Aachen University. The results of tests indicate that the date palm midrib core blockboard has excellent quality according to DIN standards (DIN 52375, 52364, 53374, 53255, 52371 and 68705) and can be used as a substitute for spruce-core blockboards: in furniture, wall and ceiling paneling, containers, equipment, etc. (Appendix 1).

The production of the blockboard pilot unit has been utilized by the UNICEF for the manufacture of furniture for 100 community schools in Asiut, Souhag and Kena governorates in 1995: The UNICEF was very satisfied with the results (Appendix 2).

2.1.2. Lumber-like Products from Palm Midribs

We were contemplating our success in the palm midrib-core blockboard. We had to use wooden veneers from both sides to manufacture the blockboard. How could we dispense with the wooden veneers? This line of thinking brought us to the idea of manufacture of lumber-like blocks from palm midribs. Therefore, a master thesis has been accomplished with the title: "An Investigation Into the Conditions of Manufacture of Lumber-Like Blocks From Date Palm Leaves Midribs" [46]. It was found that the appropriate conditions of manufacture are:

Pressure $= 0.09$ N/mm^2

Press time $= 1$ hour

The comparison between main mechanical properties of palm midrib blocks (e.g. modulus of rupture, modulus of elasticity, maximum compressive strength and nail pull through tests) and those for spruce and beech show that the results are comparable (Fig. 13). This research has been accredited the prize of the best poster in Euromat-97 conference (Appendix 3).

2.1.3. Super-Strong Materials from Palm Midribs

The study of anatomy of the palm midrib (see item 7.1.2.3) has shown that the peripheral layer greatly differs from the core layer, whereby the former is distinguished with a high density of the fibro-vascular bundles and the smallness of their diameter. This gives an indication that the palm midrib peripheral layer enjoys high mechanical properties as compared with the core material. This was the springboard of a study made on the mechanical properties of the palm midrib peripheral layer with the purpose of finding a new area of utilization of this layer being a secondary product in the blockboard industry using the palm midrib as a base material. The results of this study [28], (Fig. 14), show that the palm midrib external layer with a thickness 1.25 mm enjoys a tensile strength of ~ 250 N/mm^2. As far as the specific tensile strength (tensile strength per unit mass) is concerned this superstrong material exceeds St. 37 by more than 4 times, which opens a new area for its use in biocomposites to substitute other non-renewable materials, such as glass fibers.

2.1.4. Mashrabiah Products from Palm Midribs

How did the idea of use of palm midribs in Mashrabiah (Arabesque) come to our minds? The Mashrabiah (Arabesque) was being intensively used in houses in Egypt and the Arab World. It helps in preserving the privacy of dwellers and in ameliorating the harshness of sun rays, while allowing at the same time to have a look from windows on the outside world. The drastic increase in prices of imported beech wood has led to the shrinkage of demand on Mashrabiah (Arabesque) products. We were conducting machinability tests on palm midrib samples on metal-cutting centre lathes. The date palm midrib piece reached a diameter of ~ 2 mm without breakage. This means that palm midrib is strong enough to replace wood. Thus, the idea came to us (again the rediscovery of material resources): why wouldn't we go to remote villages possessing palm plantations and teach the populace there how to make Arabesque products that may replace wooden Arabesque products? The idea looked fantastic: reviving the traditional skills of Arabesque manufacture to support the Mashrabiah as a distinguishing component of our traditional way of life, while providing sustainable labor opportunities in remote villages, where women could work on lathes at their houses.

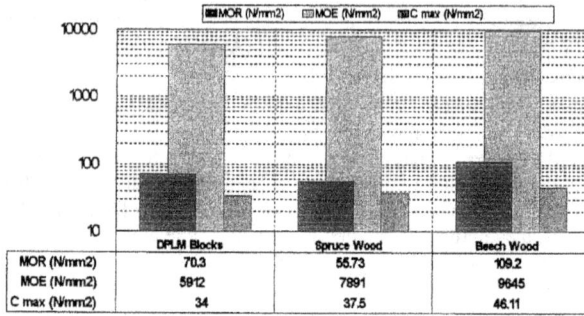

	DPLM Blocks	Spruce Wood	Beech Wood
MOR (N/mm2)	70.3	55.73	109.2
MOE (N/mm2)	5912	7891	9645
C max (N/mm2)	34	37.5	46.11

(a)

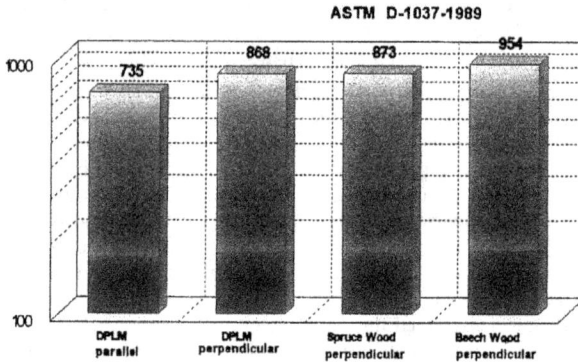

ASTM D-1037-1989

(b)

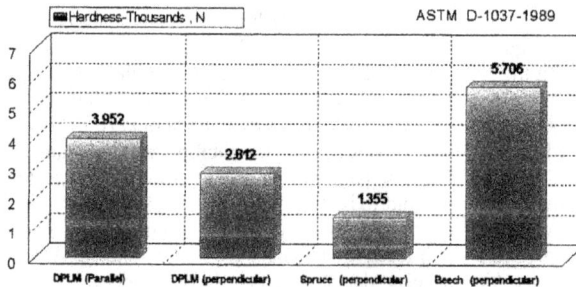

Fig. 13: A comparison between the mechanical properties of lumber like blocks from palm midribs and other species of wood: (a) bending strength, modulus of elasticity and compressive strength Cmax, (b), nail withdrawal resistance and (c) hardness [46].

Materials Research Proceedings 11 (2019) 3-61   doi: https://doi.org/10.21741/9781644900178-1

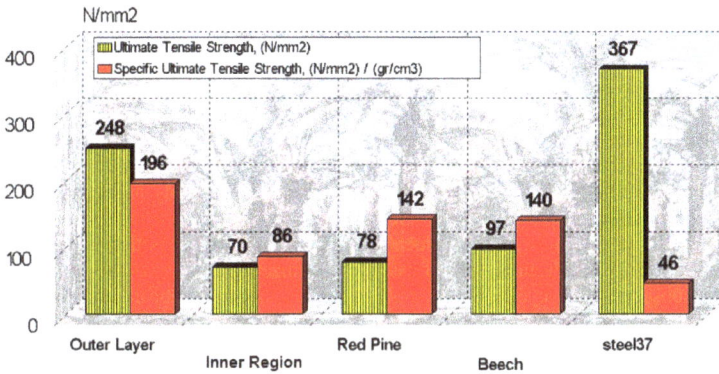

Fig.14: *A comparison between the ultimate tensile strengh and specific tensile strengh of palm midribs (peripheral and core layers), red pine and beech wood and steel 37 [28].*

The Gedeida village in El-Dakhla oases was chosen as a site of the project, which was launched by the Centre for Development of Small-Scale Industries, Ain Shams Univ. by a grant from GTZ. A new multi-purpose machine has been especially designed for this project, tailored to the needs of beneficiaries to work at home (Fig. 15). The machine could perform turning, drilling, saw cutting of palm midribs, in addition to turning solid pieces of local wood species. A training center has been established to secure palm midribs for beneficiaries, prepare palm midrib pieces and train beneficiaries (Fig. 15), as well as a permanent exhibition to help in marketing of products. Fig. (16) illustrates samples of arabesque products from palm midribs.

Fig.15: *Training of women on manufacture of Arabesque products from palm midribs.*

2.2. Organic Products from Palm Midribs

What are the arguments behind this line of research?

1. The organic products from palm midribs mean that you are making use to the maximum of the inherent structural properties of the palm midribs. The anatomical study of the palm midrib (Fig. 7) points to the increase of the intensity of the fibro-vascular bundles and fibers and consequently the strength, as we approach the peripheral zone, i.e., the strength of the palm midrib in its natural state will be certainly higher that of the core, found by research to compete with the imported wood.

2. The manufacture of organic products from palm midribs means that you are beginning the first life cycle from the top of cascade of utilization (Fig. 17). This gives the chance for subsequent life cycles for example in Arabesque, particleboard and fiberboard, etc. This is a good example of application of the full utilization principle [4].

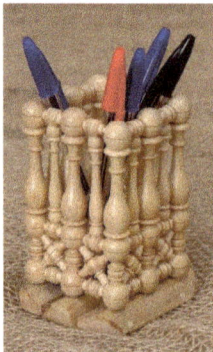

Pens holder from palm midribs

Handkerchief case cover from palm midribs.

Fig. 16: Mashrabiah (Arabesque) products from palm midribs

3. This application will provide the chance to make use of distinguishing physical properties of palm midribs. For example, the epidermal layer of the palm midrib is not porous, which makes it water resistant, in addition to its being covered by a layer of natural wax protecting it from dehydration in the conditions of aridity of the Arab region, where palms grow. This means that the organic products from palm midribs will enjoy a high water resistance without using paints or chemicals, as well as good insulation properties.

4. From a life cycle perspective [33], the organic products from palm midribs enjoy considerable environmental advantages, since the net energy requirement for their manufacture is minimal as compared with other fields of manufacture of palm midribs. In addition, these products are totally biodegradable, i.e., they will not represent any problem during the disposal stage, since they could be used as filling material to be added to the fodder, for poultry [34], and livestock [30], or burned as a CO_2-neutral material [18].

5. The organic products from palm midribs represent an eloquent example of use of the ecological specificity of our region as a strategic comparative advantage. The palm midrib carries the imprint of the local ecosystem. Therefore, the organic products will not disguise the distinguishing ecological identity of the Arabian region, which guarantees competitiveness in export (compare for example these products with those manufactured from imported wood or plastics!).

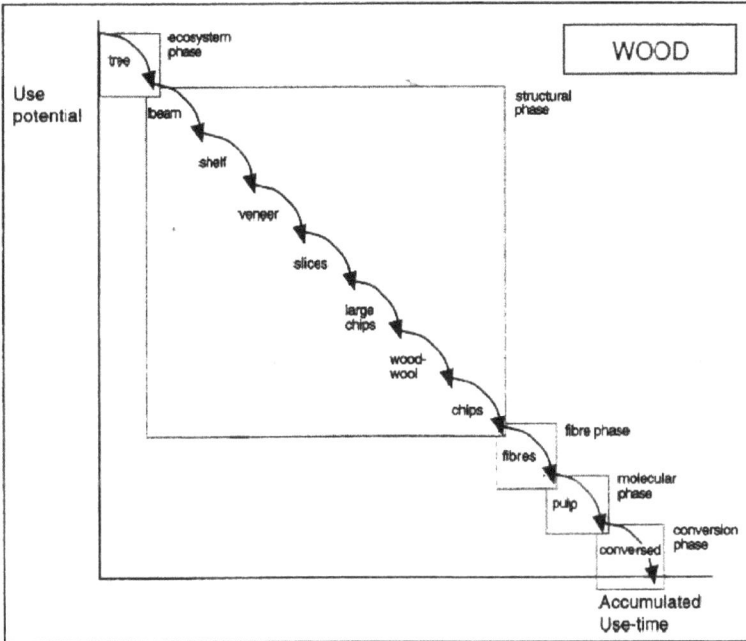

Fig. 17: A cascade of utilization of wood from a life cycle perspective [22].

6. It is expected that the organic products from palm midribs will make use of the chance that the green market offers in the international level. There are strata of environmentally conscious and socially aware consumers, who prefer those organic products, produced in socially just contexts [45]. The success of the trade outlets, such as Fair Trade Net [45], opens the potentiality of manufacture of organic products from palm midribs, as well as products of pruning of fruit trees in general. Besides, the General Preferential System of the EU gives preference for products, produced in the South in an environment friendly way. In addition, the palm midrib products are not in need of a certificate of origin, i.e., that they were manufactured from wood, obtained from sustainably managed forest. I may choose in marketing of these products the slogan: tree-free products [15].

7. The organic products from date palm midribs mean that we could realize a qualitative shift in development of uses of palm midribs simply by the innovation of new designs. This means that we could rely on the palm midrib artisans who manufacture crates in the villages in the production of these organic products using their very procedures and traditional tools via their training on the new products. Therefore, the organic products from palm midribs open wide potentiality for the endogenous development of Arabian villages utilizing – and building upon – the locally accumulated traditional knowledge and experience. This in turn gives high guarantee of success of development and makes development more sustainable.

The following organic products have been designed and manufactured from palm midribs:

- Paravane: three pieces (Fig. 18).
- Armchair (Fig. 19).
- Library.
- Chair.
- Partition.
- Basket.
- Photo frame.
- Window unit.
- Sweeper stick.

Ecotourism is one of the most appropriate areas to make use of the organic products from palm midribs and other date palm by-products.

Fig. 18: A paravane: a sample of organic products from palm midribs (Designed by Prof. Dr. Adel Y.M., Ain Shams Univ.).

Materials Research Forum LLC
doi: https://doi.org/10.21741/9781644900178-1

Fig. 19: An armchair: a sample of organic products from palm midribs.

2.3.   Space Trusses from Palm Midribs

This research has been conducted within a Ph.D. thesis [6]. The objective of this research consisted in making contribution in the field of low cost roofing using a renewable and abundantly available resource in the Arabian region like the palm midribs. As an approach to the problem space trusses were used as structural elements consisting of members, made from palm midribs and assembled by means of metal joints. During this research it was possible to design, manufacture and test a space truss (Fig. 20), composed of quadratic pyramids. Each is composed of members of equal length, made from palm midribs. The previous prototype with span 3 x 3 m was tested under load: measuring the deflections and comparing them with the theoretical values. The results of the research point to the potentiality of use of palm midrib in space trusses as an integrated system to cover the architectural spaces with different spans for permanent and temporal uses, since it is possible to disassemble the structure and reassemble it in new locations. It is possible, as well to use any appropriate locally available material to cover the truss like woven mats from palm leaflets or coir together with using the appropriate protective paints.

2.4.   Use of the Palm Midribs as Structural Elements

In this thesis, [29] the mechanical properties of the whole palm midrib (tensile, bending and compression) has been determined. Different types of connections for palm midribs have been tested: epoxy, polyester, cement mortar with additives and mechanical connections with steel bolts were investigated. From the results of testing of palm midribs, as well as the joinery exploratory tests, it was found that the truss systems are the most appropriate structural system for use of palm midribs.

After determining the most suitable connection, which was the bolt connection, a series of tests was conducted to determine its behavior and strength. Twenty specimens for single palm midrib bolted connections at different spacing and bolt number were tested. Sixty specimens for double palm midrib bolted connections were tested. It was decided that at least two palm midribs will be used in the truss at different spacing and bolt number.

Three different designs of date palm trusses with span of 3 meters and depth of ½ meter have been built and tested. Thus, trusses were built using steel plates and bolts, one using steel rods only and the last two were modification of the traditional design of crates. The results of these tests have shown that the palm midribs can be used as structural elements. A feasibility study has shown that the palm midrib trusses in light structures like canopies and sheds, where steel is dominantly used, could reduce the cost from half to one eighth that of steel.

2.5.   Structures and Sheathing by Palm Midribs

In a recent work [41], experiments have been performed on the use of palm midribs in the construction of shades, as well as in cladding. The results are encouraging. This opens wide potentialities in use of palm midribs in architectural applications.

Fig. 20: A prototype of a space truss from palm midribs [16].

2.6. Particleboards from Palm Midribs

In October 1993 the factory of the Nasr Company for Particleboards[3] and Resins in El Mansourah has been operated using palm midribs as a base material. An amount of 1.15 ton of palm midribs was used to produce particleboards of size 2240 x 1220 x 16 mm. Samples of the factory production were tested according to the Egyptian standard 906/1991 for particleboards. The results of tests showed that the average value of the modulus of rupture (MOR) for these specimens was 20.3 N/mm^2, which satisfies the requirements of the above mentioned standard. In August 1994 the factory of the Modern Arabian Company for Industry of Wood. (MATIN)[4] was operated by about 60 tons of palm midribs brought from Siwa oasis to manufacture 3-layer particleboards with melamine-impregnated paper veneer of dimensions 4.3 x 1.83 m and thickness 8 mm, whereby 20 tons were used to produce 100% palm midribs boards and 40 tons using a blend of palm midribs and casuarina wood (50% each). The experiment gave positive results, where the properties of the 100% palm midrib boards were as follows:

- Density: 0.844 gm/cm^3.

- Modulus of rupture: 21.9 N/mm^2

- Face strength: 1.07 N/mm^2.

- Internal bond: 0.9 N/mm^2

---

3 - Report of the El Nasr Company for Particleboard's and Resins, dated 4/12/1993.
4 - Report of the MATIN Company on 24/8/1994.

2.7.   Medium Density Fiber Boards from the Products of Pruning of Date Palms

Within the frame work of the project of Care for the Palms in El-Bahriah Oases[5] samples of the products of pruning of palm were collected with ratios, equal to the masses of each of these products with respect to the whole mass of products of pruning.

Item	Mass, Kg (air dried)
Palm midribs	15
Palm leaflets	14.6
Spadix stem	9
Coir	1.56
Midrib end	14
Total	**54.16**

The samples were sent to the laboratories of Naga Hammady Company of Fiber Boards. The results of tests (Appendix 4) are as follows:

Physical and chemical properties

➤ Humidity (5.2%, which falls within the limits of EN 322 standards: 4-11%);

➤ Water absorption (12.7%, which is less than the corresponding value in EN 317: 15);

➤ Formalin emission (22.54 mg/100 mg, which is less than the corresponding value in EN 120: 30).

Mechanical properties

➤ Modulus of rupture (24.4 N/mm^2, which is higher than the corresponding value in EN 310: 20);

➤ Modulus of elasticity (2911 N/mm^2, which is higher than the corresponding value in EN 310: 2200);

➤ Internal bond (0.9 Nmm^2, which is much higher than the corresponding value in EN 319: 0.55);

➤ Surface strength (1.35 N, which is higher than the corresponding value in EN 311: 1.2).

This means that it is possible to manufacture MDF boards from date palm products of pruning satisfying the international standards with respect to their physical, chemical and mechanical properties. This opens the potentiality to establish industrial projects in locations having extensive date palm plantations.

---

5 -   This project has been conducted by the Faculty of Engineering, Ain Shams Univ. in collaboration with the Ministry of Environment during the period from January to October, 2016, the project leader was Prof. Dr. Hamed El Mously.

Development of Products Using the Uniqueness of the Palm Midribs and Leaflets as a Competitive Advantage

Four designers joined our team of rediscovery of the date palm by-products. They came with their long history of interaction with consumers from the high and high-middle classes. They put their bet on the uniqueness of the date palms midribs and leaflets: their specific beauty, color and texture. Our interaction with them drove us away from using palm midribs, in a hidden way, in the core of blockboards to expose palm midribs (and leaflets) in quite new products satisfying contemporary needs of high and high-middle classes in Egypt [25]. Thus, a new market has been an opened for the use of palm by-products. Fig. 21 gives examples of these products.

Charcoal from Palm Midribs

A pyrolysis reactor has been designed and manufactured [10] to produce charcoal in laboratory conditions. This reactor may also serve as an example for productive units to be used in villages. Samples of the Baladi palm midribs have been taken to represent: the top, middle and base of the palm midrib in addition to the bent part (knee) and midrib end left after pruning on palms.

The research results have proven that it is possible to attain FAO standards concerning the calorific value by 86.3% for the whole Baladi palm midrib. Comparing the palm midrib parts, the FAO standard has been attained by ~ 96% for the top part; ~ 100% for the middle part, ~ 84% for the base part; ~ 96% for the knee and ~ 74% for the midrib end. As far as the fixed carbon is concerned, it was possible to attain 102% of the FAO standard for the whole palm midrib. Comparing the midrib parts the FAO standard has been attained by 117%, 107%, 93%, 97% and 97% for the top, middle, base, knee and midrib end respectively. The above-mentioned results show the potentiality of manufacture of charcoal for industrial and agricultural purposes from date palm midribs.

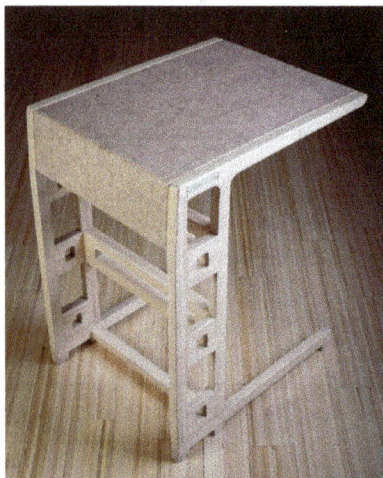

Living room table with a mosaic face   **Console Table**
Fig. 21: Examples of products, made from date palm midribs.

By-Products of Palm Trees and Their Applications   Materials Research Forum LLC
Materials Research Proceedings 11 (2019) 3-61   doi: https://doi.org/10.21741/9781644900178-1

2.8.   Impregnation of Palm Midribs to Improve Physical and Mechanical Properties

Different treatments [21] were applied to improve physical and mechanical properties of palm midribs and namely, treatment with chemicals: water-soluble phenol formaldehyde resin (PF_w), alcohol-soluble phenol formaldehyde resin (PF_a), melamine formaldehyde resin (MF), linseed oil (LD), methyl methacrylate (MMA), polystyrene (PS) and polyester (P_{est}), with different concentrations (concentration of treatment material in solution), on weight to weight basis in suitable solvent fluids for the PF_w, the PF_a and the MF: 20%, 30%, 40%, 50%, for the MMA and the LS: 100%, 90%, 80%, 70%, for the P_{est}: 10%, 15%, 20%, 30% and for the PS: 30%, 40%, 50%, 60%) in a vacuum process with suitable preconditioning before impregnation, curing and post curing after impregnation.

The studied properties included dimensional stability properties: water absorption ($WA_{2,24}$), volumetric swelling coefficient ($S_{2,m24}$), anti-shrink efficiency ($ASE_{2,m24}$) after 2 and 24 hours water soak as well as static bending properties: modulus of rupture (MOR) and modulus of elasticity (MOE), shear stress parallel to grain (S_s) and abrasion resistance (A_r).

The results of this study showed significant difference between the behavior of the treated specimens compared with the untreated control specimens. The analysis of the results points to an appropriate value of concentration to achieve the maximum retention levels of impregnation media reaching 25% with MF, LS, PF_w and P_{est}, while attaining only 15% with other polymers.

Most of the treatments showed recognized influence on dimensional stability. For example, the treatment with MF has resulted into decrease of volumetric swelling coefficient from 20.2% to 1.5% and from 71.6% to 9.6% after 2 and 24 hours water submersion test respectively as compared to the control and water absorption decreased by weight from 37.2% to 4.9% and from 165.1% to 32.5% after 2 and 24 hours water submersion test respectively as compared to the control.

The impregnation by PS, P_{est}, LS, and PF_a increased the MOR values by up to 35% for both P_{est} and PS, 30% and 23% for alcohol-soluble phenol formaldehyde and linseed oil respectively. The MOE values increased by up to 45% for PS, 35% for polyester and 33% for alcohol-soluble phenol formaldehyde.

The treatment with LS, MMA, P_{est} and MF increased the abrasion resistance by up to 61% for methyl methacrylate, 43% for linseed oil, 23% for polyester and 17% for melamine formaldehyde, while no treatment improved the shear strength of the DPLM.

Thus, selective treatments could be chosen, for the improvement of physical and mechanical properties of DPLM, based on the needed application and end use.

2.9.   Nano-Particles from Palm Midrib to Reinforce Polymers

A study [20] was devoted to the effect of nano natural particles on the mechanical properties of epoxy resin. Nano composites were prepared with 1 to 5% wt. % of palm midrib nano particles using ultrasonic dispersion method. The results show that increasing palm midrib nano particles content has no positive effect on the ultimate tensile strength, nor the tensile modulus of elasticity and bending strength compared to pure epoxy. However, an increase in impact strength (~ 300%), bending properties and hardness were detected with increasing nano particles content.

2.10.   Use of Date Palm Midribs in Poultry Feed

An experiment [34] has been conducted to determine the effects of midrib of the date palm (MDP) when included in corn-soy diets for Gemaza growing chicks on growth performance, carcass characteristics and economic efficiency. A total number of 150 one – day old Gemaza chicks (local breed) were distributed equitably into 5 dietary treatments in 3 replicates of 10 birds each. Five experimental diets in each period (starting, growing and finishing) were formulated, in which control diet was 10% wheat bran, in the other MDP were incorporated at levels 2.5, 5, 7.5 and 10% to obtain four experimental diets ($T_1 - T_4$) respectively.

The results indicated that:

1 -   There were no significant differences in body weight, body weight gain and feed conversion between chicks fed by control diets and other treatments in different growth periods.

2 -   Chicks fed by control diets or diets containing 10% MDP (T_4) significantly consumed less feed than the other dietary treatments during starter (0-6 wks) and whole experimental periods (0-12 wks).

3 -   Carcass characteristics parameters (Dressing Giblers, Breast, thight, Drumstic wing and wing %) showed insignificant figures when chicks fed by diets containing MDP compared to those fed by control diets.

4 -   Tibia characteristics, it is worth to note that birds fed by different levels of MDP ($T_1 - T_4$) reflected the lowest figures compared with control diets.

5 -   Economic evaluation, the best economical efficiency value was demonstrated when chicks were fed by 10% MDP and the value was 67% more when compared with that of chicks fed control diets.

2.11.   Use of Date Palm Midribs in Livestock Feed

This study [30] has been carried out at Regional Center Food and Feed and El-Gemmaiza experimental Station, Animal Production Research Institute, Ministry of Agriculture, Egypt. The aim of this work was to improve the nutritive values of the poor quality roughages by biological treatments with oyster mushroom and to study their effects on feed intake, digestibility, nutritive values, and performance of growing lamb. The present work was conducted to study the effect of biological treatments of palm fronds grinded (PFG) by Fungi (*Pleurotus Ostreatus*) in sheep ration. Twenty four Rahmani lambs with average about 22.50 kg live body weight were randomly chosen and divided into four groups (6 in each) to evaluate the experimental rations containing PFG. The treatments were: Control (C) 60% of allowances from concentrate feed mixture (CFM) of NRC (1986) requirements + palm fronds grinded (PGF) ad lib. Ration 1-60% of allowances CFM + palm fronds grinded treated with (*Pleurotus Ostreatus*) (PGFT) ad lib. Ration 2 – 50% of allowances CFM + palm fronds grinded treated with (*Pleurotus Ostreatus*) (PGFT) ad lib. Ration 3 – 40% of allowances CFM + palm fronds grinded treated with (*Pleurotus Ostreatus*) (PGFT) ad lib. Four digestibility trials were performed to determine the nutritive value of the experimental rations. The main results were as following: 1 – The DM intake was insignificantly increased with increasing PFGT ration; 2 – The daily body gain (kg) were 0.141, 0.149, 0.145 and 0.137 and the feed conversion were 10.23, 9.72, 10.11 and 10.58 (kg DM/kg gain) for rations C, T1, T2 and T3 respectively, which were nearly similar except for ration T3; 3 – The DM and OM digestibilities for ration T1 were significantly ($P \leq 0.05$) higher than ration T3. The CP digestibility for ration T3 was significantly lower than those of rations C and T1, which it was not significant with T2 for the other treatments; 4 – The CF digestibility for ration T2 and T3 were significantly ($P \leq 0.05$) higher than ration C, while digestibility of NFE for ration C, T1 and T2 were significantly ($P \leq 0.05$) higher than those of ration T3; 5 – The TDN of ration C and T1 were significantly ($P \leq 0.05$) higher than other treatments, while DCP of ration T3 was significantly lower than those of rations C and T1, but differences between T2 and T3 in DCP digestibility were not significant. Data also revealed that, rations containing PFG with or without Fungi treatment appeared to have higher net revenue and economical efficiency. It could be concluded that palm fronds grinded by Fungi treatment (*Pleurotus Ostreatus*) is a good quality feed in sheep ration up to 60%.

2.12. Palm Secondary Products as a Source of Organic Material for Compost Production

The Bahariah oases are located in the western desert of Egypt. The number of palms in these oases is estimated by 1.3 million palms producing ~ 70 thousand tons (air dry weight) of palm secondary products[6] within this project an experiment has been conducted to use the palm secondary products (PSP) in the manufacture of compost in Mandisha village.

The locally available poultry manure being a residue of the local poultry industry was used as a compost activator. Thus 3 tons of compost has been produced from ~ 8 tons of PSP. The physicochemical analysis of Bahariah oasis compost in as follows [12]:

Parameter	Bahariah oases compost
pH (1: 10)	7.84
EC (dsm^{-1}) (1: 10)	4.61
Organic C (%)	26.54
Organic matter %	45.66
Nitrogen (%)	1.12
C/N ratio	21.75 : 1
Phosphorous (%)	0.01
Potassium (%)	0.91
Bulls density (9 cm^{-3})	0.678
Moisture content (%)	27

---

[6] - The project of Care of Date Palms in El Bahariah oases, executed by the Faculty of Engineering, Ain Shams University in the period from January to July, 2016. The project leader is Prof. Dr. Hamed El Mously.

This opens a great developmental potentiality to use the date palm secondary products being treated as waste at the present time and thus representing a direct cause of fire in the palm gardens, in the production of compost urgently needed to reclaim desert areas in the oases.

3. Machines for the Conversion of Palm Midribs Into Uniform Cross-sections

One of the big challenges we met in our way to use palm midribs in different industrial applications was how to transform the palm midrib irregular cross-section (Fig. 5) into a regular cross-section (e.g. square, rectangular, circular or triangular, etc.). This was a precondition to obtain from palm midribs strips of regular cross-section (Fig. 22) to be further used in making boards of regular thickness (Fig. 23) or blocks of regular cross-section (Fig. 24). Thus, in embarking on designing machines for the conversion of palm midribs into pieces of regular cross-sections we took into consideration that the context, where these machines may operate may widely vary: from a small modern factory producing standard products from palm midribs, to small workshops producing palm midrib strips on a subcontracting basis to the pattern of home production in the villages with large palm plantations producing only palm midrib strips. These stripping machines could be classified according to the principle of operation in to two main categories:

I -   Stripping machines producing palm midrib strips by cutting using disc saws; (Fig. 25).

II -   Stripping machines producing palm strips by skinning or pealing (Fig. 26).

     The first family of machines using the principle of cutting could be divided into two types.

➢ Simplex stripping machine

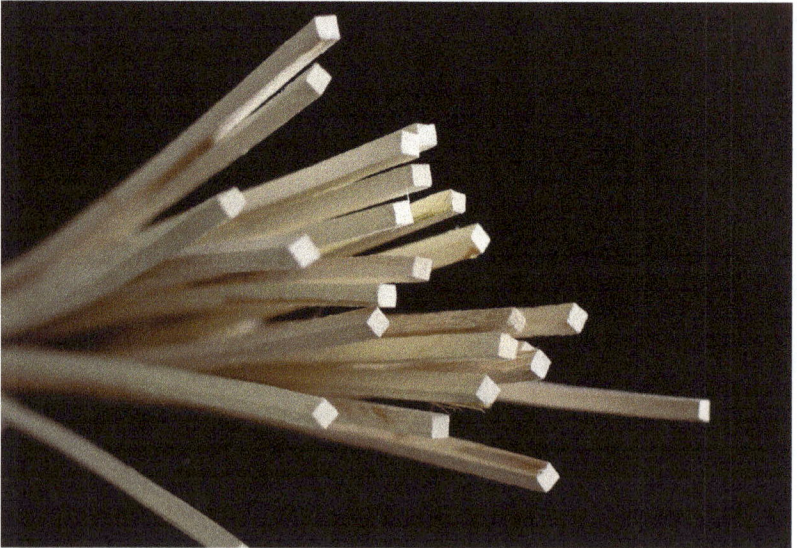

Fig. 22: Strips from palm midribs.

Fig. 23: Boards from palm midribs.

Fig. 24: A beam from palm midribs.

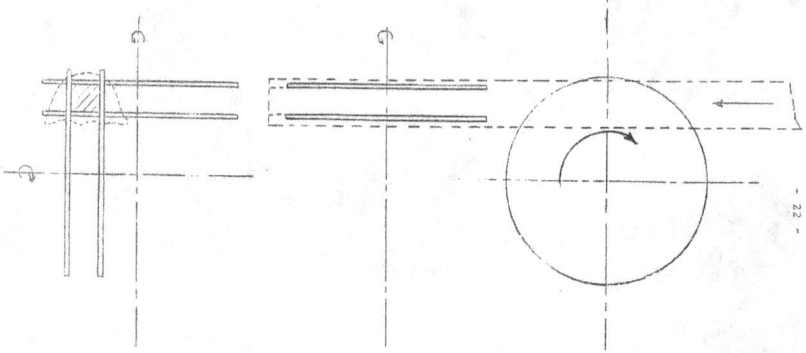

Fig. 25: A diagrammatic sketch showing the main principle of operation of the stripping machine producing palm midribs strips by cutting.

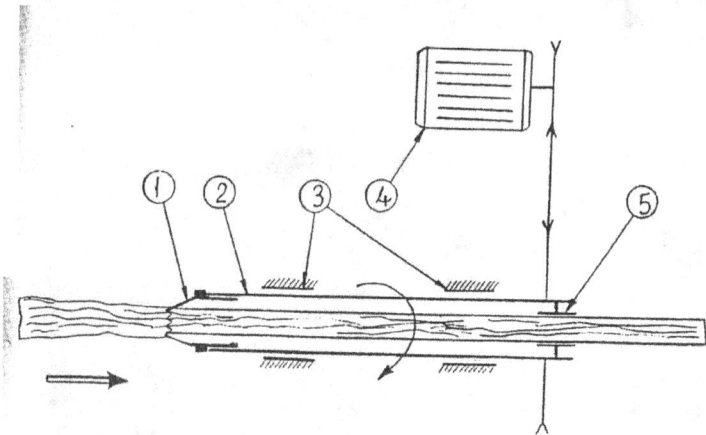

Fig. 26: A diagrammatic sketch of the machine for the producing of palm midribs strips by skinning preparation of DPLM for cutting into strips:

1. collet cutter, 2. hollow spindle, 3. bearings, 4. motor, 5. a guide collet.

This machine includes only one cutting station operating with a pair of carbide-tipped disc saws. Thus, the palm midrib strip could be produced: either using two machines operating in series, or using one machine and readjusting the distance between the disc saws: first to get the width and second to get the height of the strip from the raw palm midrib piece.

➢ Doublex stripping machine.

In this machine two cutting stations are positioned successively each having 2 disc saws set at the required space corresponding to the strip width and height. The feeding of the palm midrib piece is affected mechanically. Thus, the doublex stripping machine is a semiautomatic machine operated by one worker and of high rate of production.

The second family of machines using the principle of skinning rely on the use of collet cutters, designed to produce the whole section at once of the strip. These machines could be divided into the following models:

➢ Pull-type skinning machine

This machine is designed to convert the whole palm midrib into one strip of the required cross-section.

➢ Push type skinning machine

This machine is designed for the skinning of the palm midrib piece to the required cross-section.

➢ Lever type skinning machine

This machine is designed for the manual operation using a lever. Skinning is performed: either by using a collet-form tool or meshing blades cutter. *Different models of the skinning machines have been later designed and tested.*

Future prospects of use of date palm by-products for sustainable development

1. Date palm by-products as a substitute for wood

Egypt has imported in 2014 wood [16] and wood products for ~ 2 billion US $. This represented a big burden on the country's balance of payment. Taking into consideration the influence of inflation in wood prices in future, as well as the expected increase of population this estimate will reach ~ 177 billion US $ in 2050, which represents an unallowable burden on future generations [32] ! Besides, according to a forecast (Table 6) the World will face a huge shortage in wood supply amounting to a 484 million m^3 in 2020.

Table 6: Forecast of wood timber supply: 2010-2020

Region	1996	2010	2020
Oceana (New Zealand & Australia)	42	58	74
South America	130	158	190
North and Central America	600	503	539
Europe and the Baltics	282	330	355
Asia	252	217	288
Africa	67	66	70
Russia	67	130	160
Total supply	1,439	1,461	1,616
Forecast demand		1,801	2,100
Forecast shortfall		340	484

Source: International Forestry Report

Drawing a comparison between the date palm, short rotation forests and Aspen forest we come to the following conclusion:

➢ The products of pruning of the date palm ≅ 15.9 ton (oven dry) per hectare.

➢ The annual rate of yield of short rotation forests = 9.0 ton (oven dry) per hectare; [44].

➢ The annual rate of yield of traditional Aspens forest = 2.5 ton (oven dry) per hectare [49]

This means that quantitatively speaking the products of pruning of date palms are competitive with respect to wood.

In March 2013 EU has issued a law prohibiting the importation of any furniture or other timber product made from illegally logged timber to the EU market. This law, [31] together with the increasing concern for the preservation of natural forests world-wide provide a strong comparative advantage – and create a market niche – for tree-free products, made from palm midribs.

This means that the date may be looked to in future as an additional crop of the date palm plantations producing tree-free products!

In other terms, the date palms may be cultivated in future to obtain two crops: date crop and products of pruning as a lignocellulosic crop and a substitute for imported wood.

2. Palm Fibers for Pulp and Paper

The total world paper consumption increased from 324 million tons in 2002 to 389 million tons in 2008 and is expected to grow steadily in the next decade reaching above 500 million tons [42]. The main source of raw material for pulp and paper is soft wood and hard wood species, which are obtained from forests.

Thus, the preservation of forests necessitates the search for other alternatives as a source for pulp. The palm by-products could be an alternative. For example, the cellulose content % has been evaluated [37] for the Siwi species date palm and found as follows:

Palm Midrib	Spadix Stem	Palm Leaflet	Palm Midrib end	Palm Coir
43.34%	46.65	38.71	37.72	51.-52

This gives a very promising indicator of the potentiality of use of palm by-products as a source of pulp.

In one of the researches [38] it was possible to perform pulping of palm midrib with a yield of 45% (w/w). The physical properties of the prepared hand sheets were very similar to those displayed by other papers made of common lignocelluloses fibers. Pulps from the date palm midrib gave paper sheets with good properties without the need of refining operations. This feature can be considered as a serious advantage when looking to alternative sources of fibers for paper making. Thus, palm midribs could be considered a potential source of fibers for paper making applications. These results are confirmed by another research [36] concluding that the fiber of date palm is close to that of hardwoods with a yield of pulp (41-45%). The palm leaflets gave a low yield of (28.3%) and both pulps were bleached to good brightness and strength properties in between those for spruce and aspen.

3. Palm Fibers for the Reinforcement of Polymer Composites

Polymers have become attractive materials for various applications due to several attractive properties, including light weight, ease of processing and cost effectiveness. Hence, attempts have been significantly made to utilize polymers in different industrial applications, using various kinds of reinforcements including fibers to increase their physical and mechanical properties.

The availability of natural fibers and the ease of manufacturing have tempted researchers to study the feasibility of their application as reinforcement. Compared with the traditional reinforcements, for example, glass and carbon fibers, lignocellulosic fibers impart the composite certain benefits, such as law density and result in highly reduced wear of processing equipment. Moreover, they are readily available from natural sources at a low price [48]. A number of major industries, such as the automotive, construction and packaging industries have shown a considerable interest in the progress of new natural fiber reinforced materials [43].

In search for new fiber types a pioneering research has been conducted to develop combined multi criteria evaluation stage technique (CMCEST) to evaluate six different natural types of fibers: coir, date palm, jute, hemp, kenaf and oil palm. Utilizing the proposed technique, the date palm fibers were found to be quite promising due to beneficial characteristics revealed in the combined triple evaluation criterion, which provide a reasonable cheap and eco-friendly alternative material suitable for different applications [13].

Regarding the low compatibility of palm fibers with relatively hydrophobic polymer matrices, a research has been conducted to improve the interface and interphase interactions between date palm fibers and polyester. The combination of alkaline and silane coupling agent resulted in substantial adhesion improvement [13]. The results of this research support the use of palm fibers in composites.

The use of recycled plastics in the manufacture of natural fibers-polymer composites has high future prospects. In a pioneering work [19] date palm leaf fibers have been used to enhance the mechanical properties of recycled poly (ethylene terephthalate) (PET_r). the addition of these fibers has enhanced both the tensile and flexural strength. In addition, the impact strength was increase with higher fiber loading [19].

The increasing environmental concerns with the use of petroleum-based plastics at the disposal stage (e.g. the environmental pollution resulting from the burning of plastics or the environmental troubles associated with landfilling) point to the high future potentiality of use of biodegradable polymers, reinforced by fibers obtained from palm by-products.

4. Palm By-products as a Material Base for Bioeconomy

4.1. Date Waste

Date palm fruit, one of the most nutritive and comprehensive fruits in terms of health benefits is an ideal substrate for deriving a range of value added products in food and nutraceutical industries in the coming future employing bioprocessing technologies, which have immense scope for application in the valorization of date fruit by-products and wastes [17]. The amount of annual date waste for Egypt, Saudi Arabia, Iran, UAE and Algeria amount to ~ 840300 metric tons[(7)]. These huge quantities, annually available sustainably, provide ample scope for the emergence of new bioindustries and bioentrepreneurs in the date growing countries towards total utilization of the date palm in addition to efficient and effective date palm fruit waste management. There are high prospects, of valorization of these date fruit processing by-products and wastes employing fermentation and enzyme processing technologies towards total utilization of this valuable commodity for the production of biofuels, biopolymers biosurfactants, organic acids, antibiotics, industrial enzymes and other possible industrial chemicals [17].

The production of ethanol from dates will be an attractive option in future. Ethanol is used in a wide range of industrial products (cosmetics, perfumes, pharmaceuticals, solvents, detergents, disinfectants and organic acids). Citric acid is widely used in the food, beverage, chemical pharmaceutical and other industries. α-Amylase is used in ethanol production to break starches in grains into fermentable sugars. In one of the researches [11] ethanol, citric acid and α-amylase have been successfully produced from date waste. In another research [9], spoilage date palm fruits have been successfully used to produce: acetone – butanol – ethanol. This reveals a great potential fur furthering industrialized production of these chemicals.

4.2. Lignocellulosic Biomass

The palm lignocellulosic by-products include the palm midribs, leaflets, spadix stems, palm midrib ends, coir and palm trunk after the end of life. These represent renewable materials sustainably available in the locations of palm plantations. According to [42] and [13] the yearly amounts of products of pruning in Saudi Arabia is about 1 million metric tons. These residues contain cellulose, hemicellulose and lignin by ratios varying from 43.3 : 15.5 : 5.6 , 46.7 : 16.4 : 7.6, 38.7 : 7.9 : 11.3, 37.7 : 14.5 : 13.4 and 51.5 : 12 : 16.3 for the palm midrib, spadix stem, leaflets, midrib end and coir for the Siwi palms [26]. There is a growing interest in converting the lignocellulosic mass to respectively valuable green fuels and high value added chemicals.

---

[7] -   This estimation is based on the figures of annual date production in [17 and 11].

Various pretreatment processes are available to fractionate, solubilize, hydrolyse and separate cellulose, hemicellulose and lignin [40].

4.2.1. Cellulose

Cellulose has been widely used as a main source of paper since the beginning [14]. At the present time bioethanol is focused as a cellulose based biofuel. Bioethanol is the most widely used liquid biofuel. In 2008 worldwide production of bioethanol was over 41 billion liters. In addition, biohydrogen could be produced from cellulose. The biohydrogen could be used in chemical and oil industries. It is expected that the demand for hydrogen fuel for bitumen upgrading will increase (the capacity of upgrading bitumen is expected to be about 2045 thousand barrels in 2020). The cellulose derivatives are of great significance as well. Methyl cellulose is used on a thickener and emulsifier in various food and cosmetic products. It has medical applications to treat constipation, diverticulosis, hemorrhoids, irritable bowel syndrome, as well as to treat diarrhea. It is also used as a performance additive in construction materials. It can be used as a mild glue. It is also used in the manufacture of capsules in nutritional supplements. The ethyl cellulose-a cellulose derivative-is mainly used as a thin film coating material, as well as a food additive as an emulsifier. The cellulose acetate – another cellulose derivative – is used as a film base in photography, as a component in some adhesives, and a frame material for eyeglasses. It is also used as a synthetic fiber and in the manufacture of cigarette filters and playing cards. Nitrocellulose is used as a propellant or low – order explosive. Nitrocellulose, plasticized by camphor was used by Kodak and other suppliers from the late 1880 as a film base in photograph, x-ray films and motion picture films. Carboxymethyl cellulose is often used as its sodium salt, sodium carboxymethyl cellulose.

The hydroxypropyl cellulose is used to prepare artificial tears and to treat medical conditions characterized by insufficient tear production. Nanocellulose is a new field of research at the moment. It can be used in paper and paper board productions, in food as a low calorie replacement, in medical and pharmaceutical uses in making tobacco filter additive, in battery separators, in loudspeaker membranes, in tissue, non-woven products or absorbent structures,. It is often used as antimicrobial films, in coil recovery applications and in drilling mud [14].

4.2.2. Hemicellulose

Different biochemicals can be produced from hemicelluloses. Bioethanol as a fuel can be produced by fermentation of hemicellulose hydrolysates, as well as xglitol having important applications in pharmaceuticals and food industries due to high sweetening properties and also as sugar substitute for diabetics. Other value added products can be produced from hemicellulose hydrolysates, such as butanediol, a valuable chemical feedstock, because of its application as a solvent, liquid fuel, and as a, precursor for many synthetic polymers and resins.

Ferulic and vanillin can also be produced. There are reports about antiradical, oxidase in hibitory, antiinflammatory, antimicrobial, anticancer activities of ferulic acid and its derivatives. Vanillin is, industrially used as fragrance in food preparation, intermediate in the production of herbicides antifoaming agents or drugs, ingredient of house hold products, such as air fresheners and floor polishes and also as food preservative. Lactic acid, furfural, butanol, biohydrogen, chitosan and xylo-oligosaccharides can be produced. Lactic acid is widely used in food, pharmaceutical and textile industries. It is also used as a source of lactic acid polymers, which are being used as biodegradable plastics. Furfural is used for the production of a wide spectrum of important non-petroleum derived chemicals. It is mainly used for the production of resin. Butanol offers a number of advantages and can help accelerate biofuel adoption in countries around the world. Its primary use is as an industrial solvent in products such as lacquers and enamels. Biohydrogen production may provide a renewable more sustainable alternative. Chitosan has found numerous applications in food, cosmetics and pharmaceutical industries-Xylo-oligosaccharides are finding applications in fields related to the food and pharmaceutical industries [40].

4.2.3. Lignin

Lignin (e.g. lignosulfonates) is used to make vanillin (a flavoring for food, ice cream and bakery goods). It is also used for mud viscosity control during deep oil well drilling Lignosulfonates also help to control the viscosities of particle suspensions in making bricks, tiles, gypsum boards, in grinding and polishing, in spraying pesticides, in distributing carbon black during rubber master batching, and in textile dying. Lignin finds outlets in preparing isolation boards, linoleum and floor tile pastes, animal feed pellets, coal dust briquettes and foundry and casting form. It is also used for dust control in ceramic manufacture, synthetic fertilizer production, cement clinker milling and concrete mixing [32].

Significance of establishment of the international association for palm by-products

One of the main objectives of By-Palms conference is to establish the first of its kind in the world the International Association For Palm By-Products as a measure to guarantee the continuation of endeavors to support the use of palm by-products on the international level for the realization of sustainable development. This association will be a forum for all parties, interested and involved in the use of palm by-product in so diverse fields as: fibers for the reinforcement of polymer composites, pulp and paper manufacture, wood substitutes, fodder, biochar, compost, pharmaceutical products and food, etc. It is expected that this association will serve as a meeting area of researchers, industrialists, palm owners, as well as marketing agencies. *The roles of this association could be summarized as follows.*

1. Propagation of the culture of utilization of the palm by-products (e.g. issuing and circulation of brochures of products, success stories and demonstration material, etc.)

2. Support and coordination of R & D activities, associated with the palm by-products utilization in different applications.

3. Exchange and dissemination of knowledge, research results and experience, associated with the utilization of palm by-products for sustainable development.

4. Support of pilot projects, associated with the utilization of palm by-products (e.g. financial support, marketing support, etc.).

5. Participation in international conferences as well as holding of seminars, workshops, exhibitions and conferences on the utilization of palm by-products for sustainable development.

References

[1] حامد إبراهيم الموصلي، التكنولوجيا والنمط الحضاري، دراسة حالة من العريش، مركز بحوث الشرق الأوسط، جامعة عين شمس، 1981.

[2] حامد إبراهيم الموصلي، مشروع بحث استخدام جريد النخيل كخامة صناعية، التقرير الأول، مشروع قامت به كلية الهندسة، جامعة عين شمس بالتعاون مع أكاديمية البحث العلمي والتكنولوجيا، يونيو 1991 .

[3] حامد إبراهيم الموصلي، تجارب استطلاعية قام بها مركز تنمية الصناعات الصغيرة، 1996.

[4] حامد إبراهيم الموصلي، الموارد المادية المتجددة كمواد هندسية صديقة للبيئة، المؤتمر الثالث عشر للهندسة الميكانيكية، 28-31 مارس، 2001 .

[5] وزارة الزراعة واستصلاح الأراضي، قطاع الشئون الاقتصادية ، الإدارة المركزية للاقتصاد الزراعي، 2015.

[6] نادر حسن إبراهيم محمد ، استخدام الخامات المحلية فى صنع الأسقف منخفضة التكاليف، رسالة دكتوراه، معهد الدراسات والبحوث البيئية، جامعة عين شمس، 2001.

[7] أشرف عبد الكريم عبد المقصود، تطوير صناعة الكارينة فى مصر، مشروع تخرج تحت إشراف أ.د. حامد إبراهيم الموصلي، كلية الهندسة، جامعة عين شمس، 1995.

[8] فتحى فهيم عبد العظيم ، دراسة عن آفات جريد النخيل، مشروع بحث إمكانية استخدام جريد النخيل كخامة صناعية، مشروع بحثى قامت به كلية الهندسة جامعة عين شمس بالتعاون مع أكاديمية البحث العلمي والتكنولوجيا، التقرير الأول، يونيو 1991.

[9]   M.H. Abd-Alla, Production of Acetone-Butanol-Ethanol from Spoilage Date Palm (Phoenix dactylitera L.) Fruits by Mixed Culture of Clostridium acetobutylicum and Bacillus subtilis, Biomass and Bioenenergy 42 (2012) 172-178. https://doi.org/10.1016/j.biombioe.2012.03.006

[10]   M.M.A. Abdel Samie, Study of the Potentiality of Use of the Palm Midrib in Charcoal Production: A master thesis, the Faculty of Engineering, Ain Shams University, Cairo, 2018.

[11]   A. Acourene, A. Ammouche, Optimization of Ethanol, Citric Acid and α-Amylase Production from Date Wastes by Strains of Saccharomyces cerevisiae, Aspergillus Niger and Candida Guilliermondii, Fermentation, Cell Culture and Bioengineering 39 (2012) 759-766. https://doi.org/10.1007/s10295-011-1070-0

[12]   M.M. Ahmed, Palm Secondary Products as a Source of Organic Material for Compost Production: Applied Examples from Egypt. The First International Conference on Use of Palm By-Products for Sustainable Development, 15-17 Dec., 2018 (accepted as an oral presentation).

[13]   F.M. Al-Oqla and others, Combined Multi-criteria Evaluation Stage Technique as an Agro Waste Evaluation Indicator for Polymeric Composites: Date Palm Fibers as a Case Study, Bio Resources 9-3 (2014) 4608-4621.

[14]   D.R. Bogati, Cellulose Based Biochemicals and Their Applications Saiman University of Applied Sciences, Faculty of Technology: Bachelor's Thesis, 2011.

[15]   Brain MLEOD, Panel Source International, Dawn in the grain forest. Proceedings of the meeting of the Eastern Canadian Section of the Forest Products Society, Winnipeg, Manitoba, (1999)19-20.

[16]   Central Agency for Mobilization and Statistics, USDA; Foreign Agriculture Service, Global Agriculture Information Network, Egypt, Wood Sector Report, 29/9/2015.

[17]   M. Chandrasekaran, A.H. Bahkali, Valorization of Date Palm (Phoenix dactylitera) Fruit Processing By-products and Wastes Using Bioprocess Technology-Review, Saudi Journal of Biological Sciences, King Saud University, 2012.

[18]   Crops for Sustainable Enterprise, Design for Sustainable Development. European Foundation for the Improvement of Living and Working Conditions, Wyattville Read, Longhlinstown Co., Duplin, Ireland, 2000.

[19]   A. Dehghani and others, Mechanical and Thermal Properties of Date Palm Leaf Fiber Reinforced Recycled Poly (ethylene terephthalate) Composites, Materials and Design, EL SEVIER, 2013.

[20]   K.M. Elerian and others. Investigation on the Use of Cellulose-Based Nano Fibers for Polymer Composites Reinforcement. El Azhar University Journal, 2015.

[21]   M.S. El-Kinawy, Treatments of Palm Midrib to Improve the Mechanical Properties Compared to Imported Woods. A Ph.D. thesis, the Faculty of Engineering, Ain Shams University, Cairo, 2006.

[22]   H.I. El-Mously, M.S. Saber, Medium Density Fiberboards from the Date Palm Residues: A Strategic Industry in the Arab World. The First International Conference on the Palm By-Products, Aswan, Egypt, 15-17 Dec. 2018.

[23]   H.I. El-Mously, Date Palm Midrib Utilization Project. A research project conducted by the Centre for Development of Small-Scale Industries, the Faculty of Engineering, Ain Shams University in Collaboration with IDRC, First technical report, 1994.

[24]   H.I. El-Mously, Date Palm Utilization Project. Final report. A project conducted by the Centre for Development of Small-Scale Industries, Fac. Of Engineering, Ain Shams Univ. in collaboration with IDRC, Cairo, Oct., 1995.

[25]   H.I. El-Mously, Innovating Green Products as a Mean to Alleviate Poverty in Upper Egypt. 1st International Joint Symposium on "Product Development and Innovation, Ain Shams University 3-5 May 2016.

[26]   H.I. El-Mously, Project of Care of Palms in El-Bahariah Oases, A project conducted by the Faculty of Engineering, Ain Shams University in collaboration, with the Ministry of Environment, Final Report, Cairo, 2016.

[27]   H.I. El-Mously, Resources Searching for a Management, The Expert Consultation on Date Palm Residues Utilization. FAO Regional Office, Cairo, 27-29 Oct., 1997.

[28]   A.B. El Shabasy, H.I. El Mously, Study of the Variation of Tensile Strength Across the Cross Section of Date Palm Leaves' Midrib. Proceedings of the 5th European Conference on Advanced Materials and Processes and Applications, Maastricht, the Netherlands, 21-23 April, 1997.

[29]   T.M. El Sherbeny, Use of Palm Midribs as Structural Elements, A master thesis. Institute of Environmental Studies, Ain Shams University, Cairo, 2010.

[30]   A.A. El-Tahan and others. Upgrading Palm Fronds Through Different Treatments to be Used as Animal Feed. Egyptian J. Nutrition and Feeds 16-2 (2013) 235-242.

[31]   EU Timber Regulation 2013, European Commission.

[32]   J.M. Harkin, Lignin and its Uses, U.S. Department of Agriculture, Forest Service, Forest Products Laboratory, Madison, WIS., 1969.

[33]   L. Helen, J. Gertsakis. Design & Environment. Greenleaf Publishing, 2001.

[34]   S.A. Ibrahim and others. Effect of Using Date Palm Waste on Performance, Carcass Characteristics and Economic Efficiency of Gemaza Growing Chicks. Egyptian J. Nutrition and Feeds 16-2 (2013 309-318.

[35]   R. Khiari and others. Tunisian Date Palm Rachis Used as an Alternative Source of Fibers for Papermaking Applications, Bio Resources 6-1 (2011) 265-281.

[36]   P. Khristova and other. Alkaline Pulping with Additives of Date Palm Rachis and Leaves from Sudan, Bioresource Technology 96 (2005) 79-85. https://doi.org/10.1016/j.biortech.2003.05.005

[37]   M.S.A. Kinawy, Study of the Appropriate Conditions of the Press-Cycle for the Manufacture of Palm Midrib-Core Blockboards. A master thesis, the Faculty of Engineering, Ain Shams University, 1997.

[38]   R. Maylor, Pioneering the Production of Oil Palm MDF. Dyno Resin Technology 1 (1999).

[39]   M.M. Megahed, Hamed El-Mously, Anatomical Structure of Date Palm Leaves' Midrib and its Variation Across and Along the Midrib, IUFRO XX World Congress, Tampere, Finland, 1995.

[40]   V. Menon and others. Value added Products form Hemicellulose: Bio technological Perspective, Division of Biochemical Sciences, National Chemical Laboratory, India.

[41]   Y.m. Monsour and Others, Utilizing Palm Rachis for Eco-friendly and Flexible Construction in Egypt © SBE-Cairo, 2016.

[42]   R.A. Naser and others. Measurement of Some Properties of Pulp and Paper Made from Date Palm Midribs and Wheat Straw by Soda-AQ Pulping Process. Measurement 62 (2015) 179-180. https://doi.org/10.1016/j.measurement.2014.10.051

[43]   E. Omrani and others. State of Art on Tribdogical Behavior of Polymer Metric Composites Reinforced with Natural Fibers in the Green Materials World. Engineering Science and Technology, an International Journal, 2015.

[44]   Proceedings of the Meeting of the Eastern Canadian Section on the Forest Products Society, Winnipeg, Manitova, Canada, May 1999.

[45]   L. Ramacho. Sustainable Consumption Provides Opportunities for Developing Countries. Industry and Environment, UNEP, 22-4 (1999).

[46]   A.M. Tayssier, An Investigation into the Conditions of Manufacture of Lumber-Like Blocks from Date Palm Leaves' Midribs: A master thesis, the Faculty of Engineering, Ain Shams University, 1996.

[47]   G. Tsoumis, Science and Technology of Wood. Van Nostrand Reinhold, New York, 1991.

[48]   A.A. Waszzan, The Effect of Surface Treatment on the Strength and Adhesion Characteristics of Phoenix dactylifera L. (Date Palm) Fibers, International Journal of Polymeric Materials, 55 (2006) 485-499. https://doi.org/10.1080/009140391001804

[49]   J.A. Youngquist and others. Agricultural Fibers in Composition Panels. 1303 on: Thomas M.,ed. Proceedings of the 27th International Particleboard Composite Materials Symposium, Pullmann, 30-31 March, 1993, Washington State University, 1993.

[50]   Zaid Ab delouahhab, Date Palm Cultivation. Date Production Support Programme in Namibia, FAO, Rome, 1999.

By-Products of Palm Trees and Their Applications   Materials Research Forum LLC
Materials Research Proceedings **11** (2019) 3-61   doi: https://doi.org/10.21741/9781644900178-1

APPENDIX (1)

A Summary of the Report of
Munich Institute for Wood Research
On
Palm Midrib Blockboards
Produced in El-Kharga Factory,
The New Valley Governorate

By-Products of Palm Trees and Their Applications   Materials Research Forum LLC
Materials Research Proceedings **11** (2019) 3-61   doi: https://doi.org/10.21741/9781644900178-1

Institut für Holzforschung der Universität München

80797 München, Winzererstraße 45
Institutsleiter Prof.Dr.Dr.habil. G.Wegener
Telefon (089) 306 309 0; FAX (089) 306 309 11

Bewertung von "Date Palm Leaves' Midribs Blockboard"

nach deutschen Sperrholznormen

Gutachtliche Stellungnahme

München, 4.9.1996

Institutsleiter
Prof. Dr. G. Wegener

Verfasser
Dipl.-Holzwirt F. Tröger

By-Products of Palm Trees and Their Applications   Materials Research Forum LLC
Materials Research Proceedings **11** (2019) 3-61   doi: https://doi.org/10.21741/9781644900178-1

Quality Assessment of Palm Leaves' Midrib Blockboard
According to German Standards

Fritz Tröger, Institute for Wood Research, Munich University

Summary

Early in 1996, the Institute for Wood Research of the University of Munich was commissioned by the GTZ office in Cairo to assess the quality of blockboard, developed as part of the GTZ project program, in orientation tests using German standards.

The institute's assessment report, in its introductory section, gives a description of the cross-sectional structure of blockboard as defined in German standards. Blockboard is considered a high-quality, very well reputed wood based panel material. For cost reasons alone, applications are restricted to fields with above average quality specifications, e.g. in high-quality interior decorating (wall and ceiling panels), high-quality furniture and equipment manufacture.

Six blockboard samples measuring 1,000 mm x 500 mm x 13 mm constituted the test material supplied by the Egyptian plant El-Kharga. "Date Palm Leaves' Midribs Blockboard", as this panel with a centre layer of palm leaves' midribs is officially called, has a cross-sectional structure according to the principles of cross-banded lumber veneered board. The block core consists of slats cut from palm leaves' midribs, with a cross section of 10 mm x 10 mm. Surface veneers from poplar wood, 1.5 mm thick, are glued onto both sides of the block core using urea formaldehyde resin.

Tests involved the investigation of the following material properties: Moisture content (DIN 52 375), thickness swelling (DIN 52 364), density (DIN 53 374), internal bond strength (DIN 53 255), bending strength (DIN 52 371) and specifications regarding surface veneer and core layer (DIN 68 705, part 2).

Overall, the investigation of the material properties of Date Palm Leaves' Midribs Blockboard led to consistently positive results. Special attention is drawn to the following board characteristics:

1. The surfaces of the poplar wood veneers were evaluated class 1 and class 2 on top and underneath respectively. Some veneer surfaces proved excessively rough. This should be brought to the attention of the supplier.

2. The core of the blockboard is the centre layer which consists of slats made of the midribs of palm leaves. Emphasis is placed on the fact that this centre layer of palm leaves midribs fulfilled most specifications for centre layers made from wood. In future care should be taken to cut palm leaves' midribs in such a manner that slats with angles of 90° are achieved as well as being full edged. Grading for first and second choice could be considered.

3. The results of the internal bond test could not be better. The leverage-break tests gave excellent results regarding the gluing of surface veneers and core layer (No. 1 ranking according to DIN 53 255).

4. For surface veneers with the grain oriented in parallel to the longitudinal axis of the specimens bending strength was in the range of 37 to 43 N/mm^2. As expected, the bending strength range for veneers with grain perpendicular to the longitudinal axis of specimens was, at 27 to 30 N/mm^2, considerably lower.

In summarizing the results, it can be confirmed that Palm Leaves' Midrib Blockboard can be considered a valuable wood-based materials panel. On account of its good mechanical properties this panel is suitable for a great number of applications such as furniture manufacture, interior fitting, container and equipment manufacture as well as wall and ceiling panelling.

By-Products of Palm Trees and Their Applications
Materials Research Proceedings 11 (2019) 3-61

Materials Research Forum LLC
doi: https://doi.org/10.21741/9781644900178-1

APPENDIX (2)

A Certificate from the UNICEF Concerning
The Project of Utilization of Palm Midrib
Blockboards in the Manufacture of
Community Schools Furniture in Asiut, Sohag
And Kena Governorates
In 1995

unicef ⊛

United Nations Children's Fund
Fonds des Nations Unies pour l'enfance
Fondo de las Naciones Unidas para la Infancia
8 Adnan Omer Sidky Street, Dokki – Cairo – Egypt
Tel., 700815 – 3616346 – 710578 – 708540 – 708541
Telex: 93164 ICEF UN
Cable: UNICEF Cairo   Fax: 1605664

منظمة الأمـم المتحدة للأطفال

8 October 95

To Whom It May Concern

This is to certify that the cooperative agreement for the production of Date palm mid rib block board between UNICEF and The Center For Development Of Small Scale Industries, Faculty of Engineering - Ain Shams University and the Governorate of the New Valley was satisfactorily completed. The production was of good quality and delivered on time.

UNICEF has been very happy to support this initiative as it is in line with the development of appropriate local technology. Moreover this type of production is environmentally friendly and contributes to the creation of employment opportunities at the local level.

M. Baquer Namazi
UNICEF Representative

By-Products of Palm Trees and Their Applications   Materials Research Forum LLC
Materials Research Proceedings **11** (2019) 3-61   doi: https://doi.org/10.21741/9781644900178-1

APPENDIX (3)

A Copy of the Certificate,
Obtained from Euromat-97 Conference
On the Research, titled:
A New Lumber-like Product
From Date Palm Leaves' Midribs

Bond voor Materialenkennis

C E R T I F I C A T E

HOUWINK - PRIZE

The Board of the Netherlands Society for Materials Science (Bond voor Materialenkennis) decided to grant the HOUWINK-PRIZE 1997 to the best poster presented at EUROMAT 97.

A jury initiated by the Board consisting of Ir. M. Bliek, Ir. H. Toersen and Dr. A.P.M. van der Veek has nominated as winner:

The Ain Shams University at Cairo, Egypt
Centre for Development of Small-scale
Industries and Local Technologies

for their poster:

"A New Lumber - like Product from Date Palm Leaves' Midribs"
by: A.M. Taysseer, H.I. El Mously and M.M. Megahed

The prize of EU 1000.00 was given to Prof. Dr. Hamed El-Mously at the closing session of the meeting.

Chairman of the Jury   Netherlands Society for Materials Science

Ir. M. Bliek   Ir. H.J. van der Torren, President

Postbus 390
3330 AJ Zwyndrecht
Tel. 078 - 619 26 55
Telefax 078 - 619 57 35
E-mail bvm@worldonline.nl
Postbank 209046
Bank ABN • AMRO

By-Products of Palm Trees and Their Applications   Materials Research Forum LLC
Materials Research Proceedings **11** (2019) 3-61   doi: https://doi.org/10.21741/9781644900178-1

APPENDIX (4)

Results of testing of samples
Of MDF, made from the products
Of pruning of date palms,
Collected from El-Bahariah Oases

NAG-HAMADY FIBER BOARD CO.
(N.F.B)
E-mail : nhfiboco@intouch.com

NFB

شركة نجـع حمادى للفيبر بورد
ش . م . م

دشنا في : ٢٠١٦/٨/٢٢

<u>تقر بر فنى عن</u>

الاختبارات الفيزيائية والميكانيكية لعينات الالواح المصنعة معمليا من المنتجات
الثانوية لنخيل البلح

١- الاختبارات الفيزيائية والكيميائية

الاختبار	السمك	الكثافة	الرطوبة	التشرب	الامتصاص	ابعات الضرمالين
الوحدة	مم	كجم/م٣	%	%	%	مجم/١٠٠جم
النتيجة	١٢,٠٠	٧٥٢,٠٠	٥,٢	١٢,٧	٥٨,٠٤	٢٢,٥٤
الانحراف المعياري	٠,٠٤	٢٤,٢٨	٠,١١	١,٣	٢,٥	٠,٨٧
الطريقة القياسية	EN323	EN323	EN322	EN317	BS1142	EN120
المواصفات القياسية	-	--	١١-٤	١٥		٣٠≤

٢- الاختبارات الميكانيكية

الاختبار	السمك	معامل الكسر MOR	معامل المرونة MOE	الرابطة الداخلية IB	قوة شد السطح Surface soundness
الوحدة	مم	نيوتن/مم٢	نيوتن/مم٢	نيوتن/مم٢	نيوتن
النتيجة	١٢	٢٤,٤٠	٢٩١١	٠,٩	١,٣٥
الانحراف المعياري	٠٠,٠٤	٣,١٩	٢٨٠,٥٢	٠,٢	٠,٣٠
الطريقة القياسية	EN323	EN310	EN310	EN319	EN311
المواصفات القياسية	٢٠,٠٠	٢٢٠٠	٠,٥٥	١,٢	

مدير عام البحوث

د./عبدالباسط عبدالحميد ادم

مدير عام المعامل

كيميائي/ عبدالحميد محمد محمد

Cairo Office : 17 Gawad Hosni St.. Cairo - Egypt
Tel.: +2/02/23922109 +2/02/23902995 Fax: +2/02/23926511
P.O.Box : 432 Mohamed Farid
Factory : Desna - Qena
Tel.: +2/096/6743281 - +2/096/6743002 Fax : +2/096/6743003
website : www.geocities.com/nhfiboco/company-profile.html

فرع القاهرة :١٧ ش جواد حسنى القاهرة مصر
تليفون ٢٣٩٢٢١٠٩، ٢٣٩٠٢٩٩٥ فاكس ٢٣٩٢٦٥١١،
ص . ب :٤٣٢ محمد فريد
المصنع :دشنا محافظة قنا
تليفون :٦٧٤٣٢٨١/٠٩٦، ٠٩٦/٦٧٤٣٠٠٢ فاكس :٠٩٦/٦٧٤٣٠٠٣

By-Products of Palm Trees and Their Applications
Materials Research Proceedings **11** (2019) 62-68

Materials Research Forum LLC
doi: https://doi.org/10.21741/9781644900178-2

A Glimpse on 65 Years of Passion-driven Work for Bamboo

Walter Liese

Institute of Wood Science, University Hamburg, Leuschnerstrasse 91d, 21031 Hamburg
Germany

wliese@aol.com

Keywords: bamboo, structures, protection, utilization, international projects

Abstract. My first contact with bamboo took place in 1951, when shortage of timber for the coal mining industry in West Germany led to the idea to use bamboo as pit props. However, they failed under axial load. In 1952, pioneering the use of the electron-microscope for cell wall structures, bamboo was also tested. These photos excited a visiting Indian wood preservation expert, since structural knowledge might improve the preservative treatment of bamboo culms. So, in 1957, I went to India for 4.5 months as an FAO expert to improve methods for bamboo preservation. This mission was followed by consultancies in about 25 countries, strongly supported by laboratory research. Results were published in about 110 bamboo-related papers and 6 books as author or co-author. The latest publication from 2016 contains chapters on structures, properties and uses of bamboo. At the age of 93 I am still enjoying the discussions with bamboo colleagues worldwide.

Introduction

My first contact with bamboo took place in 1951, when shortage of timber for the coal mining industry in West Germany led to the idea to use bamboo as pit props. Culms were imported from Indonesia, but they failed because the internodes crushed under axial load without emitting any cracking sound, a warning sign for miners to escape (Fig.1, Fig.2). The following year, 1952, I pioneered the use of electron-microscopy at the "Institut für Übermikroskopie", Düsseldorf, to explore the unknown fine structure of wood. Out of pure curiosity some left over bamboo pieces were also put under the microscope to reveal structural details, (Fig 3).

Fig. 1 Dendrocalamus giganteus,
Bogor, Indonesia.

Fig. 2 Culm internode crushed under load.

By-Products of Palm Trees and Their Applications
Materials Research Proceedings **11** (2019) 62-68

Materials Research Forum LLC
doi: https://doi.org/10.21741/9781644900178-2

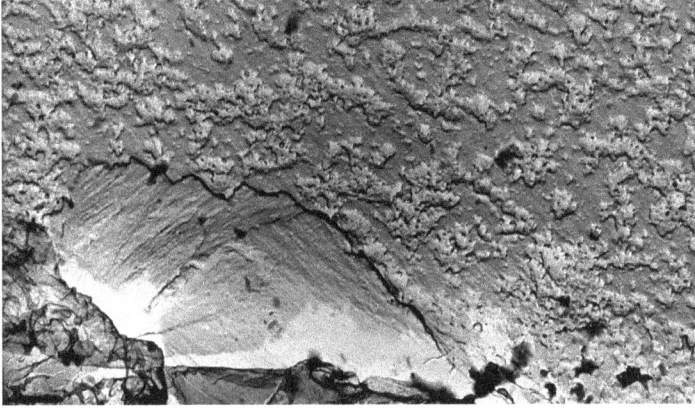

Fig 3 Parenchyma cell wall, Bambusa vulgaris, electronmicrograph,1952.

However, much later these electron-micrographs became very useful, when in 1956 an Indian wood preservation expert on an international factfinding mission paid me a visit at the University of Freiburg to discuss my earlier industrial work on the treatment of spruce. He showed little attention for this work, since his real interest were details on bamboo preservation. Happily, I showed him the slumbering bamboo electron micrographs from 1951/52. He became very excited and indicated a consultancy, since any improved knowledge of bamboo structures might help to treat bamboo culms against deterioration. So, the following year, I went to India as an FAO expert for 4.5 months, at the academically young age of 31 years. For the first time I saw a bamboo plant in its natural habitat. My task was to develop methods for the preservation of bamboo culms against deterioration by fungi and insects. Especially the frequently applied sap-replacement method (Boucherie) should be improved (Fig. 4). This could be achieved by putting an air-sucking cup on the culm end before applying pressure. The results became widely known and initiated numerous consultancies, not only on bamboo, like the one on Wood/Bamboo Preservation in Indonesia the next year and followed by electron-microsocpical work in Melbourne, Australia.

Thus, my "bamboo life" had started.

Fig. 4 Treatment of bamboo culms by the sap-replacement method, Dehra Dun, India, 1957/58.

The following reflections on bamboo activities must necessarily concentrate on some general areas such as culm structures, protection, international co-operation. After several consultancies, a general documentation about "Bamboo-Biology, silvics, properties, utilization" was published 1985 [1]. For another important monocot, the rattan palms, intensive studies about their structures and properties as well as on their fungal degradation and protection were also initiated.

Culm structures

My first studies on the general anatomical structure of bamboo started 1959 in Freiburg with two bamboo culms brought from India. Studies were continued at my following working stations at Munich and Hamburg, intensified by the use of the electron-microscope. Of the many interesting topics, only a few can be mentioned here, like the fine structure of the cell wall, the variability of fibres within a culm, the arrangement of vascular bundles and their significance for classification (Fig. 5), and structural changes during aging (Fig.6). The main tissue types of Bamboo are vascular bundles embedded in parenchyma tissue. Vascular bamboo number and distribution vary along wall thickness, giving the high strength of bamboo. The parenchyma acts as reservoir for water, plant nutrients, sugars and starch.

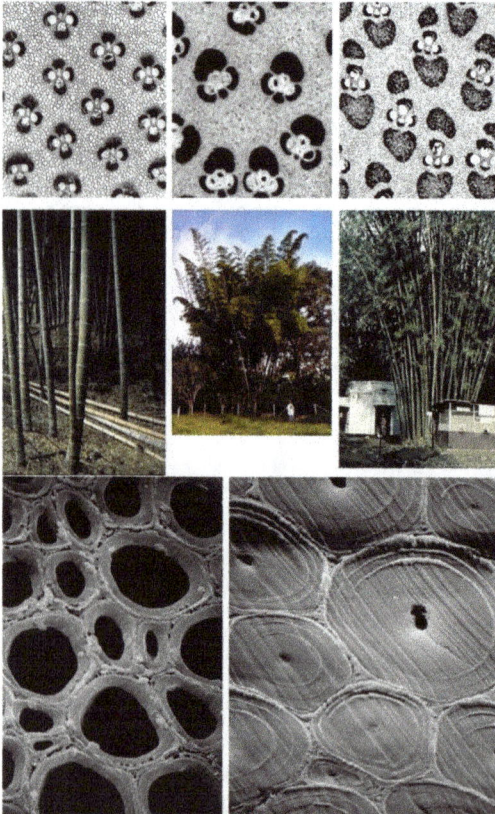

Fig. 5 Growth types of bamboo have different types of vascular bundles.

Fig. 6 Cell wall thickening from one year to six years,

Phyllostachys Viridiglaucescens.

By-Products of Palm Trees and Their Applications   Materials Research Forum LLC
Materials Research Proceedings **11** (2019) 62-68   doi: https://doi.org/10.21741/9781644900178-2

The state of knowledge on "the anatomy of BAMBOO CULMS" 1998 was documented in the INBAR Technical Report No. 18, [2].

Bamboo, rattan and palms, all belonging to the monocotyledons, show many similarities in structural aspects defined by vascular bundles and parenchyma. This results in typical property variation of the tissue according to the vascular bundle density and number as well as the age of parenchyma. Looking at the literature on bamboo, rattan and palms one can learn from each other. My college Johannes Welling will have a presentation on the comparison of bamboo and palms.

Preservation

Bamboo culms are easily attacked by insects, especially termites, moulds, blue-stain fungi, white-, brown-and soft fungi, as well as bacteria under suitable conditions. A number of laboratory experiments dealt with the basic factors of degradation of bamboo and the effects of physiological and chemical conditions for protection. In field tests at various locations the natural durability in soil contact was tested as well.

The good results with the sap-replacement treatment in India became widely known and led to a number of consultancies. Thus, a wider spectrum of applied treatment methods could be critically reviewed. These are the non-chemical methods, like storage conditions, clump curing, water storage, boiling, lime washing, traditional smoking and heat treatment. Great attention was paid to the various chemical treatment methods, like brushing, spraying, dipping and especially the ones for longer sustainability, as sap-displacement and the pressure methods. While being on site, a number of methods could be applied. Of special significance was the further improvement of the sap replacement method by inventing a special cap on the culm end for removing the air before the preservative is pushed in (Fig. 7).

Fig. 7 The sap-replacement method requires an air suction/pressure cap at the culm end.

In cooperation with the Environmental Foundation in Bali, the Vertical Soak Diffusion System (VSD) was developed. For its application all diaphragms of a fresh culm are punched through, except the lowest one. The culms are then placed vertically in a basin and filled with the preservative up to the top. The preservative diffuses into the wall tissue for a given period of time after which the lowest internode is punched through as well, so that the preservative can flow out and can be used for the next treatment after adjusting the concentration (Fig 8).

The "Bamboo Preservation Compendium" from 2003 presents a comprehensive overview (Fig. 9). 1992 a general documentation on "Wood Protection in Tropical countries was provided as background knowledge [3].

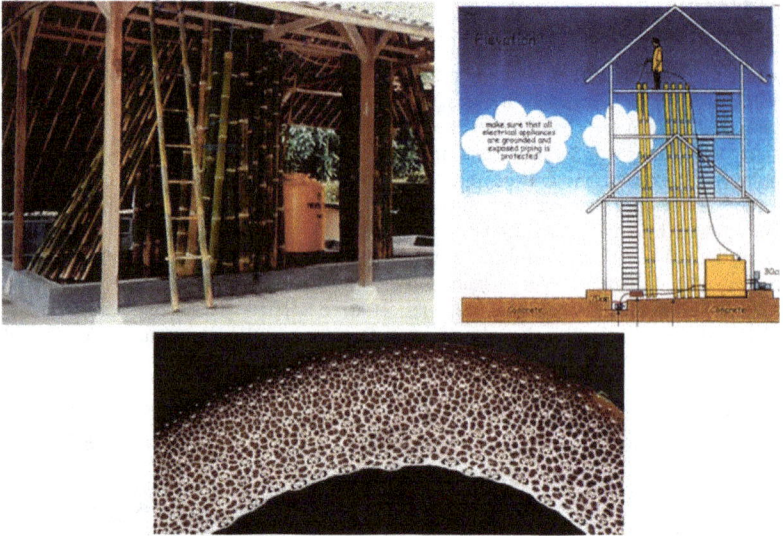

Fig. 8 The Vertical Soak System, Environmental Bamboo Foundation, Bali.

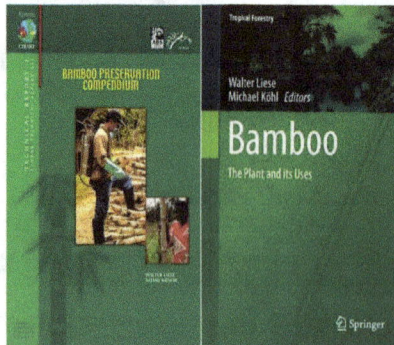

Fig. 9 Bamboo Preservation Compendium, 2003.   Fig. 10 Bamboo. The Plant and its Uses, 2015.

By-Products of Palm Trees and Their Applications   Materials Research Forum LLC
Materials Research Proceedings **11** (2019) 62-68   doi: https://doi.org/10.21741/9781644900178-2

Our activities on preservation methods became widely known, resulting in a number of foreign guests to our institute in Hamburg, travelling from countries like India, Indonesia, Thailand, Iran, Ghana, Nigeria, Tanzania, South Africa, USA, Canada Costa Rica, Cuba, Mexico, Chile, Colombia, Brasilia and Chile. They all came to learn and cooperate, taking home with them knowledge, international contacts and personal memories.

International co-operation
The first mission in India was followed by many consultancies in about 25 countries for national and international organizations, as GTZ, FAO, INBAR, ITTO, EU, SES, Kolping and others. Since a number of projects were arranged by the GTZ, an overview report no.180 on "Bamboo-biology, silvics, properties, utilization" was published 1985 [1]. My last projects were conducted 2009 in Thailand, 2010 in Korea and 2012 for a key lecture at the XVIIth World Bamboo Congress, Antwerp, Belgium.

The international activities were strongly supported by intensive laboratory research in collaboration with colleagues, thesis students and guests from various countries. About 110 bamboo-related scientific papers and six books were published as author or co-author. The last book in 2015 "Bamboo. The Plant and its Uses" contains chapters on structures, properties and uses of bamboo (Fig.10) [4].

Since I was engaged in various international organizations, two examples must suffice. I was instrumental in getting the International Development Research Centre (IDRC) of Canada interested in bamboo for the creation of the International Network for Bamboo and Rattan, INBAR, consisting now of 43 Member States countries. I am sometimes referred to as the "Grandfather of INBAR". A similar welcome as "The Father of Bamboo" was offered to me at a consultancy in 1999 at the Kolping Society of the Philippines (Fig.11). During my presidency of the International Union of Forest Research Organizations IUFRO 1978-1981 the activities for bamboo were much strengthened. A special highlight was the XVII IUFRO World Congress in Kyoto, Japan 1981, where at a memorial ceremony *Phyllostachys pubescens var.aureosulcata* was planted by the Prince and Princess of Japan (Fig.12).

Fig. 11 Visit of a bamboo project led by the Kolping Society, Phillipines, 1999.
Fig. 12 Royal Prince and Princess planting bamboo at the IUFRO Congress, Kyoto, 1981.

At the age of 92 my passion for bamboo continues with enjoying the international contacts and discussions (Fig.13).

Fig. 13 Walter Liese in his bamboo garden, 2018.

References

[1] Liese W., Bamboos-Biology, silvics, properties, utilization, GTZ Schriftenreihe, Eschborn, 180 (1985).

[2] Liese W., The anatomy of BAMBOO CULMS, Technical report 18, INBAR, Beijing, 1998.

[3] Willeitner H., Liese W., Wood Protection in Tropical Countries, GTZ Schriftenreihe Eschborn, 227 (1992).

[4] Liese W. and Satish Kumar, Bamboo Preservation Compendium, New Delhi, 2003.

[5] Liese W., Köhl M. (eds), Bamboo - The Plant and its Uses, Springer, Heidelberg, 2015, Chapters 8,9,10: pp. 227-364

By-Products of Palm Trees and Their Applications
Materials Research Proceedings 11 (2019) 69-80

Materials Research Forum LLC
doi: https://doi.org/10.21741/9781644900178-3

The Use of Oil Palm Trunks for Wood Products

Fruehwald Arno[1,a*], Fruehwald-Koenig Katja[2,b]

[1]PalmwoodNet Detmold, Germany and Palmwood R+D, Engelbergerstr. 19, 79106 Freiburg, Germany

[2]University of Applied Sciences Ostwestfalen-Lippe, Department 7: Production and Management, Liebigstr. 87, 32657 Lemgo, Germany

[a]arno.fruehwald@gmx.de, fruehwald@palmwood.de, [b]katja.fruehwald@hs-owl.de

Keywords: oil palm trunks, palm wood, processing, palm products

Abstract. Worldwide, oil palms cover an area of nearly 25 million ha with over 75 % located in Asia. After 25 years of age, the palms are felled and replaced due to declining oil production. The average annual total volume of trunks from plantation clearings amounts to more than 100 million m³. Like all other biomass, the trunks remain on the plantation site for nutrient recycling. But this leads to increased insect and fungi populations causing problems for the new palm generation. Many regions where oil palms grow currently suffer from a decline in timber harvested from their tropical forests. An extensive project, involving partners from both R+D and industry, is studying the possibility of improving the use of oil palm trunks to manufacture marketable timber products. The consortium consists of some 20 partners mainly from Germany, Malaysia, and Thailand. Areas of development are: harvesting and storage of trunks, sawmilling, drying, processing into various products like solid wood-based panels (block-board), flash doors, furniture elements as well as CLT and gluelam for the building sector. All sectors have shown remarkable success.

Introduction

The availability of timber from tropical forests is steadily declining due to over logging and measures taken towards sustainable forest management and conservation of tropical forests. In Asia the demand for wooden products is rising due to a growing population and greater economic development. The declining wood supply from tropical forests in Southeast Asia is partly being compensated for by imported timber (i.e. from North and Latin Americas, Australia, New Zealand, Europe), and new fiber sources are also being developed. Rubberwood from Indonesia, Malaysia, and Thailand has found its way into the markets and the use of bamboo is rapidly increasing. Rubberwood, however, is limited in quantity, because rubber plantations are being converted into oil palm plantations due to improved economy. Fast growing forest trees like albizzia (*Albizia falcataria* (L.) Fosberg) are being promoted but can hardly fill the increasing supply shortage.

Palms have long been a source of fiber for manufacturing products, but mainly fibers from husks (i.e. coconut fibers) or, to a lesser extent, from palm fronds or fruit bunches. Also nut shells are often used as fillers of (activated) coal. The trunks of coconut palms are widely used as building material, for furniture and crafts. A good example is in the Philippines were coco-wood has an important market share. But utilization is performed more locally in small workshops with partly inferior processing techniques and tools, resulting in low quality and more simple products. Processing is difficult as density of coconut trunks is high and hard vascular bundles, ash and silica causes high tool wear. The trunks from date palms generally have lower and evenly distributed density making processing easier. Nevertheless, date palms are less available, because as their growing area is much smaller (Table 1 and Table 2) and the average age of a palm is high, resulting in less felled palms.

Oil palms (*Elais guineensis* JACQ.) were introduced in Asia around 100 years ago. With initially limited distribution, the plantation areas have grown steadily since around the 1970s, first in Malaysia, later in Thailand and Indonesia. Table 1 shows the estimates for the plantation area, which is worldwide above 25 million ha with a growing tendency, especially in Indonesia and some Latin-American countries. Experts estimate global coverage will range between 30 to 40 million ha in the year 2030.

Table 1: Palms with potential for industrial conversion of trunks into products.

palms	world area [million ha]	number of palms [million]	rotation period [years]	number of available palms [million]	available million m³ [palm trunks per year]
oil palm	25	3,000	25	120	180
coconut palm	12	1,200	50	24	40
date palm	0.8	110	55	2	3

Table 2: Main growing countries for palms and areas in million ha (various sources).

oil palm		coconut palm		date palm	
Indonesia	13.0	Indonesia	4.0	Iran	0.22
Malaysia	5.0	Philippines	3.5	Iraq	0.21
Nigeria	3.5	India	2.0	UAE	0.16
Thailand	1.0	Brasil	0.5	S. Arabia	0.04
World	>25.0	world	~12.0	world	~1.0

Past attempts to use oil palm trunks (OPT) as a supplement or substitute for tropical timber in product manufacturing failed due to the palms' different material properties and processing behavior compared to traditional wood species. Intensive R+D, especially in Malaysia [i.e. 1, 2, 3, 4], has provided a clearer understanding of the material (structure, mechanical, and chemical properties) and worked to test manufacture of products. With the exception of plywood manufactured in Malaysia (some 50.000 m³/y), all efforts towards semi industrial or industrial use proved unsuccessful. Product quality, processing of the material (i.e. sawing, planing, drying) and logistics / supply did not meet high enough standards to make to a manufacturing break through.

Yet, given the tremendous supply of OPT (180 million m³/y, see Table 1) and the rapid decline in common timber stocks, the need to launch a "new start in OPT utilization" is obvious. After several years of scientific oriented material research in various German and Asian universities and research centers, a consortium was founded consisting of five industrial core partners and some 20 associated partners from academia and industry in Germany resp. Europe, Malaysia and Thailand. Information about the consortium can be found on the project website, www.palmwoodnet.com.

By-Products of Palm Trees and Their Applications   Materials Research Forum LLC
Materials Research Proceedings 11 (2019) 69-80   doi: https://doi.org/10.21741/9781644900178-3

OPT Harvesting: Potentials and Logistics

After 25 years of age, oil palms are felled and replaced due to declining oil production (remarkably less than 5 t/ha palm oil). Plantations are cleared on plots ranging from only a few hectares to up to 100 ha (or more) depending on ownership, age distribution, and site conditions. Generally, clearing starts at the beginning of the dry season, the sites are prepared for replanting at the beginning of the following wet season. Traditionally, most of the biomass from the clearing was piled up in rows and burned (with the help of sprayed diesel) at the end of the dry season. The main aim was to avoid pests caused by fungi (Ganoderma) and beetles. Today, most countries have introduced a zero-burning-policy so now the trunks are chipped and evenly distributed (together with fronds and leaves) on the site ore piled up in rows of 10 or 20 m distance (in-between the rows for re-planting) to let the material rot.

For the OPT volume to be harvested for use, PalmwoodNet developed a concept for removing some 70 m³ of OPT per ha (from 150 – 180 m³/ha in total) for reasons of nutrient management (among others K, P, Ca, N, Mg) and soil quality. Normally, the palms are felled by "push-felling" where an excavator pushes the palms to the ground and chips it into pieces between 20 – 40 cm in length. The felling technique will be modified to secure less or no damage to the trunks. The discrepancy between felling within two dry periods per year of 2 – 3 months only and a continuous supply of processing mills must be bridged by either extension of the plantation clearing periods and / or appropriate storage techniques for the trunks. Intensive laboratory research followed by field tests has led to storage techniques and conservation of the trunk cross cut sections with "green chemicals" in order to avoid rapid and intensive manifestation of mold followed by stain of the wood. The results have shown little damage to the wood even after storage of 2 – 3 months; the techniques are also very cost effective.

Properties of Oil Palm Wood

A large number of publications describe the basic properties of oil palm wood, but quite often as secondary literature. Experimental research is not always systematic in terms of material selection and methods applied. In the following, a general overview of the properties is given, some references are made. The partners of PalmwoodNet have dealt with material properties relevant for processing and use of the palm wood.

Density variation: As a monocot, the density varies remarkably along the trunk diameter and along the trunk length. The outer peripheral zones (at the trunk base) show dry densities of 0.5 – 0.7 g/cm³ caused by high density fibers / fiber cups of the vascular bundles (VB) and high share of the VB of the wood volume. The more inner / central part of the trunk show densities between 0.2 – 0.3 g/cm³ (less VB). Fig. 1 and 2 show typical density distribution. Along the trunk axis, the density decreases to 0.4 – 0.5 g/cm³ at the periphery and 0.15 – 0.25 g/cm³ in the inner zone due to younger age of the cells. The cell walls show "secondary growth" of their thickness by additional cell wall layers with the age. This might be one of the reasons for the higher density of coconut wood compared to oil palm wood as coconut palms are only harvested at the age of 50+.

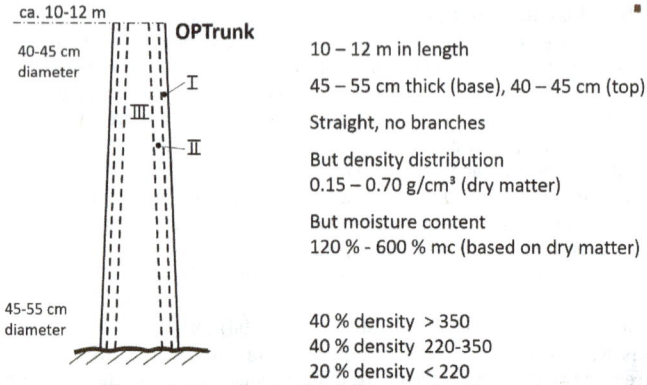

Figure 1: Density distribution in an oil palm trunk.

Figure 2: Cross section of an oil palm trunk.

Moisture content: For reasons of physiology (to bridge water shortage during dry seasons) the tissue of OPT contains a high percentage of water – generally the parenchyma cells show almost maximum moisture content (which depends on the density). Fig. 3 shows moisture contents (based on dry density) of between > 100 % (peripheral zone, base of trunk) and 600 % (inner zone, top of the trunk). The high moisture content results in the high weight of the trunks (logistics), risks of mechanical damages and difficult, long and expensive drying. OPT shows an average moisture content (whole trunk) of 200 – 250 % whereas coconut trunks show only 50 % resp. 350 % moisture content (densities 0.75 resp. 0.35 g/cm³) [5].

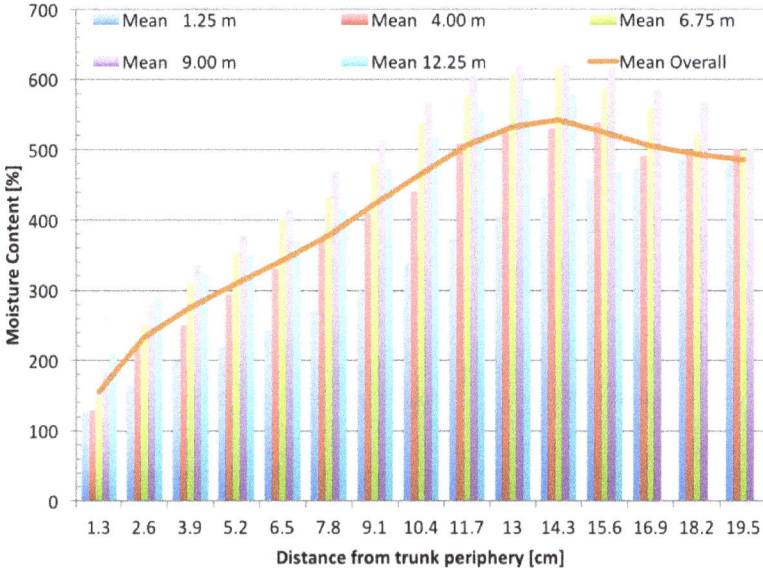

Figure 3: Moisture content distribution [13].

Ash and silica content: All palms show higher ash and silica contents compared to most common timber species (Table 3). This results in increased tool wear during processing. Oil palm wood contains ash up to 2.0 % of dry mass (especially potassium content is high) and silica of up to 0.7 %. For silica these values are higher compared to coconut palms and date palms [6]. All palms (as well as bamboo) require special tools and tool materials for high quality processing (i.e. sawing across fiber direction) and sufficient tool life. PalmwoodNet places strong emphasis on tool development (together with adopted processing and process parameters) and has achieved remarkable results [7, 8].

Table 3: Mineral contents and pH-values (average values) of date palm and oil palm wood [6].

	oil palm	date palm	pine	oak
pH	4.2	5.1	3.4	3.4
ash [%][1)]	1.8	2.2	0.3	0.5
silica [%][1)]	0.56	0.18	0.04	0.01
density [g/cm³][2)]	0.43	0.64	0.52	0.70
[1)] Based on dry density [2)] dry densities				

Hygroscopicity: As with all timber species, oil palm wood shows hygroscopic behavior which means that the moisture content is in equilibrium with the moisture conditions of the surrounding air. Bound water level depends on the composition of the main chemicals. As [1] and [2] have shown, the level of cellulose, hemicellulose and lignin is in the same order compared to many common timber species. For moisture content between 15 % and 30 % the cell wall structure is important as "liquid water" is already captured in small caverns. Generally, oil palm wood should have similar sorption isotherms as common timber species. One "big problem" is the content of sugars in the wood, like glucose and sucrose, which absorb additional water from the air. This results in increased equilibrium moisture content at higher relative humidity levels. PalmwoodNet partners have studied this (well known) phenomenon and have developed measures to cope with this problem.

Swelling/shrinkage: In the past little systematic research was done on the shrinkage of oil palm wood. What is described in literature [i.e. 9] are values for maximum shrinkage from green to bone dry. The values indicate that under these test conditions cell collapse can occur which must be excluded from "normal shrinkage". [10] has given values for volume shrinkage between 10 % and 23 % for the various wood densities. He also mentioned the problem of cell collapse. For the use of oil palm wood-based products the shrinkage/swelling factors related to the variation of humidity should be known. PalmwoodNet partners have set up a comprehensive test program to study this aspect. Generally, oil palm wood is isotropic in radial and tangential directions of the trunk. This means that shrinkage and swelling is similar in both directions.

Mechanical properties: The mechanical properties elasticity and strength are very much influenced by the structure of oil palm wood. Along the trunk axis, the vascular bundles (having high-density fiber caps) "dominate" the properties, especially tension. Across trunk axis, the low-density parenchyma is responsible for much lower property values compared to along the trunk axis. Mechanical properties play a significant role in the design and use of products as well as for processing of the wood. In a joint effort, PalmwoodNet partners have analyzed "areas of necessary knowledge for product design for use and material processing". Related to the intended products (see below) and the developed processes for product manufacture (sawing, planning, moulding, sanding, glueing, drying etc.) the necessary set of properties were studied. With this, products and processes can be modeled and designed. To give a rough idea about mechanical properties, the information in Table 4 was compiled from literature.

Table 4: Mechanical properties of oil palm wood and date palm wood in comparison with other wood species [11]

	oil palm	coconut palm	date palm	spruce	beech	poplar	rubber-wood
density [g/cm³]	0.22-0.55	0.25-0.85	0.41	0.30-0.64	0.49-0.88	0.36-0.56	0.53
MOE [MPa]	800-8,000	5,300	1,719-2,745	11,000	16,000	8,300	8,800
MOR [MPa]	8-45	36	11-23	66	105	76	58
compression strength [MPa]	5-25	24	6-10	43	53	36	26
hardness [N]	350-2,450	4,230	2,000	2,140	5,650	2,500	4,320

By-Products of Palm Trees and Their Applications   Materials Research Forum LLC
Materials Research Proceedings **11** (2019) 69-80   doi: https://doi.org/10.21741/9781644900178-3

Products made from Oil Palm Wood

Since about 1980, several attempts have been made to use oil palm wood for manufacturing products. Especially in Malaysia with its large plantation areas and reduced logging in natural forests (particularly in Peninsula Malaysia) OPW became an obvious source for bridging the demand and supply gap for wood and wood products. It was inspired by the market success achieved with rubberwood-based products. The main players were the Forest Research Institute Malaysia (FRIM), University Putra Malaysia (UPM) and Malaysia Palm Oil Board (MPOB) and the former Malaysian Oil Palm Research Centre. The work was twofold: on one side the research focused on material properties, on the other side the development of processes and products.

Various conferences in Malaysia and Southeast Asia aimed to distribute the findings. Examples of state-of-the-art reports are [1, 2, 3, 4]. Most effort was directed towards products like furniture, plywood, chipboard, and fiberboard. When evaluating this work, it becomes clear that it was extremely difficult to achieve progress similar to the rubberwood success. The material properties of oil palm wood differ so greatly from common timber species (including rubberwood), and the methods of processing material also varies significantly due to structure, density variation, drying, inhomogeneity of mechanical properties etc. The product quality therefore was much below market expectations; quality in the sense of appearance, durability, design and working quality. Various attempts have been made in Malaysia to support the plywood and composite panels industry. Reports like [12] and [4] describe achievements, but also many hurdles towards broad commercialization. The main drawback certainly was processing of the oil palm wood. Because of the lack of appropriate processes and tools and the focus on "traditional products" most of the attempts failed or had very little success (plywood with the use of oil palm veneer as core layers).

After 2005, several research institutions, universities and engineering companies in Germany with knowledge in coconut R+D started looking deeper into material properties, possible product design and identified areas for process and tool development. In 2013 discussions started aimed at establishing a network with members of R+D institutions and engineering companies (woodworking machines) for comprehensive development work. In 2015 PalmwoodNet was officially founded and started development activities. The products in focus are panels based on solid wood and veneer design, multilayer solid wood composites, door panels, acoustic panels, building elements like glued laminated timber and solid wood multi-layer panels. Fig. 4 – 8 show some examples of the developed products for which processes are already being developed as well.

Figure 4: Blockboard, core from low density material (overall density 0.22-0.30 g/cm³).

Figure 5: Door panel.

Figure 6: Sandwich Panel (middle layer vertical).

Figure 7: 3-Layer-Panel from mixed density.

Figure 8: left: 3-Layer Solid Wood Panel (CLT) with core from low and medium density material; right: Glued-Laminated Timber (Gluelam) from high density material.

PalmwoodNet consists of five core partners: Moehringer Sawmill Technology, Leitz Woodworking Tools Technology, Minda Engineering (Panels and Beams), Jowat Adhesives and Palmwood R+D Engineering. More than 20 associated partners from academia in Germany, Malaysia, and Thailand as well as industry partners from Germany, Italy, and Malaysia have contributed to development in all relevant sectors. Fig. 9 shows a rough outline of the areas of work.

availability of resource	basic properties bio, chem, phys, tech	primary, secondary processing	environmental, energy aspects
harvesting, locistics, storage	use related properties - customer, standards - building regulations	product design + engineering	market issues, certification

Figure 9: Areas of concern for R+D in palm wood utilization.

Fig. 10 shows an overview on the key processes in oil palm wood utilization.

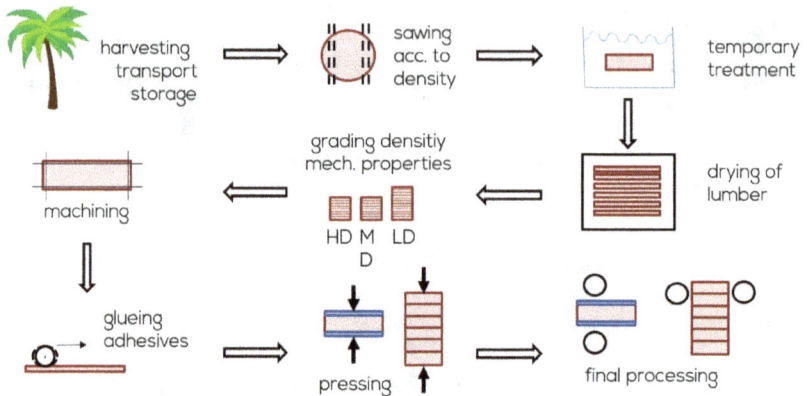

Figure 10: Key processes in palm wood utilization

The working principle is "easy": identify development needs, move the relevant experts towards solutions, test the new technologies, combine and optimize single processes and product design to complete manufacturing strategies.

Concepts for Products and their Manufacturing

Based on market studies for products, on technical product design, process development, supply with raw material and cost/revenue analysis several manufacturing strategies are being developed. These strategies include various product families, capacities, mix of products (or product families resp.) for improved raw material and capacity utilization, local aspects for supply and production sites as well as cost/benefit evaluation. The concepts developed are ready for implementation, adoption for local conditions and individual request is possible through computer-aided planning tools. The network will continue to develop industrial solutions on specific request which can include market and product specification, process design, integrated manufacturing together with existing wood processing mills etc.

Socio-economic aspects: Depending on the products in focus, capacity of production and standard of mechanization and automatization (PalmwoodNet favors the latest state of the art of technology) for each 200 – 400 m³ of OPT utilization one work-place for one year (1 WPy equals 200 – 400 m³ OPT) can be calculated. Including supply with other necessary material (glue etc.) and local supplies as well as sales chain to the final customer an equivalent of 1 WPy equals 100 – 150 m³ OPT is realistic. For a potential use of OPT in Malaysia (total availability some 200,000 ha/y replanting) of 10 – 15 million m³ OPT per year this results in some 100,000 jobs. A certain number of these jobs will replace jobs in the timber industry using common species from tropical forests. 10 million m³ OPT may result in 3 – 4 million m³ final products (material equivalents) having a production value of around 1.5 – 2 billion Euro (including residue valorization).

Ecological aspects: There are several aspects with ecological relevance:
a) OPT provides a substitute for timber from tropical forests or wood from plantations stocking on tropical forestland.
b) Using OPT for products (and partly for energy) reduces the high volume of trunk biomass being chipped on the replanting site. High biomass volume can cause a risk for fungi and beetle spread on the re-planting sites.
c) The use of wood-based products is known to have CO_2-positive effects. Compared to the use of alternative materials, the use of wood results in less CO_2-emissions – some 1 t CO_2 per 1 m³ wood used. A rough calculation of the effects is given for Malaysia in Table 5.
d) OPT use contributes to the development of an increased sector bio-economy.
e)

References

[1] S. Khozirah, K.C. Khoo, A.R.M. Ali, Oil Palm Stem Utilization - Review of a research, Forest Research Institute Malaysia (FRIM) (1991).

[2] I. Wan Asma, K. Wan Rashidah, J. Rafidah, J. Khairul Azmi, As Good As Wood, Forest Research Institute Malaysia (FRIM), Kepong (2012).

[3] S.P. Chandak, I. Wan Asma, converting Waste Oil Palm Trees into a Resourve, FRIM-UNEP Collaborative Project Report, Osaka, Japan (2012).

[4] MTIB, Palm Timber: A New Source of Material with Commercial Value (2015).

[5] A. Frühwald, R.D. Peek, M. Schulte, Utilization of Coconut Timber, Federal Research Centre for Forestry and Forest Products, Institute of Wood Physics and Mechanical Technology & Institute of Wood Biology and Wood Protection, Hamburg (1992).

[6] T. Tufashi, Physical-Mechanical and Chemical Properties of the Wood of Oil palm and Date palm Trees, Bachelor Thesis at the University of Hamburg, Zentrum Holzwirtschaft (2013).

[7] S.P. Sukumaran, A. Kisselbach, F.Scholz, C. Rehm, Investigation of machinability of kiln dried oil palm wood with regard to production of solid wood products, in: M. Zbiec, K. Orlowski, Proceedings of the International Wood Machining Seminar, Warsaw, Poland, 2017, pp. 131-142.

[8] S.P. Sukumaran, A. Kisselbach, F. Scholz, C. Rehm, J. Graef, Investigation of machinability of kiln dried oil palm wood with regard to production of solid wood products, Holztechnologie, 59-3 (2018) 19-31.

[9] E. Bakar, D. Hermawan, S. Karlina, O. Rachman, N. Rosdiana, Utilization of oil-palm trees as building and furniture materials (I): Physical and chemical properties, and durability of oil-palm wood, Journal Teknologi Hasil Hutan (Indonesia) 11-1 (1998) 1-12.

[10] S.H. Erwinsya, Improvement of Oil Palm Wood Properties Using Bioresin. PhD-Thesis at the Technical University Dresden, Fakultät für Forst-, Geo- und Hydrowissenschaften, Institut für Forstnutzung und Forsttechnik, Dresden (2008).

[11] W. Killmann, S.C. Lim, Anatomy and Properties of Oil Palm Stem, in: Proceedings National Symposium of Oil Palm By-Products, Kuala Lumpur, Malaysia (1985).

[12] L.Y. Feng, A. Mokhtar, P.Md. Tahir, C.K. Kee, H.W. Samsi, Y.B. Hoong, Handbook of Oil Palm Trunk Plywood Manufacturing, Malaysian Timber Industry Board, Kuala Lumpur, Malaysia (2014).

[13] N. Kölli, Density and Moisture Distribution in Oil Palm Trunks from Peninsular Malaysia, BSc-Thesis at Hamburg University, Zentrum Holzwirtschaft (2016).

By-Products of Palm Trees and Their Applications
Materials Research Proceedings **11** (2019)

Materials Research Forum LLC
doi: https://doi.org/10.21741/9781644900178

Wood Alternatives and Panels

By-Products of Palm Trees and Their Applications
Materials Research Proceedings **11** (2019) 83-87

Materials Research Forum LLC
doi: https://doi.org/10.21741/9781644900178-4

Wood, Bamboo and Palm Wood - Similarities and Differences in Research and Technology Development

Johannes Welling[1,a*], Walter Liese[2,b]

[1]Thünen-Institut of Wood Research, Leuschnerstr. 91c, 21031, Hamburg, Germany
[2]Institute of Wood Science, University Hamburg, Leuschnerstr. 91d, 21031 Hamburg, Germany

[a]johannes.welling@thuenen.de, [b]wliese@aol.de

Keywords: palm wood, wood, bamboo, technology development, fundamental knowledge, renewable resources

Abstract. Wood science has a history of several hundred years, bamboo research started in the of the last century and palm wood research is even younger. Consequently, there are differences not only in depth and width of knowledge, but also in the state of the art of conversion technologies and utilization options. There are considerable wood resources all over the world, but bamboo and palm resources are restricted to certain regions. Similarities and differences in research and technology development related to the three raw materials will be examined and expected future developments will be discussed. Technological progress needs time for a) development based on fundamental knowledge and practical experience, b) diffusion of knowledge into industry, and last but not least c) consumer acceptance and commercial breakthrough. Policy interaction may accelerate development and diffusion of knowledge, however in some cases may also impede or hinder the utilization of a specific raw material resource. While wood science and wood technology have reached a mature stage, research on bamboo and bamboo utilization is progressing rapidly; however, research on palm wood and, especially, the processing of palm wood and the utilization of palm products is still at an early stage. Existing knowledge and expertise around wood/bamboo science and technology should be used for speeding-up the development and realization of palm wood utilization options.

Introduction
In the evolution process Mother Nature designed dicotyledon plants with a lignified cellulose matrix some hundred million years ago. This was the time when the success story of trees started. At that time, the monocotyledon ferns, grasses and palms did already exist for quite a long period for time. Mankind occurred only some millions of years ago. Considering the age of our planet, the evolution of science happened during the very recent few seconds of our planets history.

Wood is one of the oldest raw and building materials used by mankind. But most probably, early men have used also bamboo and palm wood wherever this was available. Looking back in the history of science, it becomes obvious that the early researchers concentrated on investigating the structure, function and behavior of wood, and not that of bamboo or palms. The success of wood as the basic material for construction was initially based on experience and tradition gained in regions where natural materials with high natural durability existed, which could stay in service for long periods of time. Only later, wood science explained why and how this could occur. Nowadays, wood science has reached a mature stage, but bamboo and palm wood science are still at an early stage.

Key developments for progress in wood science and bamboo / palm wood research

Wood has been used by mankind since thousands of years and for many different purposes. Selection of wood species for specific fields of application were mainly based on experience. The development of the microscope in the 16th/17th century by the Dutch lens maker Zacharias Janssen firstly allowed an insight view into the microstructure of bio-based materials [1]. This invention led to a better understanding of why certain wood species have better properties and how it comes that some species are more suitable for certain applications than other species. A variety of test methods had to be developed in order to describe and compare the properties and behavior of wooden materials in an objective manner. As a result of the abundance of wood all over the world, highly sophisticated wood processing techniques were developed and introduced worldwide. Due to its regional distribution, bamboo and palm wood utilization naturally occurred only in certain regions.

Sustainable forestry was firstly introduced by Carlowitz in 1713 [2]. This led to man-made managed forests not only in Germany but in many other countries worldwide. At that time and for long after, bamboo was considered a natural resource, which did not need management by men. Palms were cultivated mainly because of its fruits. Bamboo was used as building material and its sprouts as foodstuff. The palm fruits (e.g. dates, coconut and many others), in addition to its regional importance, were important articles of trade because they could be transported over wide distances without deterioration.

Wood research has been focusing from the very beginning on the woody tissue. Since several hundred years wood scientists have been deepening and widening the knowledge on wood, its formation, properties, processing, and its application in form of a large variety of wood products.

Research on bamboo started in the 20th century in India and in China. In the Western world Walter Liese in the early 50ies of the last century tested the suitability of bamboo species for substituting wooden poles in German coal mines to overcome the serious raw material shortage after World War II. At that time some science-based knowledge on bamboo did already exist, but it was widely unknown to the Western research community. Research on palm wood only started some decades ago when it became obvious that millions of palm trees in large coco, date and oil palm plantations had to be replaced. Solutions for elimination of the palm trunks (worst case) or for its transformation into usable products (best case) had to be found. This shows that the drivers for development of research on wood, bamboo and palm wood are quite different.

Comparison of the utilization pathways of trees, bamboo and palms

The tables 1, 2, and 3 provide an insight view into the various utilization pathways of trees, bamboo and palms. Some of these pathways had considerable relevance in the past, but nowadays hardly do exist anymore. Other utilization pathways had disastrous consequences. Huge amounts of wood were burnt for potash production, a chemical needed for the glass manufacturing, and for energy needed by the glass and porcelain manufacturing industries. In Germany this led to an almost complete depletion of the wood resources of the Black Forest and the Bavarian Forest. The wood demand for ship building activities led to deforestation and karstformation in many Mediterranean countries. Oil palm plantations in the tropics are nowadays one of the major reasons for destruction of rain forests.

However, some pathways show a great potential for future development. For wood a growing demand is foreseen for the construction of prefabricated houses in Central Europe. Many experts also see wood as an important feedstock for future bio-refinery. Pulp and paper industry are still relying on wood as its main resource. In this sector, bamboo will play an important role in the future. The rising demand for wood and bamboo as a resource for textile fibers will lead to the substitution of cotton fibers. A possible consequence will be a reduce of the area of agricultural land needed for cotton production, which then can be used for other crops.

Palm plantations normally are established for the utilization of the palm fruits (dates, coconut, oil palm nut). Millions of hectares of coconut plantations exist since many decades. The area for oil palm production is increasing rapidly. But only recently the older coco nut and oil palm plantations have reached an age where replacement of the palms has become necessary. Currently the trunks must be burnt or disposed to avoid pests that might also infest the palms of the plantation. Here an integrated utilization concept for the palm trunks is urgently needed. This concept should focus on the unique properties of the palm trunks. The Palmwood R+D Net is working in this field.

Palm fronds are used locally, the fibers originating from oil palm fruit bunches and the coco mesocarp fibers have found many applications (ropes, floor mats and carpets, filling material). Due to its high starch content certain palm species are used for alimentation (sago starch).

Table 1 Tree utilization pathways

	Currently established	Options with high potential
Fruits, seeds	Alimentation; chemicals; handicraft, ornaments	
Wood	Energy; ashes (mainly potassium carbonate); building material; underground engineering material; pile foundation; ship building; interior design; source for fibers; barrels; handicraft; weapons, resource for engineered products; resource for textile fibers	Energy; building material; interior design; resource for engineered products; resource for biorefinery; resource for textile fibers
Bark	Energy; chemicals; medicine; cork stopper; mulching; soil improvement	Medicine
Sap, resin	Alimentation; chemicals, medicine	
Leaves	Humus formation	

Table 2 Bamboo utilization pathways

	Currently established	Options with high potential
Fruits, sprouts	Alimentation; chemicals; handicraft; ornaments	
Culm	Building material; interior design; handicraft; resource for textile fibers, charcoal, medical application	Resource for fibers; resource for engineered products; resource for biorefinery; resource for textile fibers
Sap	Beverage	
Leaves	Shelter; animal fodder	

By-Products of Palm Trees and Their Applications   Materials Research Forum LLC
Materials Research Proceedings 11 (2019) 83-87   doi: https://doi.org/10.21741/9781644900178-4

Table 3 Palm utilization pathways

	Currently established	Options with high potential
Fruits, seeds	Alimentation; pharmaceutical industry; handicraft; ornaments; energy	Pharmaceutical industry
Culm	Handicraft	Building material; interior design; resource for engineered products; resource for biorefinery
Sap	Beverage; starch (sago)	
Leaves, fronds	Shelter; material for braiding; animal fodder	

Reasons for different industrial developments in the wood, bamboo and palm sector
Wood from forests is available all over the year. It can be harvested in the forest which represents a standing stock for the woody resource. Once harvested, many wood species can be stored in form of round wood either in the forests, at the forest road or in the factory. In the past wood harvesting was seasonal in many regions, but nowadays wood for large scale industrial application is harvested all over the year. This kind of raw material availability was the reason for the development of huge factories and integrated industrial sites, where sawmilling, wood-based panel production, glulam production and energy generation can be combined. This allows a very effective use of all parts of a tree, which leads to high added value.

Annual agricultural crops, which are also considered a valuable fiber resource, normally are harvested during a very short period of the year. For industrial processing of agricultural fibers this means that the raw material has to be stored to ensure a continuous production all over the year. Up to now this has been a major obstacle for industrial utilization of e.g. cotton stalks, bagasse, oil palm fruit bunches and coco mesocarp. Not only the missing disposability, but also the often-found high sugar or starch, content hinder the storage bagasse, bamboo, and palm biomass. While growing in tropical and subtropical regions bamboo and palm biomass deteriorates easily due to fungal and insect attack. Only when solutions for storage and preservation of palm biomass are found, an industrial utilization of palm biomass will become feasible. Managing the felling of the palm stems in a plantation in such a way that allows a continuous supply of raw material all year long, would be an important factor for establishing a palm wood industry. Another important factor for industrial investors is the quantity of material available within a certain transport distance. In addition, a good infrastructure and roads usable during all seasons are needed to supply a palm wood-based industry with sufficient raw material.

In case of wood and bamboo the lignocellulose biomass is considered the main renewable raw material and all the other parts are by-products. This is different in cultivation of palms. Here the fruit is the main product and all other plant parts, including the stem and fronds, have to be considered as by-products. The main product always comes first while by-products often stay behind. This changes only when the profits made with the by-products start exceeding the profit made with the main-product.

An extensive source of literature is available in wood science. Handbooks summarizing the knowledge collected over decades do exist and are in use [3] [4] [5]. In bamboo research the publications of INBAR comprise a huge source of knowledge and information [6] [7], but in the palm sector only some examples [8] of such literature exist.

By-Products of Palm Trees and Their Applications   Materials Research Forum LLC
Materials Research Proceedings **11** (2019) 83-87   doi: https://doi.org/10.21741/9781644900178-4

How can we accelerate progress in palm utilization?

The rapid progress achieved in wood science has been and still is based to a great extent on knowledge exchange between scientists, which are organized in associations and networks. Very well-known are organizations such as IUFRO and InnovaWood (both having institutional members), and IAWS (personal members). In the bamboo world INBAR plays a key role. In the palm wood sector scientists, consultants, entrepreneurs, industry representatives, and developers have just recently started to organize themselves by establishing Palmwood R+D net.

Even though the three raw materials (wood, bamboo culm and palmwood) are different, palm researchers can use the knowledge and consider many of the ideas developed in the wood or bamboo sector. Especially when it comes to processing palm-based woody matter many techniques and processes are already available, which can be used after minor modification and adaptation for converting palm stems and palm fronds into valuable products.

Obstacles which have led to disappointment or financial flops in the past with wood, bamboo, sugarcane, cotton stalks or other agricultural lignocellulose crops should be considered and analyzed carefully in order to avoid repeating mistakes and unprofitable investments.

References

[1] Ilse Jahn: Janssen, Sacharias. In: Werner E. Gerabek u. a. (ed.): Enzyklopädie Medizingeschichte, De Gruyter, Berlin/ New York , ISBN 3-11-015714-4, 2005, pp. 688.

[2] Sylvicultura oeconomica, Leipzig, Braun 1713 mit dem Zitat "nachhaltende Nutzung" auf Seite 105 (Digitalisat der SLUB Dresden, Digitalisat der BSB München); Reprints: Freiberg, TU Bergakademie Freiberg und Akademische Buchhandlung (bearb. von Klaus Irmer und Angela Kießling), ISBN 3-86012-115-4, 2000.

[3] F. Kollmann, Côte, Principles of Wood Science and Technologie, Part 1 Solid Wood. Springer-Verlag Berlin Heidelberg New York, 1968. https://doi.org/10.1007/978-3-642-87928-9

[4] F.W. Jane, The structure of wood, 2nd edition, completely revised by Wilson K and White D J B. A. & C., London, Black Ltd, 1970, pp. 478.

[5] D. Fengel, G. Wegener, Wood-chemistry, ultrastructure, reactions, Walter de Gruyter, Berlin and New York, 1984, pp. 613. https://doi.org/10.1163/22941932-90000910

[6] Ziang Zehui eds. Bamboo and Rattan in the World, China, Forestry Publishing House, Beijing, ISBN 978-7-5038-5109-4, 2007, pp 360.

[7] W. Liese, Köhl M. (eds), Bamboo-The Plant and its Uses, Springer, Heidelberg, 2015, Chapters 8,9,10: pp 227-364

[8] P.B. Tomlinson, The structural biology of palms. Harvard University, Harvard Forest, Petersham, Massachusetts, USA, 1990, pp. 477.

By-Products of Palm Trees and Their Applications
Materials Research Proceedings 11 (2019) 88-98

Materials Research Forum LLC
doi: https://doi.org/10.21741/9781644900178-5

Innovative Bio-composite Sandwich Wall Panels made of Coconut Bidirectional External Veneers and Balsa Lightweight Core as Alternative for Eco-friendly and Structural Building Applications in High-risk Seismic Regions

González Oswaldo Mauricio[1,a*], Barrigas Hua Lun[2,b], Andino Nathaly[2,c],
García Andrés[2,d], Guachambala Marcelino[3,e]

[1]Graduate Studies Centre, Universidad de las Fuerzas Armadas ESPE, Av. Gral. Rumiñahui
s/n, Sangolquí 171-5-231B, Ecuador

[2]Construction and Earth Sciences Department, Universidad de las Fuerzas Armadas ESPE, Av.
Gral. Rumiñahui s/n, Sangolquí 171-5-231B, Ecuador

[3]Research and Development Department, 3A Composites Core Materials – Plantabal, Quevedo
Industrial Complex, Ecuador

[a]omgonzalez@espe.edu.ec, mauricio.gonzalezmosquera@griffithuni.edu.au,
[b]hlbarrigas@espe.edu.ec, [c]nsandino@espe.edu.ec, [d]magarcia19@espe.edu.ec,
[e]marcelino.guachambala@3acomposites.com

Keywords: engineered wood products, biomaterial mechanical characterization and mechanical efficiency, *Cocos nucifera L*, *Ochroma pyramidale*

Abstract. The research that constitutes this paper is based on a series of publications that aimed at understanding, from an engineering perspective, the optimised mechanical efficiency of senile coconut palm stem-tissues as foundation for non-traditional building applications. Particularly, this study aims at determining, evaluating and analysing the mechanical properties of lightweight bidirectional sandwich-like structure wall panels made of balsa core material and coconut external veneers. To achieve these objectives, 10 test specimens cut from prototype panel 1 (1200 mm high, 600 mm wide and 124 mm total thick) and 10 test specimens cut from prototype panel 2 (1200 mm high, 600 mm wide and 74 mm total thick) were investigated under mechanical and seismic behaviours in accordance to the current American Society for Testing and Materials (ASTM) building standards. Preliminary results show that the proposed wall panels are up to two and three times more efficient, in terms of mechanical high-performance, than equivalent sections of solid wall bricks and concrete block walls, respectively. Therefore, the innovative panels constitute a feasible alternative to reduce/replace typical construction materials (e.g. steel, concrete and bricks) with a significant positive environmental impact that fully address current engineering requirements. These bio-panels are meant to be used as important non-traditional elements during the rebuilding process of low-rise and mid-rise residential buildings that were dramatically affected during the 2016 Ecuador earthquake.

Introduction

Building collapse or damage is one of the major causes for earthquake injuries and fatalities. The catastrophic Ecuador earthquake in April, 2016, left approximately 35,300 affected dwellings, out of which about 19,500 resulted totally destroyed or demolished. Tragic result of it, around 670 people died and 6,300 individuals were injured [1, 2]. Despite some advantages (e.g. fire

By-Products of Palm Trees and Their Applications   Materials Research Forum LLC
Materials Research Proceedings **11** (2019) 88-98   doi: https://doi.org/10.21741/9781644900178-5

resistance and durability) offered by traditional building structures made of typical materials (e.g. steel, concrete, bricks) [3], their partial failure or total collapse during extreme seismic events can lead to critical consequences as hereinabove mentioned. It has been estimated that during the 2016 Ecuador earthquake, many casualties occurred, not only by the structural framing collapse effect, but greatly by the overbalance masonry effect as shown in Fig. 1. Moreover, typical manufactured structural materials all involve very substantial use of energy during their production process, which in turn involves high generation of CO_2 to the atmosphere. Indeed, building with steel or concrete is 20 and 9 times, respectively, more CO_2 emissions intensive (i.e. compared on mass basis) than structural timber [4, 5].

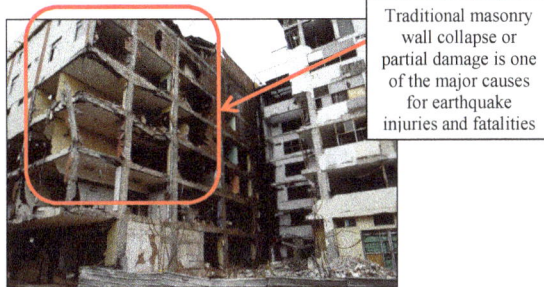

Traditional masonry wall collapse or partial damage is one of the major causes for earthquake injuries and fatalities

Fig. 1. Overbalanced brick masonry recorded during the 2016 Ecuador earthquake occurred on April 16, with a moment magnitude of 7.8 and a maximum VIII severe Mercalli intensity. Adapted from [6]

Unfortunately, part of the Ecuadorian area affected by the earthquake is currently being rebuilt using the same traditional building methods and materials. The curious aspect of the rebuilding process is that huge amounts of concrete and steel are daily transported to the construction project sites whereas massive plantations of biomaterials surrounding the zone (e.g. coconut palms and balsa trees) are totally disregarded. These observations were the driven force behind the work in this investigation, which aims at addressing the hereinabove stated problems by proposing innovative bio-composite structural wall panels as alternative for masonry construction that makes the most of both fundamentals: (1) the enhanced performance of engineering wood products, cross laminated timbers, specifically, and (2) the optimal mechanical efficiency [7-9], in terms of mechanical performance (i.e. high strength versus moderate stiffness) per unit mass; the optimal mechanical efficiency that is best represented in biomaterials by either a sandwich-like structure (e.g. coconut stem tissues) or a tubular-like structure (e.g. bamboo culms) [10].

Materials and equipment
Two wall panel types were built in this study: prototype panel 1 (1200 mm high, 600 mm wide and 124 mm total thick) and prototype panel 2 (1200 mm high, 600 mm wide and 74 mm total thick). The prototype panels resemble a complex sandwich-like structure (see Fig. 2) that is made of two different biomaterials: (1) Ecuadorian balsa hardwood (*Ochroma pyramidale*) as core material [11], and (2) Ecuadorian coconut palmwood (*Cocos Nucifera L*) veneers as external boards. The balsawood core material was used in the form of the BALTEK® SB.100 product due to its high level of stiffness to weight ratio [i.e. Avg. Moduli of Elasticity (MOE) perpendicular to the plane of 2,526 MPa for an equivalent basic density of 148 kg/m^3 at an Avg.

moisture content of 12.6%). BALTEK® core material was acquired from the local supplier 3AComposites. Each external board (i.e. one board per external side of each panel as shown in Fig. 2) comprises three coconut veneers glued each other bidirectionally with acrylic vinyl resin following the same principle of cross laminated timbers (CLT) that are used for wall building purposes [12]. Coconut veneers were obtained by peeling process [13] of the peripheral section (Avg. MOE parallel to the fibers of 8,920 MPa for an equivalent basic density of 900 kg/m^3 at an Avg. moisture content of 12.6%) of three mature coconut palm stems. 2-component Polyurethane adhesive (Pur 2C) was used to glue the external coconut boards with the BALTEK® core material. Once fully assembled and glued, each prototype panel were hot-pressed at 400 psi and 100°C for about 30 minutes.

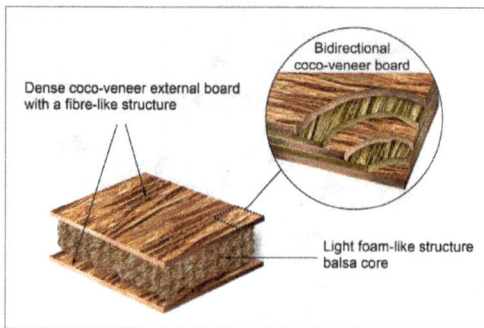

Fig. 2. Sandwich-like structure wall panel made of Ecuadorian balsa lightweight core and coconut bidirectional external veneers.

Methods

The research scope of the whole investigation includes the following tests: compression, bending, shear, tension, cyclic assessment, hardness, fire resistance, acoustic isolation, resistance to pathogens, glue and ply-delamination. Yet, only the first two mechanical modes with the corresponding determination of basic density and moisture content properties are included as part of the present paper. Specifically, this paper presents results from (1) axial stiffness and strength in compression and (2) bending strength in flat-wise four-point loading.

The mechanical tests were all carried out in an AGS-X Shimadzu universal testing machine (UTM) 300 kN capacity equipped with a non-contact digital video extensometer to measure deformations. Moreover, the acquired results for each mechanical mode were double-checked by pilot testing on selected samples using 5 mm long single-element strain gauges glued on the longitudinal-radial (L-R) external faces (refer to Fig. 3a) of each sample using adhesive cyanoacrylate ester and coated with instant repair epoxy resin/tertiary amine. The experimental equipment was complemented with Wheatstone bridge circuits to connect the strain gauges, a data logger (National Instruments NIcRIO-9074) and a computer for data processing.

Before testing and after sanding, each sample was labelled according to the mechanical mode to be investigated. Experimental tests were performed at room temperature and humidity.

Axial stiffness and strength in compression

According to the ASTM C364/C364M-16 Standard Test Method for Edgewise Compressive Strength of Sandwich Constructions, a total of 10 compressive tests were carried out on 5 small-

By-Products of Palm Trees and Their Applications
Materials Research Proceedings **11** (2019) 88-98

Materials Research Forum LLC
doi: https://doi.org/10.21741/9781644900178-5

clear panels cut from prototype 1, nominal size of 250 mm × 250 mm × 124 mm, and on 5 small-clear panels cut from prototype 2, nominal size of 150 mm × 150 mm × 74 mm (refer to Fig. 3).

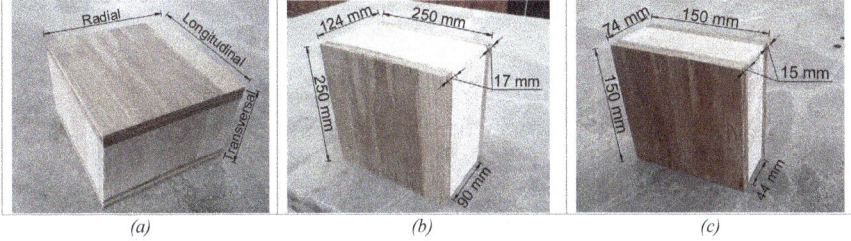

(a) *(b)* *(c)*

Fig. 3 *Bio-composite sandwich panel-samples for compression tests, a) loading directions, b) nominal panel-sample from prototype 1, and c) nominal panel-sample from prototype 2.*

The UTM lower platen was fixed while the upper platen was mounted on a half sphere bearing which could rotate, so as to provide full contact between the platen and the panel-samples (see Fig. 4). Between the platens and the panel-specimens, 10 mm thick acrylic plates were inserted. To minimise friction, dry lubricant (graphite powder) was used between the panel-samples and the testing platens. Each panel-sample was then loaded in the longitudinal (L) direction (see Fig. 3a) up to failure at a cross-head speed of 0.5 mm/min to reach failure between 8 to 10 minutes.

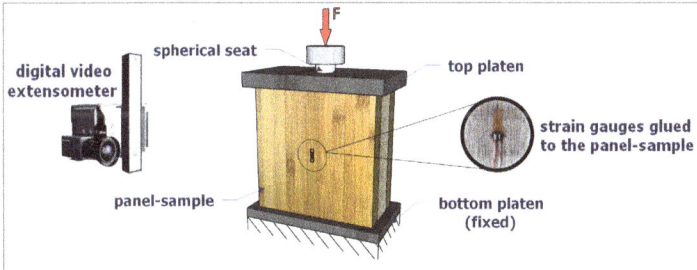

Fig. 4 *Compressive panel-sample test set-up.*

As friction was limited between the panel-samples and the platens, no stress developed in the plane perpendicular to the loading direction i and the Hooke's law [14] applied. The stress σ_i - strain ε_i relationship was therefore given as,

$$\sigma_i = MOE_i \cdot \varepsilon_i. \tag{1}$$

The elastic stiffness of the sandwich wall panel (i.e. the MOE) was then calculated for each test by performing a linear regression on the linear part (i.e. the proportional limit) of the stress-strain curves (please refer to the Results section).

The ultimate edgewise compressive strength [i.e. the Moduli of Rupture (MOR)] that reflects the maximum load carrying capacity of the sandwich construction in the L direction of the applied load was also determined herein and calculated by applying Eq. 2 given in the ASTM C364/C364M-16 as,

$$MOR_L = F_{max} / (w \cdot t_{fs}),$$

(2)

where MOR_L is given in MPa, F_{max} is the ultimate force prior to failure (N), w is the width of the panel-sample (mm) and t_{fs} is the thickness of a single facesheet (mm).

Bending strength in flat-wise four-point loading
According to the ASTM C393/C393M Standard Test Method for Core Shear Properties of Sandwich Constructions by Beam Flexure, a total of 10 flexural tests were carried out on 5 small-clear panels cut from prototype 1, nominal size of 500 mm × 250 mm × 124 mm, and on 5 small-clear panels cut from prototype 2, nominal size of 300 mm × 150 mm × 74 mm (please refer to Fig. 5).

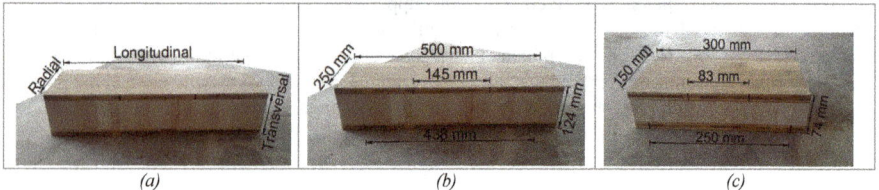

Fig. 5 *Bio-composite sandwich panel-samples for bending tests, a) loading directions, b) nominal panel-sample from prototype 1, and c) nominal panel-sample from prototype 2.*

A 4-point loading configuration was carried out as shown in Fig. 6. The panel-sample was placed onto two lower supporting pins as set distance apart (S). The UTM top platen was mounted onto two loading pins placed equidistantly from the centre as set distance apart of 1/3 S. To minimise friction and prevent local damage between the panel-sample facings and set of upper/lower pins, 3 mm thick rubber pressure pads were used. Each panel-sample was then loaded in the transversal (T) direction (see Fig. 5a) up to failure at a cross-head speed of 6 mm/min to reach failure between 4 to 6 minutes.

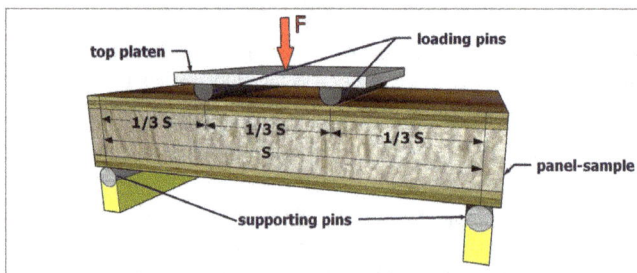

Fig. 6 *Bending panel-sample test set-up.*

By-Products of Palm Trees and Their Applications
Materials Research Proceedings **11** (2019) 88-98

Materials Research Forum LLC
doi: https://doi.org/10.21741/9781644900178-5

The panel-sample facing bending maximum stress (σ_{max}) that reflects the maximum load carrying capacity of the sandwich construction in the direction parallel to the applied load was determined by applying Eq. 3 given in the ASTM C393/C393M -11 as,

$$\sigma_{max} = (F_{max} \cdot S) / [\, 3t_{fs}(d + c)w\,],$$ (3)

where σ_{max} is given in MPa, F_{max} is the maximum force carried by test specimen before core-failure (N), S is the span length between lower supporting pins (mm), t_{fs} is the thickness of a single facesheet (mm), c is the thickness of the core material (mm), w is the width of the panel-sample (mm) and d is the sandwich panel-sample total thickness (mm).

The core shear ultimate strength (τ_{max}) that reflects the maximum load carrying capacity of the core sandwich construction in the longitudinal-transversal (LT) plane was calculated as (ASTM C393/C393M -11),

$$\tau_{max} = F_{max} / [\,(d + c)w\,],$$ (4)

where τ_{max} is given in MPa, F_{max} is the ultimate force prior to core failure (N), d is the total thickness of the panel-sample (mm), c is the core thickness (mm) and w is the panel-sample width (mm).

Results and Discussion

Axial stiffness and strength in compression
Detailed results of the complete set of tested panel-samples under compression are given in Table 1 and Fig. 7.

Table 1. Results from the compressive tests carried out on panel-samples cut from prototype panels 1 and 2.

	PROTOTYPE PANEL 1				PROTOTYPE PANEL 2			
Panel-sample	Weight [kg]	Density at 11% of M.C. [kg/m³]	MOE_L [MPa]	MOR_L [MPa]	Panel-sample	Weight [kg]	Density at 11% of M.C. [kg/m³]	MOE_L [MPa]	MOR_L [MPa]
PCE1-1	2,31	300,86	8928,10	35,09	PCE2-1	0,62	405,23	15044,00	39,92
PCE1-2	2,39	308,47	9346,90	37,04	PCE2-2	0,70	422,16	15586,00	39,50
PCE1-3	2,40	313,35	10293,00	37,86	PCE2-3	0,69	411,58	15542,00	39,39
PCE1-4	2,35	303,13	9244,80	35,92	PCE2-4	0,68	414,11	15291,00	42,11
PCE1-5	2,41	311,37	10173,00	35,94	PCE2-5	0,69	416,79	15575,00	37,00
Avg.	2,37	307,43	9597,16	36,37	Avg.	0,68	413,97	15407,60	39,59
CoV	0,02	0,02	0,06	0,03	CoV	0,05	0,02	0,02	0,05

The results in Table 1 reflect a prototype panel 1 that is in average 1.6 times more elastically efficient (i.e. more deformable) than prototype panel 2 for similar range of densities (i.e. in between 307,43 and 413,97 kg/m³) at 11% of moisture content. The higher elastic performance is likely produced by the greater amount of lightweight balsawood core material in prototype panel 1 (i.e. almost double compared with the core material of prototype panel 2). Thus, the MOE_L in the proposed bio-composite sandwich wall panels varies proportionally to the light foam-like structure balsa core. Yet, from the point of view of the mechanical performance per unit mass (i.e. mechanical efficiency) of both sandwich constructions, the results herein reveal highly efficient prototype panels 1 and 2 with average performance indexes (*PI*) equal to 10.07 GPa$^{\frac{1}{2}}$ m^3 Mg^{-1} and 9.48 GPa$^{\frac{1}{2}}$ m^3 Mg^{-1}, respectively. The performance index for this specific case denotes the material's performance for undergoing deformations when compressive stresses are acting over/on the panels.

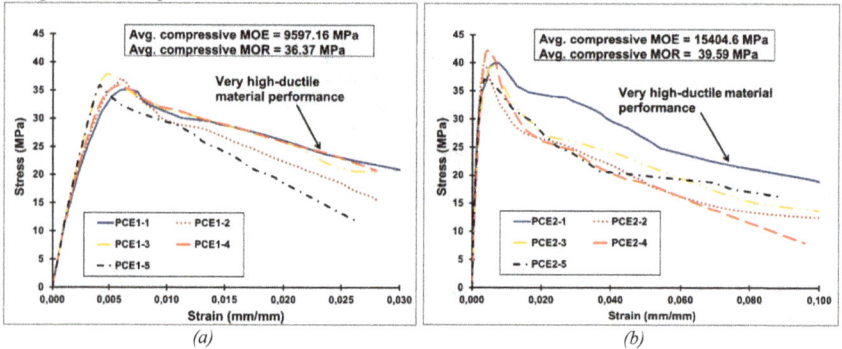

Fig. 7. Compressive stress – strain relationship for (a) five specimens cut from prototype panel 1, and (b) five specimens cut from prototype panel 2.

As shown in Table 1 and Fig. 7, the ultimate edgewise compressive strength (MOR$_L$) appears to be in the same order of magnitude (i.e. in between 36 MPa and 39 MPa) for both prototype panels. These compressive MOR$_L$ results were expected as the thickness of the coconut external boards for both prototype panels varies by just 2 mm. The coco-veneer external boards with a fibre-like structure [15] are the denser part of the sandwich construction and, therefore, are meant to fully resist the progressive generation of compressive stresses. Accordingly, among the variables in Eq. 2 for the MOR$_L$ calculations, it is not consider the total thickness of the panel-sample but only the thickness of a single facesheet (t_{fs}). It can also be inferred from Table 1 and Fig. 7 that the average compressive strengths of both prototype sandwich wall panels are up to three times more efficient, from the point of view of structural mechanics, than conventional concrete masonry units (CMU) with an average compressive strength of 12.5 MPa for the best CMU type according to the ASTM C-90/91 Standard Specifications for Load-Bearing Concrete Masonry Units. Similarly, the average compressive strengths of both prototype sandwich wall panels are up to two times more mechanically efficient than conventional solid wall bricks with an average compressive strength of 20 MPa for the best brick-type according to the ASTM C-55 Standard Specifications for Concrete Building Brick.

Fig. 7 also shows a very high-ductile material performance (i.e. the material's ability to undergo significant plastic deformation before failure) for both prototype panels, which makes them suitable to be used in eco-friendly and structural building applications located in high-risk seismic regions. It technically means the progression of compressive stresses within the bio-composite sandwich wall panels (see Fig. 8a) allows the whole building structure to gradually resist the cyclic and seismic forces before total collapse (see Fig. 8b).

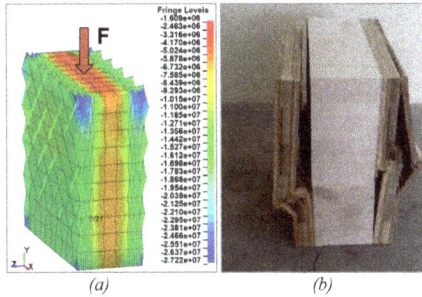

(a)   *(b)*

Fig. 8. *Experimental test post-analyses given by (a) the progression of normal stresses shown in a finite element panel model undergoing compression, and (b) the panel-sample facesheet compressive failure.*

Bending strength in flat-wise four-point loading

MOE values are not calculated herein as, theoretically, this property would not significantly vary regardless the mechanical mode under investigation. What considerably varies is the material's capacity to resist progressive stresses (see Fig. 10b), which depends on the mode of loading, e.g. columns carry compressive axial loads, shafts carry torques, and beams carry predominantly bending moments. Therefore, this part of the paper focuses its analyses on both (1) the panel facing bending maximum stresses, and (2) the panel core shear strengths. Table 2 and Fig. 9 give the complete set of results for the bending tests carried out on panel-samples cut from prototype panels 1 and 2.

Table 2. *Results from the bending tests carried out on panel-samples cut from prototype panels 1 and 2*

PROTOTYPE PANEL 1					PROTOTYPE PANEL 2				
Panel-sample	Weight	Density at 11% of M.C.	Facing bending max. stress	Core shear strength τ_{max}	Panel-sample	Weight	Density at 11% of M.C.	Facing bending max. stress	Core shear strength τ_{max}
	[kg]	[kg/m³]	σ_{max} [MPa]	[MPa]		[kg]	[kg/m³]	σ_{max} [MPa]	[MPa]
PFE1-1	4,77	312,37	14,43	1,70	PFE2-1	1,07	322,61	12,51	2,25
PFE1-2	5,04	328,79	17,00	2,00	PFE2-2	1,18	355,06	13,18	2,37
PFE1-3	4,69	306,09	11,72	1,38	PFE2-3	1,18	352,35	12,92	2,33
PFE1-4	4,88	321,57	13,06	1,53	PFE2-4	1,24	372,45	12,45	2,24
PFE1-5	4,69	305,60	10,15	1,19	PFE2-5	1,22	367,41	14,05	2,53
Avg.	4,81	314,88	13,27	1,56	Avg.	1,18	353,97	13,02	2,34
CoV	0,03	0,03	0,20	0.20	CoV	0,06	0,05	0,05	0,05

As shown in Table 2, the average facing bending maximum stress for both prototype panels are in the same order of magnitude (i.e. in between 13,02 MPa and 13,27 MPa) due to the similar thickness configuration of the coco-veneer external boards in both sandwich constructions. It is worth noting that the facing bending strength (i.e. the MOR_T) could not be calculated herein as, for this specific case, the light balsawood core shear failure always preceded bending failure of the dense coco-veneer external boards with a fibre-like structure (i.e. the panel-sample facings). Within this context, the core shear strength of prototype panel 2 is in average 1.5 times greater than prototype panel 1, which in theory is correct as the lighter core section of the sandwich prototype panel 1 is about 1.2 times greater in volume than prototype panel 2, and consequently, it makes the latter prototype less vulnerable to shear stresses. Moreover, it confirms one important finding in [8-10, 15-18] that states the mechanical properties in biomaterials are all quasi-linearly proportional to density. On the other hand, it is unfortunately not possible to establish any comparison between the resulting bending performances acquired in this study and conventional wall building elements (e.g. concrete masonry units and solid wall bricks) as they only mechanically perform under compression. Yet, the optimised bending performance of the proposed bio-composite sandwich wall panels clearly denotes herein a big plus over the limited capacity of conventional wall building systems (i.e. they do not hold the capacity to mechanically perform under bending stresses). It gives another reason to consider the proposed bio-composite wall panels as a feasible alternative to be used in eco-friendly and structural building applications in high-risk seismic regions.

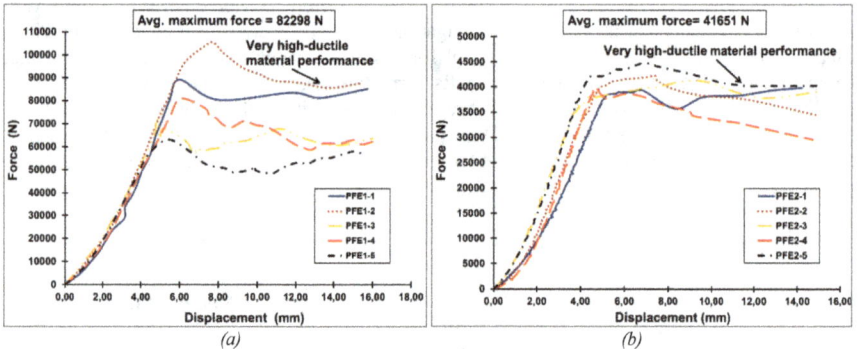

(a)   *(b)*

Fig. 9. *Bending force – displacement relationship for (a) five specimens cut from prototype panel 1, and (b) five specimens cut from prototype panel 2.*

Similar to the panel-sample's compressive performance shown in Fig. 7, the bending performance of both prototype panels (see Fig. 9) shows a remarkable ductility that allowed the core-panel reach failure only after having suffered a large plastic deformation (see Fig. 10). Moreover, it was also observed during the bending tests a high flexibility (i.e. the deflection due to a unit value of the applied load) of the panel-sample coco-facings with a high mechanical resilience (i.e. the material's capacity to spring back into shape). It simply reflects, from an engineering perspective, an optimal design of the bio-composite sandwich-like structure that efficiently performs its mechanical functions.

Materials Research Forum LLC
doi: https://doi.org/10.21741/9781644900178-5

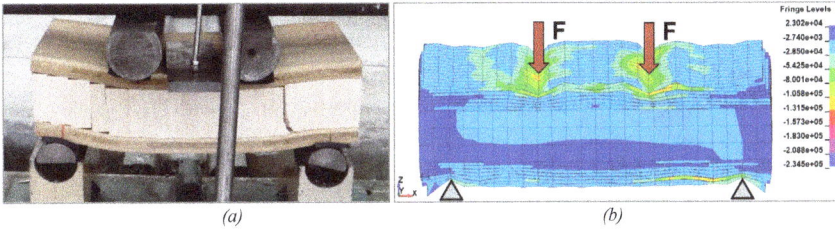

(a)   *(b)*

Fig. 10. *Experimental test post-analyses based on (a) the balsawood core material shear failure, and (b) the progression of normal stresses shown in a finite element panel model undergoing bending.*

Conclusion

Significant findings have been achieved from this investigation that show the bio-composite structural wall panels fully address current engineering and environmental requirements like high structural performance, sustainability, design flexibility, low construction costs, short construction timelines, efficient and low embodied energy, durability, light weight, readily availability, easy transport and assembly, and minimum environmental impacts. The results thereof show that the proposed wall panels are up to two and three times more efficient, in terms of mechanical high-performance, than equivalent sections of solid wall bricks and concrete masonry blocks, respectively. More notably, the optimal cocowood mechanical efficiency [9] biomimicked into the innovative bio-composite sandwich wall panels, makes them suitable to be used in building projects located in high-risk seismic regions as its remarkable ductility, flexibility and resilience, could significantly reduce the overbalance masonry effect (typical for conventional construction wall materials) during high-intensity seismic events.

References

[1] SENPLADES, Secretaría Nacional de Planificación y Desarrollo, Evaluación de los costos de reconstrucción: Sismo en Ecuador, April 2016.

[2] S.G.R., Secretaría de Gestión de Riesgos, Informe de Situación N°71– 19/05/2016 (20h30) Terremoto 7.8 Pedernales, 2016.

[3] C. Beall. Masonry and Concrete, ISBN: 978-0070067066, 2000.

[4] J.S. Gregg, R.J. Andres, G. Marland, China: Emissions pattern of the world leader in CO2 emissions from fossil fuel consumption and cement production, Geophysical Research Letters, ISSN: 1944-8007, 35-8 (2008). https://doi.org/10.1029/2007gl032887

[5] J.G. Vogtländer, N.M. van der Velden, P. van der Lugt, Carbon sequestration in LCA, a proposal for a new approach based on the global carbon cycle; cases on wood and bamboo. The International Journal of Life Cycle Assessment, ISSN: 0948-3349, 19-1 (2014) 13-23. https://doi.org/10.1007/s11367-013-0629-6

[6] C. Sim, E. Villalobos, J.P. Smith, P. Rojas, S. Pujol, A. Puranam, L. Laughery, Performance of Low-rise Reinforced Concrete Buildings in the 2016 Ecuador Earthquake, Purdue University Research Repository, DOI: R7ZC8111, 10 (2017).

[7] U. Wegst, M. Ashby, The mechanical efficiency of natural materials. Philosophical Magazine, ISSN: 1478-6435, 84-21 (2004) 2167-2181. https://doi.org/10.1080/14786430410001680935

[8] O.M. González, K.A. Nguyen. Influence of density distribution on the mechanical efficiency of coconut stem green tissues, World Conference on Timber Engineering, Austria, Vienna, ISBN: 978-390303900-1, 2016.

[9] O.M. González Mosquera. The Ingenious Tree of Life - A Biomechanical Approach to Cocowood Science, Lambert Academic Publishing, Germany, ISBN: 978-613-9-87000-4, 2018.

[10] O.M. Gonzalez, K.A. Nguyen, Senile coconut palms: Functional design and biomechanics of stem green tissue, Wood Material Science & Engineering, ISSN: 1748-0272, 12-2 (2017) 98-117. https://doi.org/10.1080/17480272.2015.1048480

[11] M. Osei-Antwi, J. De Castro, A.P. Vassilopoulos, T. Keller, Shear mechanical characterization of balsa wood as core material of composite sandwich panels, Construction and Building Materials, ISSN: 0950-0618, 41 (2013) 231-238. https://doi.org/10.1016/j.conbuildmat.2012.11.009

[12] E. Pulgar, P. Soto, E. Flores, S. Vavra, J. Conde, S. Parada, Mechanical characterization and seismic behaviour of cross laminated timber panels made of Chilean radiata pine, World Conference on Timber Engineering, WCTE, ISBN: 978-390303900-1, 2016.

[13] H. Bailleres, L. Denaud, J.-C. Butaud, R. Mcgavin, Experimental investigation on rotary peeling parameters of high density coconut wood, BioResources, DOI: 10.15376, 10-3 (2015) 4978-4996. https://doi.org/10.15376/biores.10.3.4978-4996

[14] A.C. Ugural. Mechanics of Materials, John Wiley & Sons Inc. New Jersey Institure of Technology, United States of America, ISBN: 978-0-471-72115-4, 2008.

[15] O.M. González, K.A. Nguyen, Cocowood Fibrovascular Tissue System—Another Wonder of Plant Evolution, Frontiers in Plant Science, ISSN: 1664-462X. DOI: 10.3389/fpls.2016.01141, 7-1141 (2016). https://doi.org/10.3389/fpls.2016.01141

[16] O.M. Gonzalez, B.P. Gilbert, H. Bailleres, H. Guan, Senile coconut palm hierarchical structure as foundation for biomimetic applications, Applied Mechanics and Materials - Advances in Computational Mechanics, DOI: 10.4028, 553 (2014) 344-349. https://doi.org/10.4028/www.scientific.net/amm.553.344

[17] O.M. Gonzalez, B.P. Gilbert, H. Bailleres, H. Guan. Compressive strength and stiffness of senile coconut palms stem green tissue. 23rd Australasian Conference on the Mechanics of Structures and Materials, ACMSM23. Byron Bay, Australia, ISSN: 978-0-994152-00-8, (2014) 881-886.

[18] O.M. Gonzalez, B.P. Gilbert, H. Bailleres, H. Guan, Shear mechanical properties of senile coconut palms at stem green tissue, 4th International Conference on Natural Sciences and Engineering (2015ICNSE), Kyoto Research Park, Japan, ISSN: 978-986-87417-8-2, (2015) 529 - 543.

By-Products of Palm Trees and Their Applications
Materials Research Proceedings **11** (2019) 99-112

Materials Research Forum LLC
doi: https://doi.org/10.21741/9781644900178-6

Medium Density Fiberboards from Date Palm Residues a Strategic Industry in the Arab World

Hamed El-Mously[1,a] *, M.Saber[2]

[1]Faculty of Engineering, Ain Shams University, 1 Elsarayat St., Abbaseya, 11517, Cairo, Egypt

[2]Expert in economy and project feasibility studies, Cairo, Egypt

[a]Hamed.elmously@gmail.com

Keywords: wood availability, date palm products of pruning, medium density fiber boards

Abstract. The success of the environmental movements world-wide has led to the decrease of wood availability in the world market, and hence to the soaring of wood prices .This in turn has led to the increase of the burden on the balance of payment of the Arab countries, relying on the importation of wood to satisfy the needs of their populations in shelter, furniture, etc. Meanwhile, the Arab World includes the palm belt extending from Morocco in the far West to Iraq in the Far East. Therefore, it makes sense to look to the date palm residues, mainly resulting from the palm pruning, as a sustainable renewable material base to locally manufacture wood substitutes as, for the example, the composite panels including the medium density fiber boards (MDF), particle boards, block boards, etc. Within a research project, conducted by the Faculty of Engineering, Ain Shams University with the collaboration of the ministry of environment, samples of the date palm secondary products have been collected in proportion with the available products of palm pruning (palm midribs, leaflets, spadix stems and coir), threshed and sent to the laboratory of Deshna MDF factory in Kena governorate .The result of tests confirm that the MDF samples, manufactured from the date palm secondary products, satisfy the mechanical and physical requirements of international standards of MDF. A technical and economic feasibility study has been conducted on a suggested industrial project to manufacture MDF boards in EL-Bahariah oases. The results of this study show that the profitability indicators of this project are high: the return rate on invested capital is (39.4%), the revenue to cost rate is (1.43:1), the payback period is 3.6 years and the internal rate of return is 36.2%.

Introduction

The success of the environmental movements world-wide has led to the decrease of wood availability in the international market. This has led to the soaring of wood prices, which has increased the burden on the balance of payments in the Arab countries, relying basically on importation to satisfy people's need, of wood for shelter, furniture, etc. Meanwhile, the Arab countries are distinguished with the date palm belt extending from Morocco in the West to Iraq in the East including ~ 102.4 million palms [5]. Thus, it is logic to look to the products of pruning of date palm as a renewable and sustainable resource for the manufacture of wood substitutes, such as the medium density fiberboards (MDF), particle boards (pb), blockboards, etc.

Present Status of Wood Market: A Case Study from Egypt

Fig. 1 illustrates the value of Egypt's wood imports until 2011 [8]. It is clear that beginning from 2000 the value of wood imports is steadily ascending reaching ~ 2.5 billion US$ in 2011. Fig. 2 [10] illustrates the change of value of Egypt's wood imports during the period from 2000 to

2016. It is noteworthy that the increase of the financial burden of wood imports is tremendous, especially as weighed by the local currency. As is clear from this figure, the value of wood imports has doubled over the ten years from 2005 to 2015 and the average rate of growth in Egypt's imports of wood during this period was about 6.8% annually. This represents a huge burden on the shoulders of the furniture manufactures, as well as the consumers leading to the collapse of furniture industry in Egypt and the tendency of the consumers to rely on imported furniture probably coming from china. Fig. 3 [7] represents a five-year forecast of Egypt's wood sector under a 6 percent GDP growth rate. It is clear from this figure that both the low and high endforecasts show a tremendous increase of the value of wood imports, representing a huge burden on the balance of payments of Egypt. Fig. 4 [11-20] illustrates the change of prices of wood species, as well as wood products due to the big devaluation of the Egyptian pound in November 2016. This has had a drastic influence on the small-scale furniture establishments, leading to the closure of many of them.

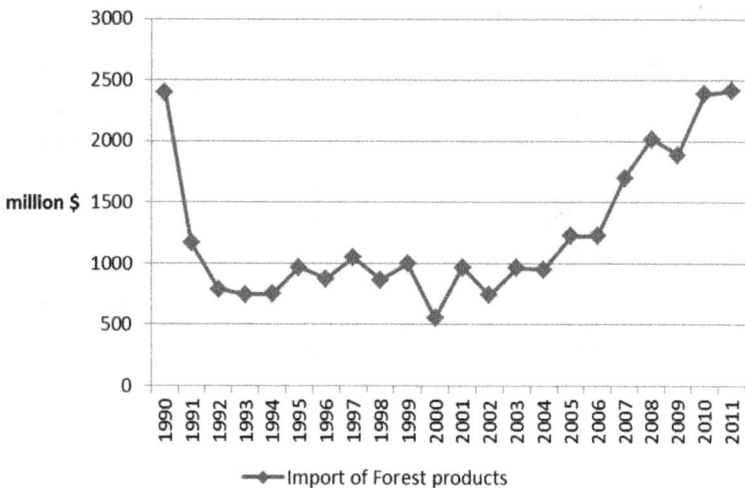

Fig. 1: The value of Egypt's wood imports until 2011 [8].

Materials Research Proceedings **11** (2019) 99-112

doi: https://doi.org/10.21741/9781644900178-6

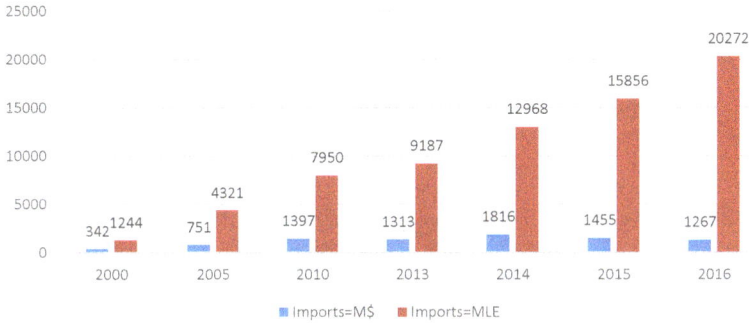

Fig. 2: The value of Egypt's imports of wood in US dollars and the equivalent value in local currency [10].

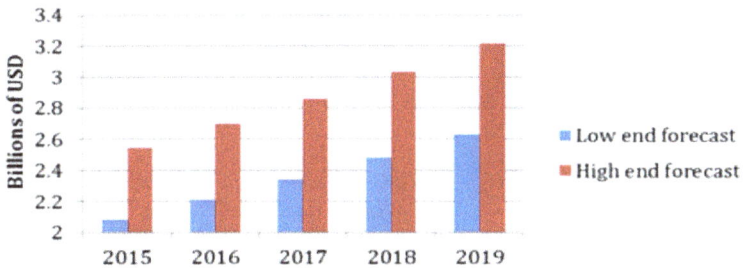

Fig. 3: Five-Year forecast of Egypt's wood sector value under a 6 percent GDP Growth Rate [7].

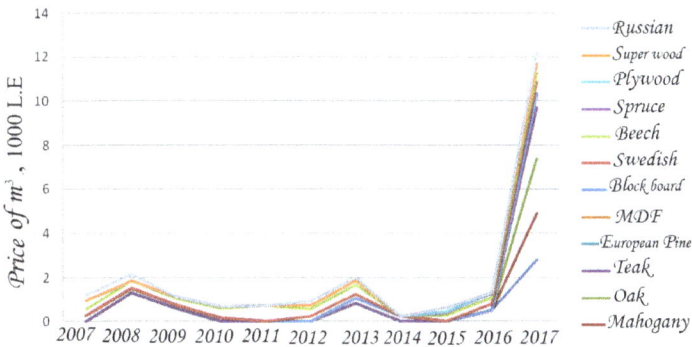

Fig. 4: Change of value of prices of wood species and wood products during the period from 2007 to 2017 [11-20].

Strategic Necessity of Reliance on Agricultural Residues as a Substitute for Imported Wood

Fig. 5 illustrates the value of Egypt imports of wood and wooden products during the period from 2010 to 2014 [6]. It is clear from this figure that the value of Egypt's imports in 2014 is approaching $\cong$ 2 billion US $. Assuming the same previous average rate of wood imports (6.8%), the future estimates suggest that the cost of imports of wood and wood products will exceed ~22.8 billion US $ in 2050! This represents an unacceptable burden on the future generations. *Thus, there is a strategic necessity to rely on the agricultural lignocellulosic residues (ALR) as an alternative to imported wood.* The annual amounts of these ALR amount to about 80 million tons (oven dry weight): ~ 76 million field crops residues and ~ 4 million products of pruning of fruit trees [1]. The industrial utilization of ALR will lead to the building of endogenous scientific and technological capacities, as well as the rising of successive waves of innovation beginning from rural areas and reaching urban areas in the country.

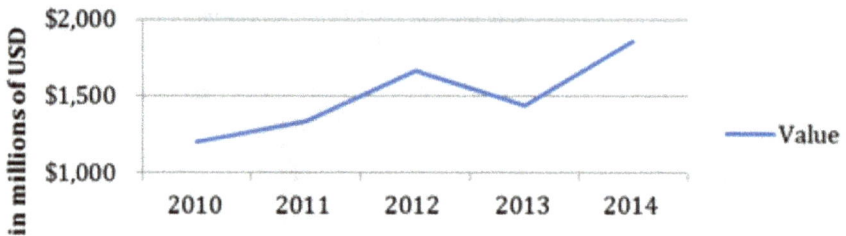

Fig. 5: The value of Egypt imports of wood and wooden products during the period from 2010 to 2014 [6].

The Medium Density Fiber Board Industry in the World: Present Status and Future Prospects

The medium density fiber boards (MDF) are classified among the lignocellulosic composite panels. This industry relies basically on wood resources. There is a growing trend at the present time to utilize the agricultural residues as an alternative to wood in view of the environmental concerns of cutting of wood trees in forests, as well as the increase of wood demand for the MDF products.

The MDF industry came to existence during the ends of the sixties of the last century and witnessed a considerable growth during the seventies and eighties of that century. The MDF industry in Egypt has begun by the establishment of Naga-Hammady Fiberboard Company in 2001. This is the sole company in Egypt producing MDF boards from bagasse. The bagasse is one of the first ALR to be used in particle board production since the First World War. Then it was used in pulp production. Afterwards other ALR were used for the production of MDF, such as cotton stalks, wheat straw, maise stalks, rice straw, rice husk and palm midribs.

The MDF boards are distinguished with acceptable mechanical and physical properties. That is why MDF products found wide applications in indoor use (e.g. furniture and flooring) and outdoor use. The international statistics (Fig.6) illustrate the progressive increase of the production capacities of MDFduring the period from 2007 to 2017 reaching ~ 100 million m³ [9].

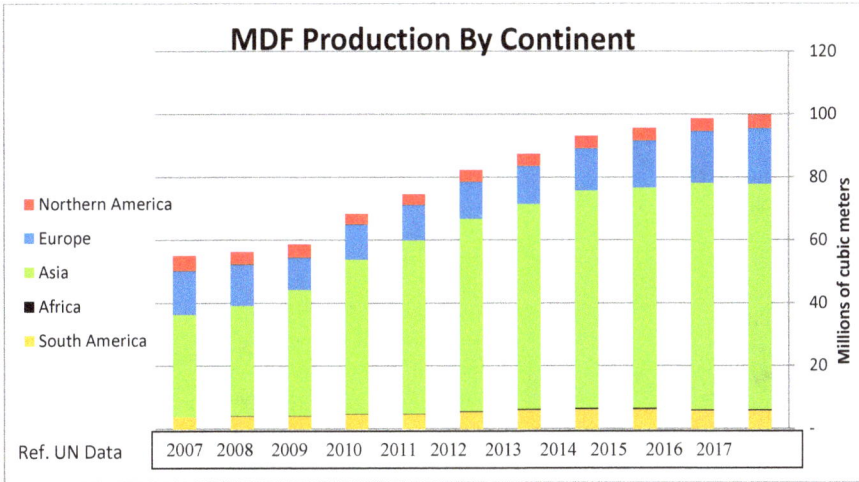

Fig. 6: The progressive increase of the production capacities of MDF [9].

MDF from Date Palm Products of Pruning

The Date Palm: An Essential Component of the Flora in Egypt

Table (1) illustrates an estimation of the number of palms in Egypt in 2015. It is clear from this table that Egypt possesses ~15 million productive palms, distributed among Egypt's 28 governorates. Proceeding from the data of this table the following governorates may be considered leading in palm plantations: Aswan, Giza, Beheira, New Valley and Sharkia.

N	Governorates	Number of Palms (million)	Order
1	Aswan	2.48	First
2	Giza	1.81	Second
3	Beheira	1.37	Third
4	NewValley	1.26	Fourth
5	Sharkia	1.21	Fifth

Table 1: The number of palms in Egypt in 2015 [4]

Governorates	Area	fruitful palm (palm)	Productivity (Kg\ palm)	Production (ton)
Alexandria	436	82563	83.706	6911
Beheira	14327	1371794	168.436	231060
Garbiya	315	48368	103.560	5009
Kafr El Sheikh	5159	343427	128.062	43980
Dekhalia	674	216716	110.010	32841
Damiatta	15	866216	99.166	85899
Sharkia	260	1211196	171.536	207764
Ismailia	1327	670532	131.809	88382
port said	-	11195	84.413	945
Elsuez	456	93879	94.590	8880
Monofia	75	163339	100.656	16441
Qalyubia	547	203469	125.729	25582
Cairo	810	37586	53.424	2008
Lower Egypt(Total)	24392	5320280	140.350	746702
Giza	21089	1813322	130.798	237178
BeniSuef	61	320783	92.324	29616
Faiyum	1158	643074	133.832	86046
Minya	586	337608	110.000	37137
Middle Egypt (Total)	22894	3114787	125.208	389995
Asyut	400	462501	95.431	44137
Sohag	799	414071	93.291	38713
Qena	1039	361346	61.368	22175
Luxor	552	192360	70.254	13514
Aswan	24840	2477458	90.840	225054
Upper Egypt (Total)	27630	3908663	87.906	343593
Inside Valley(Total)	74916	12343730	119.922	1480290
New Valley	18482	1262475	81.681	103120
Matruh	7207	330674	90.001	29761
Red Sea	134	39528	42.856	1694
North Sinai	9076	320650	53.429	17132
South Sinai	-	91304	39.998	3652
Nubariya	5795	567970	86.744	49268
Outside Valley (Total)	40694	2612601	78.323	204627
Total	**115610**	**14956331**	**112.656**	**1684917**

Evaluation of the Available Quantities of Products of Prunning: A Case Study from El-Bahariah Oases

Within the framework of the Project of Care of Date Palm in El-Bahariah oases [3], the mass of the products of pruning has been determined for the Siwi, Freihi and Gaga palm species. Taking as an example the Siwi species (a dominant species in Egypt), it is possible to give the following estimations for the annual products of pruning of one palm.

N	Secondary products	Mass per palm kg (air dry weight)
1	Palm midribs	15
2	Palm leaflets	14.6
3	Spadix stems	9
4	Coir	1.56
5	Midrib end	14
Total		54.2 kg

Taking the Siwi palm as a basis for estimation, the total annual available mass of the products of pruning of date palm in Egypt amounts to 810,000 tons, which represents a considerable sustainable material base for the establishment of a wide spectrum of industrial activities.

Study of the Technical and Economic Feasibility of Manufacture of MDF from the Products of Pruning of Date Palms

  Stages of Manufacture of MDF

Fig. 7 Illustrates the stages of manufacture of MDF, whereas Fig. 8 shows a process chart of the manufacturing operations.

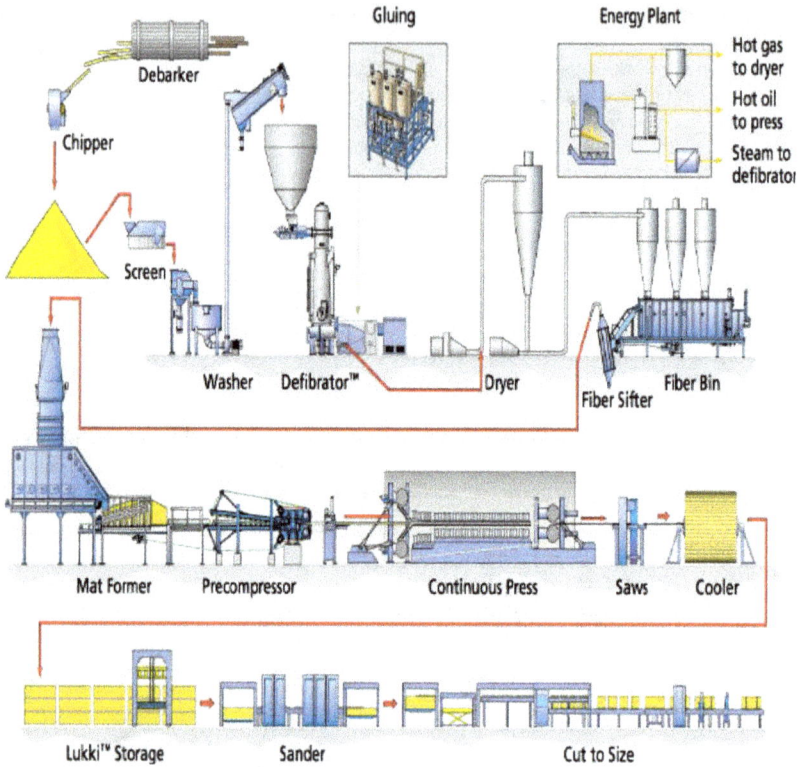

Fig. 7: Stages of manufacture of MDF [3].

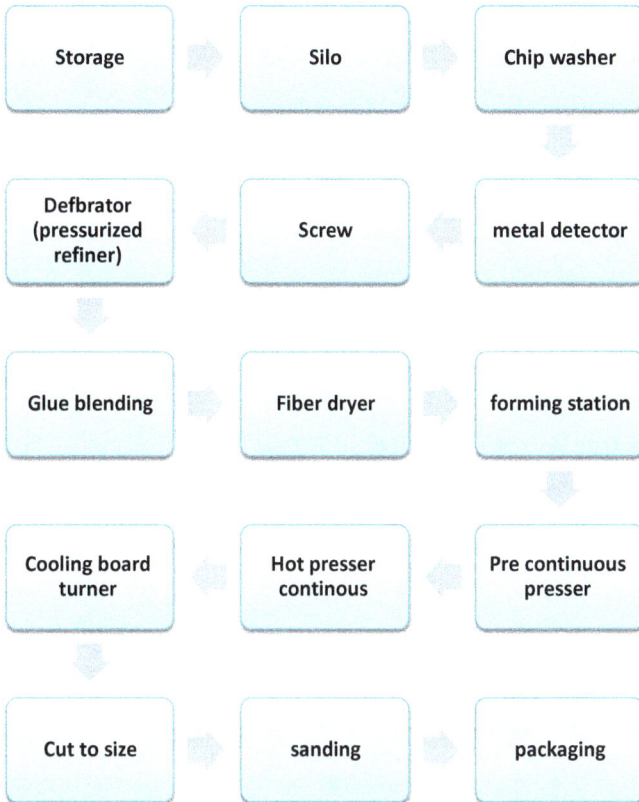

Fig. 8: Flow process chart of the sequence of manufacturing operations of MDF.

Demand Estimates and the Market Gap of Medium Density Fiber Boards in Egypt
Table (2) shows the demand and local production of MDF from 2016 to 2026. It is known that there is currently only one MDF factory in Naga-Hammadi in Qena governorate (working with bagasse); its expansion in production will not reach more than 97 thousand m³ in 2026.

It is clear that the market gap of MDF in Egypt ranges from 80% to 90%. This represents a promising opportunity to invest in the MDF industry, based on palm by-products. In addition, local investment in MDF production will reduce the burden on Egypt's balance of payments.

*Table 2: Estimated demand for MDF and the market gap during the period from 2016 to 2026
(thousand cubic meters)*

Years	Demand estimates	Local production	Market gap	Market gap ratio
2016	300	30	270	90%
2017	325	65	250	79%
2018	331	97	234	71%
2019	347	97	250	72%
2020	365	97	268	73%
2021	383	97	286	75%
2022	402	97	305	76%
2023	422	97	325	77%
2024	443	97	346	78%
2025	465	97	368	79%
2026	489	97	392	80%

Results of the Financial Study of the Project

Net Annual Revenue
According to estimates of the annual revenues of the project for the construction of a MDF plant operating with palm by-products in E--Bahariah oases as well as annual operating expenses; according to the 2015 estimates, the project achieves net pre-tax revenues of about LE 33.4 million annually, which has an annual average of 22.5% (in accordance with the Egyptian tax system at the time of conduction of study: 2016); the net after-tax income of the project (table 3) is estimated at LE 25.9 million per year.

Table 3: Net annual revenue of MDF panel production project

	Value (million pounds)	Relative importance (%)
Project income	82.3	100%
Operating expenses	48.8	59.4%
Net profit before tax	33.4	40.6%
Tax payable	7.5	9.1%
Net profit after tax	25.9	31.5%

By-Products of Palm Trees and Their Applications   Materials Research Forum LLC
Materials Research Proceedings **11** (2019) 99-112   doi: https://doi.org/10.21741/9781644900178-6

Profitability Indicators
The financial feasibility study for the project of MDF production using palm by-products in the El-Bahariah oases is discussed below. The following are a set of indicators to guide the feasibility of the project, as shown in Table 4:

Table 4: Profitability indicators for MDF production project

The project profitability	Value of the indicator
Return rate on invested capital	39.4%
Revenue/rate: costs	1:1.43
Payback period	3.6 Year
Net present value	92.9
Internal rate of return	36.2%

Result of Testing of Samples of MDF Manufactured from Palm Products of Pruning
Samples of products of pruning of date palms have been sent from El-Bahariah oases to the laboratories of Naga-Hammadi Company for Fiber boards. The quantities of palm midribs, palm leaflets, spadix stems, coir and palm midrib ends were proportional to the real masses of these products per palm. Appendix (1) illustrates the results of testing of the MDF specimens. The results of test are as follows:

Physical and chemical properties
- Humidity (5.2%, which falls within the limits 4-11% of EN 322).
- Water absorption (12.7%, which is less than the corresponding value of 15% in EN 317).
- Formalin emission (22.54 mg/100 mg, which is less than the corresponding value 30 mg/100 mg in EN 120).

Mechanical properties
- Modulus of rupture (24.4 N/mm^2, which is higher than the corresponding value 20 N/mm^2 in EN 310:)
- Modulus of elasticity (2911 N/mm^2, which is higher than the corresponding value 2200 N/mm^2 in EN 310)
- Internal bond (0.9 Nmm^2, which is much higher than the corresponding value 0.55 N/mm^2 in EN 319)
- Surface strength (1.35 N, which is higher than the corresponding value 1.2 N in EN 311)

All previous profit indicators for the MDF production from palm by-products in El-Bahariah oases reflect, according to the 2016 estimates, high financial returns and economic viability.

Conclusion
The establishment of MDF projects in Egypt, as well as in countries having extensive date palm plantations, have good economic returns. In addition, the economic success of these projects will increase the added value of date palm plantations, which provides a stimulus for the augmentation of planting of date palms in Egypt and the whole Arab region. In addition, there are several developmental returns: providing labor opportunities in rural and urban areas, building of endogenous scientific and technological capabilities and the improvement of the

balance of payments via substituting a portion of imported wood products by local MDF production.

References

[1] حامد إبراهيم الموصلي، دراسة عن استخدام البواقي الزراعية بمنطقة الشرق الادنى في تحقيق التنمية المستدامة، منظمة الصحة العالمية، المكتب الاقليمي لشرق المتوسط، 2006.

[2] حامد إبراهيم الموصلي، مشروع تحسين الاوضاع البينية ورعاية النخيل والاستخدام الاقتصادي لمنتجاتة الثانوية في الواحات البحرية، التقرير المرحلي، مارس، 2006.

[3] حامد إبراهيم الموصلي، مشروع تحسين الاوضاع البينية ورعاية النخيل والاستخدام الاقتصادي لمنتجاتة الثانوية في الواحات البحرية، والتقرير النهائي، يونيو 2016.

[4] وزارة الزراعة و استصلاح الاراضي، قطاع الشؤن الاقتصادية، الادارة المركزية للاقتصاد الزراعي، نشرة الاقتصاد الزراعي، 2015.

[5] سعود عبد الكريم الفدا ،رمزي عبد الرحيم ابو عيانه، المنتجات الثانوية للنخيل، اوقاف الراجحي، 2017.

[6] Central Agency for Mobilization and Statistics, USDA, Foreign Agriculture Service, Global Agriculture Information Network, Egypt, wood sector Report, 29/9/2015.

[7] Egypt wood sector report, USDA Foreign Agriculture Service, 2015.

[8] UNECE/FAO Timber data base, July 2018.

[9] United National Data, 2018.

[10] WTO and CAPMAS: 2000-2016 Central Agency for Mobilization and Statistics.

[11] Information on http://www.masress.com.

[12] Information on http://www.ahram.org.eg/Index.aspx.

[13] Information on https://www.nmisr.com.

[14] Information on https://www.alnharegypt.com.

[15] Information on http://hanooon.yoo7.com.

[16] Information on http://elbashayeronline.com.

[17] Information on https://alwafd.org.

[18] Information on http://www.almalnews.com/Default.aspx.

[19] Information on http://live-match-stream-online.blogspot.com.eg.

[20] Information on http://prices.hooxs.com/contact.

By-Products of Palm Trees and Their Applications   Materials Research Forum LLC
Materials Research Proceedings 11 (2019) 99-112   doi: https://doi.org/10.21741/9781644900178-6

Appendix (1):
Results of tests of samples of MDF,
manufactured from the products
of pruning of date palms in
El-Bahariah oases
The tests have been conducted
in the laboratories
of Naga-Hamadi Fiberboards Company.

Materials Research Proceedings 11 (2019) 99-112   doi: https://doi.org/10.21741/9781644900178-6

NAG-HAMADY FIBER BOARD CO.
(N.F.B)
E-mail : nhfiboco@intouch.com

شركة نجـع حمادى للفيبربورد
ش . م . م

دشنا في: ٢٠١٦/٨/٢٢

تقرير فني عن

الاختبارات الفيزيائية والميكانيكية لعينات الالواح المصنعة معمليا من المنتجات الثانوية لنخيل البلح

١- الاختبارات الفيزيائية والكيميائية

انبعاث الفرمالين	الامتصاص	التشرب	الرطوبة	الكثافة	السمك	الاختبار
مجم/١٠٠جم	%	%	%	كجم/م٣	مم	الوحدة
٢٢,٥٤	٥٨,٠٤	١٢,٧	٥,٢	٧٥٢,٠٠	١٢,٠٠	النتيجة
٠,٨٧	٢,٥	١,٣	٠,١١	٢٤,٢٨	٠,٠٤	الانحراف المعياري
EN120	BS1142	EN317	EN322	EN323	EN323	الطريقة القياسية
٣٠ ≤		١٥	١١-٤	--	-	المواصفات القياسية

٢- الاختبارات الميكانيكية

قوة شد السطح Surface soundness	الرابطة الداخلية IB	معامل المرونة MOE	معامل الكسر MOR	السمك	الاختبار
نيوتن	نيوتن/مم٢	نيوتن/مم٢	نيوتن/مم٢	مم	الوحدة
١,٣٥	٠,٩	٢٩١١	٢٤,٤٠	١٢	النتيجة
٠,٣٠	٠,٢	٢٨٠,٥٢	٣,٤٩	٠٠,٠٤	الانحراف المعياري
EN311	EN319	EN310	EN310	EN323	الطريقة القياسية
١,٢	٠,٥٥	٢٢٠٠	٢٠,٠٠		المواصفات القياسية

مدير عام البحوث

د.عبدالباسط عبدالحميد ادم

مدير عام المعامل

كيميائي / عبدالحميد محمد محمد

Cairo Office : 17 Gawad Hosni St., Cairo - Egypt
Tel.: +2/02/23922109 +2/02/23902995 Fax: +2/02/23926511
P.O.Box : 432 Mohamed Farid
Factory : Desna - Qena
Tel.: +2/096/6743281 +2/096/6743002 Fax : +2/096/6743003
website : www.geocities.com/nhfiboco/company-profile html

فرع القاهرة: ١٧ ش جواد حسني - القاهرة - مصر
تليفون : ٢٣٩٢٢١٠٩ - ٢٣٩٠٢٩٩٥ - فاكس ٢٣٩٢٦٥١١
ص . ب : ٤٣٢ محمد فريد
المصنع : دشنا : محافظة قنا
تليفون : ٠٩٦/٦٧٤٣٢٨١ - ٠٩٦/٦٧٤٣٠٠٢ - فاكس ٠٩٦/٦٧٤٣٠٠٣

By-Products of Palm Trees and Their Applications   Materials Research Forum LLC
Materials Research Proceedings **11** (2019) 113-132   doi: https://doi.org/10.21741/9781644900178-7

Evaluation of Palm Fiber Components as Alternative Biomass Wastes for Medium Density Fiberboard Manufacturing

Abdel-Baset A. Adam[1], Altaf H. Basta[2,a*], Houssni El-Saied[2]

[1]Nag-Hamady Fiberboard Company, Quena, Egypt

[2]National Research Centre Cellulose & Paper Dept., El-Behoos Street, Dokki-12622, Cairo, Egypt

[a]altaf_basta2004@yahoo.com

Keywords: medium density fiber-board, date palm components, defibration process, thermal behaviour, UF-fibers interaction, strength properties, water resistance property

Abstract. This work deals with assessing the date palm component wastes as alternative lignocellulosic material for production of Medium density fiberboards (MDF), in order to establish economic and balance between production/consumer ratio at different provinces rather than Upper Egypt. Palm leaves and palm frond was used as MDF precursors. Different urea formaldehyde (UF) levels (10-14%/fiber) and pressing pressure (25-35 bar) were applied in this evaluation. The acceptable interaction of palm fibers component with UF was optimized by characterizing its DSC & TGA, in comparison with commercial used sugarcane bagasse fibers. The promising MDF Panel is obtained from palm frond fibers and its mechanical and water resistance properties fulfill the ANSI standard for high grade MDF wood products, especially on applying UF level 12-14%, and pressing pressure, 35 bar. It is interesting to note that, applying higher pressing pressure together with 12% UF level provided palm frond-based MDF with static bending properties, higher than commercial Bagasse-based MDF. The insignificant effect of pressing pressure was noticed on water swelling property and free-HCHO of MDF panels. Where, both type of fibers have the same water swelling property (reached ~ 10%), and free-HCHO (~ 27 mg/100g board).

Introduction

In Egypt agricultural wastes accumulate in huge quantities, it reaches about 35 MT/year. Part of this amount is used as animal fodder and to produce energy; as well as in production of paper and engineered wood products (particle and fiber boards). Still large amount of agro- wastes remains unused and it is burned in the open atmosphere causing environmental pollution. In laboratory scale some of wastes were used as filler for rubber composites, carbon materials, and hydrogels for water purification [8,10-14,20,25]. Sugar-can bagasse (SCB) regards the main residue available and used as precursor for production of engineered wood products (particle-boards and Medium density fiberboards; MDF) and paper, in Upper Egypt. Our previous work was focused on examining the ability of controlling the steam digestion step; to improve the strength of SCB-based MDF produced [12].

Medium density fiberboards (MDF) and particleboards have replaced the natural wood, and plywood in many furniture applications. MDF is superior to particleboards due to its properties including strength, homogeneity and machining performance [4, 26]. They are also appropriate for interior and exterior construction, as well as in industrial applications. The preparation of particleboards and MDF in mill scale started from about four decades. The possibilities of utilizing the available agricultural wastes, (e.g., wheat, bagasse, rice straws, peanut husks, and hazelnut shells), using urea-formaldehyde, melamine-modified urea-formaldehyde, and soybean-pMDI adhesive systems in the production of engineered wood products has sparked attention with many authors [1,3,11,13,14,16,17,19,21,24,27,32,33,46,47]. Other trials in producing green artificial wood were carried out by using HCHO-free adhesive and changing the surface properties of natural lignocellulosic fibres by different grafting techniques [34,38-41], or by in-situ grafting of agro-fibres with the free styrene containing polyester [22]. To improve the water resistance (WA and TS) of wood fibers-based MDF together with limitation of formaldehyde emission of promotion of the quality of the resins, wollastonite and its nano-particles were used in internal and surface treatments [42-43].However, surface treatment of MDF by Calcite (100%), clay (100%) or mixture of clay/calcite did not cause a significant difference on surface quality [28].

Date palm is a multi-purpose tree; it regards a highly national heritage in many countries. It provides food, shelter, timber products. Because of these qualities, and its tolerance to harsh environmental desert conditions, areas under cultivation have increased tremendously in last decades (Mahmoudi et al, 2008). Egypt is the largest date producing country in the world. In 2012, it produced 1.47 million tones that make 19% of world dates production. Cultivation of date Palms in Egypt dates backs to thousands of years. Approximately seven million fruiting palm trees are grown in Egypt in the Nile Valley, Sinai and similar areas.

Based on the availability of palm date, as well as the environmental and economic impacts of sugar-cane bagasse from storage processes, the objective of this study is focused on evaluating the interaction of palm component wastes (date palm pruning mixed products, leaves & frond), with commercial UF to be alternative substrate to manufacture of MDF. The success of this alternative material will be supported by comparing the MDF properties resulted from promising Palm component with that produced from bagasse fiber and the standard specifications [5].

Materials and Methods

MDF Fibers and board preparation
Date palm rachises (DPR) were collected from date palm trees grown up at Sinai. The used species is " EL BARMATODA ". DPR were air dried in sun light for 48 hours, and then cut to 25x25x3 mm chips. These chips were softened by steam in a horizontal digester. The steam pressure was maintained at 7.8 bars for 6 min. then defibrated through ANDRITZ refiner parallel experiments, sugarcane bagasse (SCB) samples were also subjected to the same condition of digestion and refining. The softening fibers were sunlight dried and mixed through laboratory blender with different levels of urea formaldehyde (UF) adhesive (based on oven dry basis) ratios as shown in Table 1. The UF was delivered from Speria Co., with free-formaldehyde (HCHO) 0.18 %. The ammonium sulfate (1%based on UF) was added as hardener, followed by 1 % paraffin wax (based on dry fibers). The mats were formed and pre-pressed using 400 mm x400 mm box. Medium density fiberboard (MDF) boards with 12 mm thickness were made by hot pressing at different specific pressures condition as shown on Table 1, at 165 –170°C press temperature. For feasibility application such waste, the average weight of palm fronds and palm leaves in palm waste are estimated and recorded in Table 2.

After open air-conditioning for 24 h, these boards were cut according to standard sawing pattern into test specimens serving for the subsequent determination of basic mechanical and physical properties. Each result recorded in the following Figures is the average of five replicate measurements.

Materials Research Forum LLC
doi: https://doi.org/10.21741/9781644900178-7

Fiber analysis and MDF tests

Chemical analysis
Parts of Date palm rachises were subjected to mill, using sieve 250 μm and 400 μm, followed by conditioning in polyethylene bags for 12 hours, and labeled to be ready for work. The chemical constituents (e.g., extractives, hollocellulose, α-cellulose, lignin and pentosans) were determined by standard methods [30,36,37,44].

Thermogravimetric Analysis(TGA)
The non-isothermal TGA of the selected representative adhesive systems, was carried out using Instrument SDT Q600 V20 Build 20 module (made at US), under nitrogen atmosphere at a heating rate of 10 °C/min. The analysis was carried out on adhesive casting films (~7.032- 9.40 mg), and their subjected to heat at temperature range from ~ 30 °C to 1000 °C,

TG-curve analysis
Kinetic studies, based on the weight loss data, were obtained by TG curve analysis. The activation energy against the appropriate order of degradation was evaluated by applying an analytical method proposed by Coats and Redfern (1964) [18] and Basta & El-saied 2008 [9].

Differential Scanning Calorimetry (DSC)
Differential scanning calorimetry (DSC) is the thermoanalytical technique used to measure the thermal properties of the investigated adhesive systems, as phase transition and glass transition temperature. This analysis was carried out by employing DSC on the same previous Instrument SDT Q600 V20 Build 20 DSC module (made at US), on the dynamic run, also under nitrogen atmosphere at a heating rate of 10 °C/min.

MDF Tests
Different parameters of the MDF making involved different UF level (10-14%) and pressing pressure (25-35 Kgf/cm$^{2)}$ were carried out, as shown in Table 1. The density of MDF was fixed by changing the amount of pressed fibers. The levels of UF were selected to provide MDF of E2 type with free-HCHO not exceed 27 mg/100 g board.
Three-point static bending (modulus of rupture; MOR), modulus of elasticity (MOE), and internal bond strength (IB) tests were performed in conformance with ASTM D1037 and ANSI A208.2 standards [5,6], for MDF panel's requirements for interior uses], using an IMAL IB500 testing machine.
For evaluating the low toxicity of the resulted composites, the perforator method [23] for determination of free-HCHO in composites was carried out.

Statistical Analysis
The data of the mechanical and physical properties of MDF samples manufactured from bagasse, frond, and mixed frond fibers were subjected to statistical data handling through the TWO WAY (ANOVA) variance analysis by using IBM SPSSV20 software in order to evaluate the significance of the board properties. The statistical analyses were carried out separately for both glue additions and pressing pressure.

Table 1. Experimental parameters

| sample code | Additives | | | sp. Press. kg/cm^2 |
	UF % / substrate	Had % / UF	Wax % / substrate	
fa1				25
ba1				25
fa2	10	1	0.5	30
ba2				30
fa3				35
ba3				35
fb1				25
bb1				25
fb2	11	1	0.5	30
bb2				30
fb3				35
bb3				35
fc1				25
bc1				25
fc2	12	1	0.5	30
bc2				30
fc3				35
bc3				35
fd1				25
bd1				25
fd2	14	1	0.5	30
bd2				30
fd3				35
bd3				35

fxyf: date palm fronds, b : sugarcane bagasse x= a: resin level 10 % x= b: resin level 11 % x= c: resin level 12 % x= d: resin level 14 % y =1 pressed at specific pressure 25 kg/cm2y =2 pressed at specific pressure 30 kg/cm2y =3 pressed at specific pressure 35 kg/cm2.

Table 2. UF resin specification.

Spec. Weight g/cm3	Solid content %	Viscosity cp	Gel time Sec.	Free HCHO %
1.275	60.55	378	68	0.18

By-Products of Palm Trees and Their Applications   Materials Research Forum LLC
Materials Research Proceedings **11** (2019) 113-132   doi: https://doi.org/10.21741/9781644900178-7

Table 3. Chemical analysis of date palm frond and leaves.

Test/Material	DPR Frond	DPL Leaves	SCB Bagasse
H.Cell %	69.96	42.44	74.14
α- Cell %	38.99	22.66	42.58
Hemi %	30.97	18.78	31.56
lignin %	18.55	25.65	18.22
Pentosan %	24.79	16.65	29.22
Water Ex. %	14.16	24.24	3.97
Solvent Ex. %	1.6	4.95	1.82
Total Ex. %	15.76	29.19	5.79
NaOH %	29.26	57.79	34.5
Ash %	5.82	5.36	2.85

H. Cell % = holocellulose %, α-cell % = α-cellulose %, Hemi % = Hemicellulose %,Water Ex % & Solvent Ex % = water & solvent extractives % respectively, Total Ex. % = total Extractives %, NaOH % = solubility in 1 % NaOH, C.W % = Cold water solubility %.N.B. Hemicellulose content % & total Extractives can be calculated as follow   Hemicellulose % = (H.Cell %) – (α-cell %) and Total Extractives % = Water Ex % + Solvent

Table 4. Average weight of fresh cut date palm components.

Test/Material	Avg. Weight	DPR Frond	DPL Leaves
Frond with leaves	2.35	1.45	0.9
Moisture content	50.00	54.00	45 %
Dry weigh	1.18	0.67	0.50
%		56.8	43.2

By-Products of Palm Trees and Their Applications   Materials Research Forum LLC
Materials Research Proceedings **11** (2019) 113-132   doi: https://doi.org/10.21741/9781644900178-7

Results and discussion

Characterization of fibers

The chemical analysis and average weight of palm components were studied to examine their chemical constituents in comparison with sugar-cane bagasse, which is a commercial substrate used in local MDF mill; as well as, to estimate its available as alternative biomass instead of bagasse. Thermal analyses (DSC and non-isothermal TGA) were carried out to examine the interaction of UF adhesive with the foregoing agro-fibers. The importance of chemical constituents depends on the fact that, the relatively higher cellulose content may impart the fiber strength; while the extractives will represent serious influence on the steam digestion process, and consequently MDF produced. Because, steam digestion process regards important step operation in manufacturing process, which facilitates the individualization of the fibers, and enhanced MDF formation. Moreover, the available amount of this waste for MDF production and the positive interaction between fibers and UF during resinification and curing processes are also very required for production acceptable MDF.

The results obtained are illustrated in Tables 3- 6 and Figures 1 and 2. Table 3 shows that, leaves of date palm included higher extractives and lignin content together with lower cellulose and hemicellulose contents compared with palm frond component and SCB. The higher cellulose content together with hemicellulose in case of palm frond will play profound effect on fiber strength and self resinification of fiber during exposing to high temperature and pressure. This observation persuades us to recommend this waste in further work as precursor for production of MDF. Moreover, to exclude the serious problem effect of extractives included the leaf on the machines of steam and fibrilization processes. Also, this view is emphasized from the relatively higher average weight percentages of palm fronds (~ 57 %) than palm leaves (43%); Table 4.

For application in industrial scale and to reduce the effect of extractives, further study was carried out on possibility of using frond and leaves in blend, as a precursor fibrous of MDF production. To recommend the possible using blend-based fibers, its interaction with UF should be studied, in comparison with both palm frond and bagasse fibers. In this respect DSC and TGA of Fibers-UF samples were studied.

For thermal analysis of resinated fibers, the differential scanning Calorimetry (DSC) and the non-isothermal thermogravimetric (TGA) analyses are shown in Figures 1 and 2; while their calculated parameters are recorded in Tables 5 and 6.

With regard to DSC, Figure 1 shows that, the resinated fibers of palm leaves and frond blend with UF exhibits three peak temperatures like bagasse- UF. While, its DSC profiles exhibit an endothermic peaks at relatively higher temperature (248.99°C, 340.01°C and 563.22°C), than bagasse-UF (226.9°C, 272.0°C and 359.2°C). The $\triangle T$ value of each peak, which calculated from the peak temperature and onset temperature (T_P-T_o), is a measure of curing rate and recorded in Table 5. Table 5 shows that the thermal curing of palm fibers- UF behaves at lower temperatures than that of the commercial bagasse-UF. The $\triangle T$ value of 1st peak is about 10.68 °C for the whole palm fibers; while in case of bagasse fiber is 35.1°C. The relatively lower value of $\triangle T$ in the case of blending from palm components (palm mix), indicates a higher rate of curing. While, in the case of bagasse-UF, curing is started at lower temperature and gives a higher value of $\triangle T$, which means lower rate of curing. This trend is probably accompanied by shorting the time of penetrating the UF adhesive through whole palm fibers, and weakness the adhesion between the fibers, due to fast the condensation of UF adhesive on fiber surface. This unaccepted trend is reduced on using palm frond, whereas the difference values of onset temperature and peak temperature ($\triangle T$) are also increased than those observed in bagasse-UF.

118

Where, $\triangle T$ for the main peaks are 47.15 °C and 76.29°C; while for bagasse are 35.1 and 36.7°C.

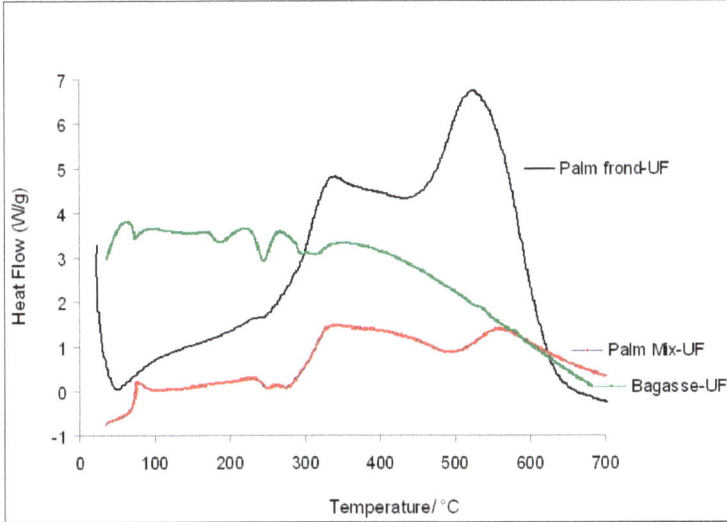

Figure 1. DSC of palm frond fibers with UF in comparison with Bagasse fibers-UF.

Table 5. DSC analysis of the main Endothermic event of Bagasse- and palm components-UF samples.

Sample	First step of curing reaction			Second step of curing reaction		
	T_o (°C)	T_P (°C)	$\triangle T$ (°C)	T_o (°C)	T_P (°C)	$\triangle T$ (°C)
Bagasse -UF	191.8	226.9	35.1	322.5	359.2	36.7
	246.9	272.0	25.1			
Palm mix-UF	238.31	248.99	10.6 8	506.62	563.22	56.60
	288.84	340.01	51.17			
Palm Frond-UF	287.39	334.54	47.15	458.41	534.70	76.29

Materials Research Forum LLC
doi: https://doi.org/10.21741/9781644900178-7

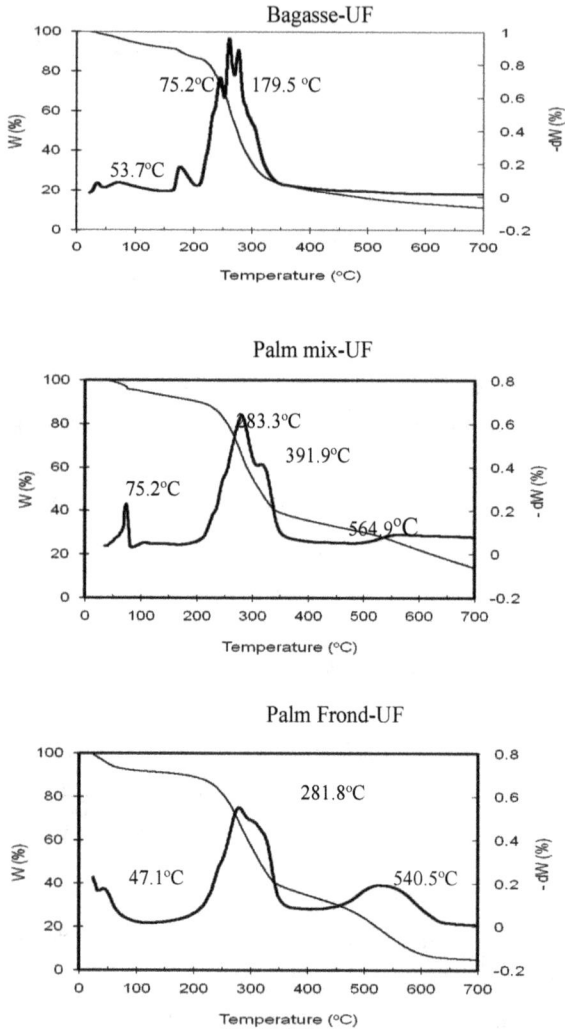

Figure 2. Thermo-gravimetric weight loss and derivative curves of palm frond fibers with UF in comparison with Bagasse fibers-UF.

Table 6. TGA kinetic parameters of Bagasse- and palm components-UF samples and their

Sample code in chart	stage	Temp. range $^{\circ}$C	DTG peak Temp., $^{\circ}$C	"n"	R^2	SE	E_a kJ/ mole	Wt. remain %
Bagasse-UF								
	1st	r.t.-100.4	-	-	-	-	-	94.46
	2nd	157.2-211.6	179.5	1.0	0.985	0.168	173.01	86.24
	3rd	211.6-390.5	261.5	2.0	0.980	0.230	127.25	20.82
Palm Mix-UF	1st	r.t- 106.5	75.2	-	-	-	-	94.57
	2nd	164.8-302.7	283.3	1.0	0.988	0.1087	84.82	53.33
	3rd	302.7-328.4	391.9	1.5	0.985	0.146	117.90	40.22
	4th	461.3- 700.0	564.9	1.5	0.968	0.156	94.30	6.86
Palm Frond-UF	1st	r.t- 104.33	33.7	-	-	-	-	91.97
	2nd	150.4 - 398.5	281.8	1.0	0.989	0.125	64.18	34.74
	3rd	398.5-651.9	540.5	0.5	0.982	0..124	124.06	5.53

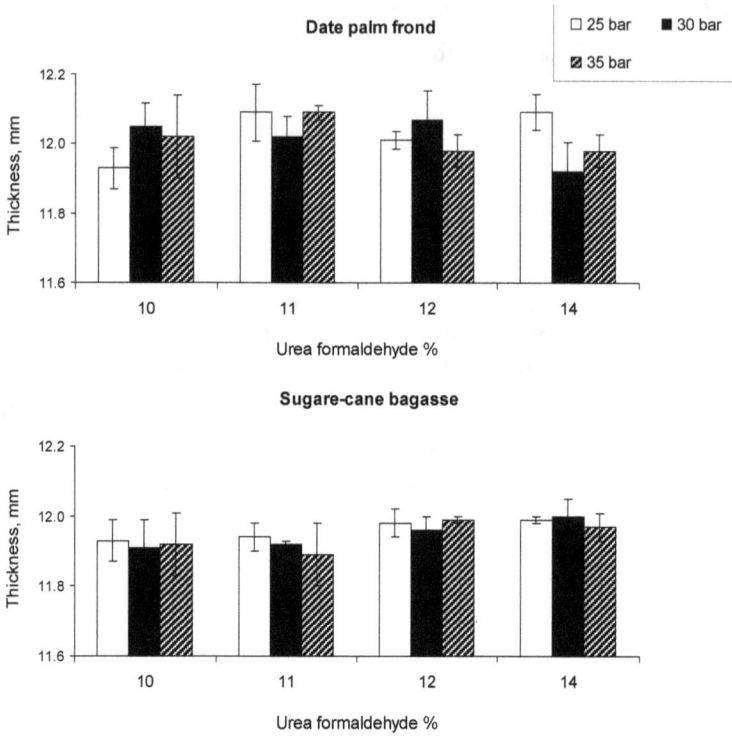

Figure 3. Variation of Thickness of MDF of palm frond and Bagasse fibers versus UF %.

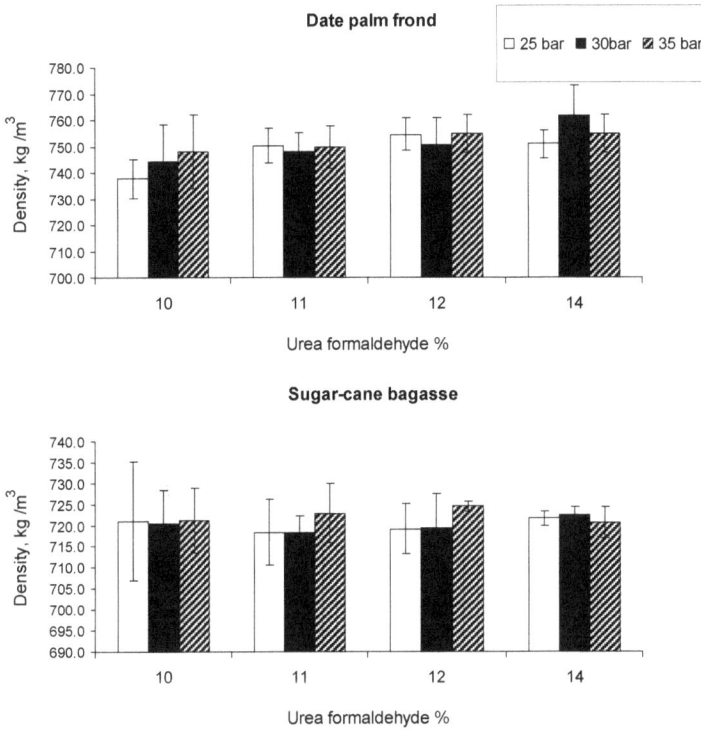

Figure 4. Variation of Gross density of MDF from palm frond and Bagasse fibers versus UF % and specific pressure.

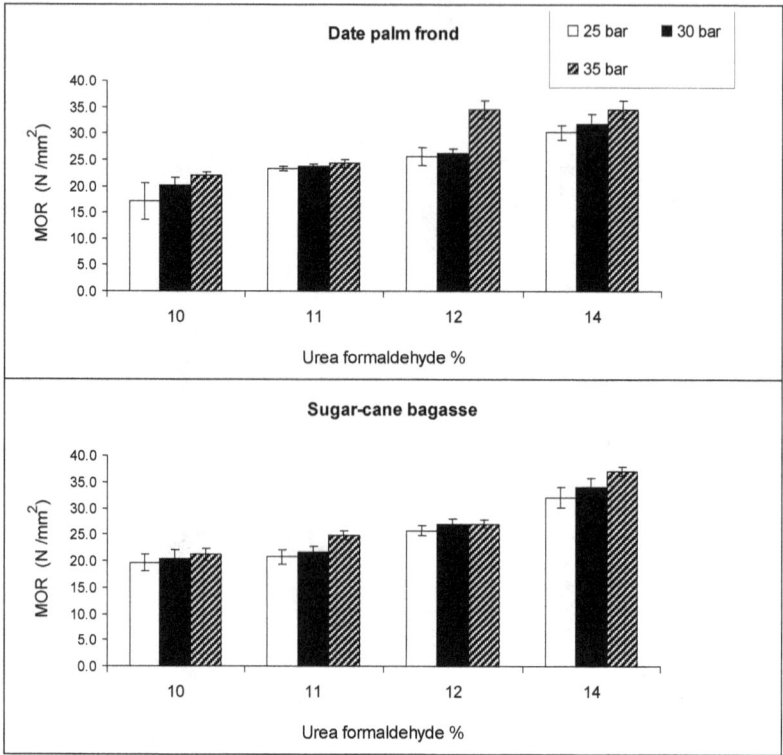

Figure 5. Variation of MOR of MDF from palm frond and Bagasse fibers versus UF % and specific pressure.

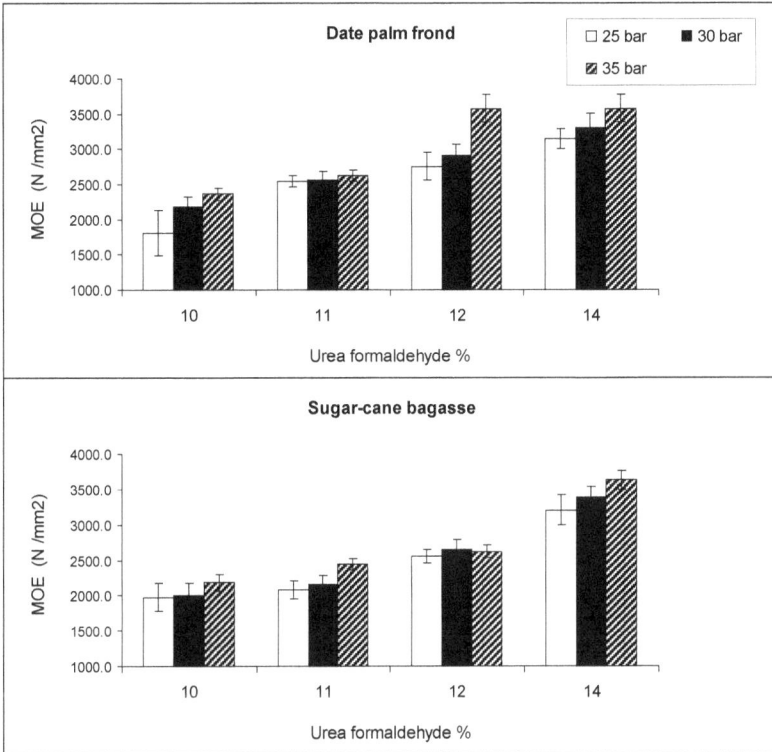

Figure 6. Variation of MOE of MDF from palm frond and Bagasse fibers versus UF %. and specific pressure.

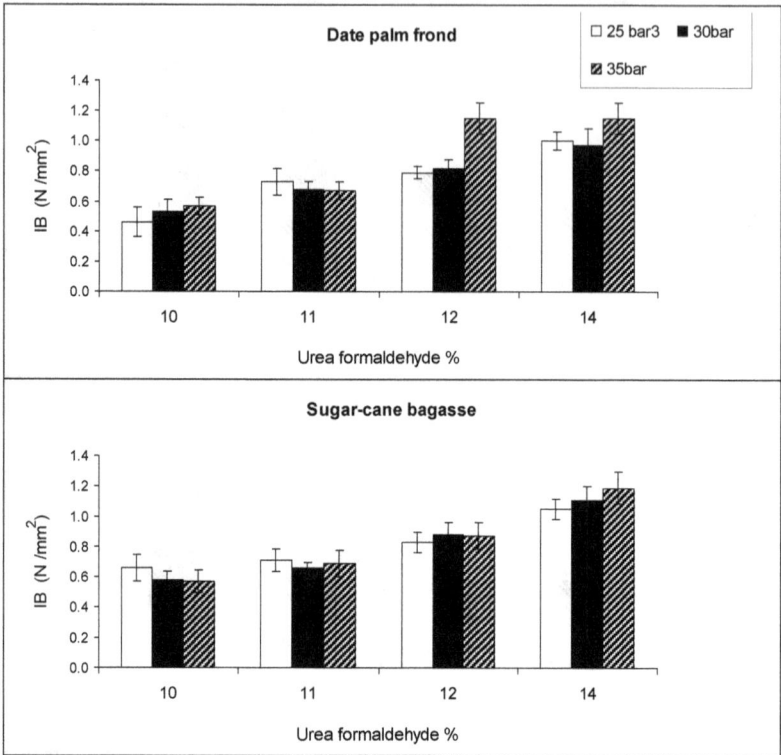

*Figure 7. Variation of IB of MDF from palm frond and Bagasse fibers versus UF %
and specific pressure.*

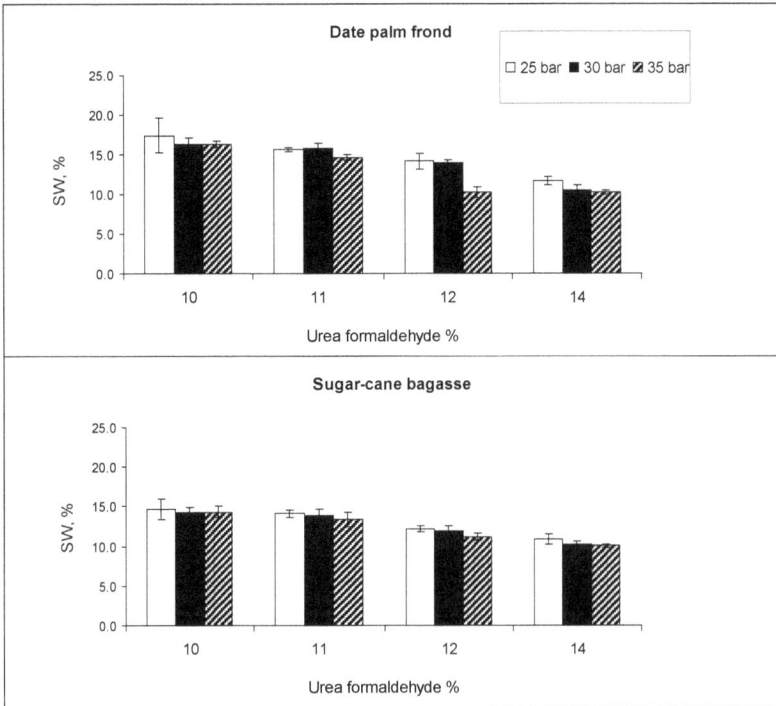

Figure 8. Variation of water swelling of MDF from palm frond and Bagasse fibers versus UF % and specific pressure.

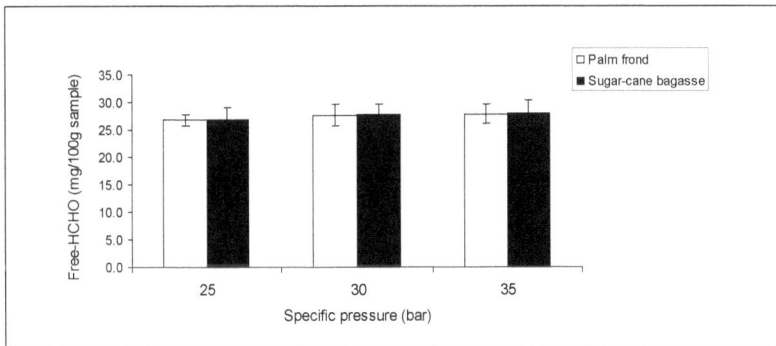

Figure 9. Free-HCHO of MDF palm frond and Bagasse fibers versus specific pressure

With regard to non-isothermal thermogravimetric analysis, Figure 2 illustrates the TGA and DTG curves of the blend of palm components (palm mix) and palm frond resinated fibers with UF, in comparison with bagasse fibers-UF. Table 6 summarizes their kinetics parameters. Figure 2 shows that, there are three decomposition steps for bagasse-UF and Palm frond-UF resinated fibers. The first one at temperature lower than 100 °C, indicates the evolution of adsorbed water, in addition to second and third steps, in the ranges from 150-399 °C and 211-652 °C corresponding to volatilization and carbonization (degradation of selected adhesive systems. The ranges of degradation stages of palm frond-UF (150.4 - 398.5°C and 398.5-651.9 °C) are greater than bagasse-UF (157.2-211.6 °C and 211.6-390.5 °C). Moreover, the peak temperature of these degradation stages in case of Palm frond (281.8°C and 540.5 °C) is higher than bagasse (179.5 °C and 261.5 °C). This indicates the slowness degradation of resinated palm frond than bagasse (higher thermal stability). This observation is in agreement with that found by DSC which supports the curing of UF on palm frond fibers is occurred at longer time.

With regard to the blend of palm components (leaves + frond), it is observed that the volatilization is occurred in 2 stages with DTG peaks 283.3 °C and 391.9 °C; while the carbonization stage with DTG peak at 564.9 °C. This is probably related to the relatively higher extractives included palm leaves, which resist to some extent the role of steam and digestion process, and consequently it may affect the individualization of the fibers and interaction with UF. This extractive leads to increase the activation energy required for volatilization stage (ΣEa = 202.7 kJ/mole) than the case of palm frond (64.2 kJ/mole and bagasse 173.01kJ/mole, respectively).

Based on the foregoing results of chemical constituents and thermal analysis, the palm frond was candidate as substrate for MDF production. Till it possible to use as alternative to commercial bagasse, the properties of MDF made from both fibers are compared.

Properties of MDF
The influence of preparation parameters, e.g., UF level and pressing pressure on the performance of MDF produced from palm frond and SCB fibers are illustrated graphically in Figures 3-9.

Figures 3 and 4 show that, changing the UF % together with specific pressure provided board from palm frond and bagasse with thickness ~ 12 mm; while for board density in case of palm frond fibers is around 750 kg/m^3. This value is higher than that made from bagasse (~720 kg/m^3); Figure 4. Higher board density is observed in case of board made from palm frond fibers at relatively higher UF% (14%) and applied specific pressure (35 bar) (it reached 762 kg/m^3). Static bending (MOR and MOE), and IB are greatly affected by changing both UF% and specific pressure (Figures 5-7). At relatively lower UF%, the specific pressure has a profound effect (improvement) on strength properties of MDF than higher ones. Where, for boards made from palm frond fibers, at 10%UF, changing the specific pressure accompanied by increasing in MOR, from 17 MPa to 22 MPa, MOE from 1811.6 MPa to 2368.8 MPa, and IB from 0.46 MPa to 0.57 MPa. While, on applying 14%, the changes in MOR, MOE and IB with increasing the specific pressure from 25 bar to 35 bar are from 30 to 34 MPa, from 3153 to 3579 MPa, and from 1.0 to 1.15 MPa, respectively. A similar improvement is observed in case of SCB-based MDF. It is interesting to note that, both palm frond and bagasse fibers at 12- 14% UF, and different specific pressure provided MDF boards fulfill the requirement of high grade MDF reported according to ANSI standard. In other words, the increasing in UF level was more significant on producing high quality MDF than specific pressure.

Results for the thickness swelling (SW) of the various MDF produced under the foregoing parameters are illustrated in Figure 8. It is clear that, the SW property of MDF decreased

Materials Research Forum LLC
doi: https://doi.org/10.21741/9781644900178-7

(improved) considerably as UF % increased from 10% to 14%. Greater reducing in this property is observed at relatively lower specific pressure (25 bar), where SW decreased from 17 % to 11% and from 15% to 10%, in case of palm frond- and Bagasse-based MDF, respectively. Increasing the applied specific pressure to 35 bar, together with UF% up to 14% provides SW value ~10%, in both types of fibers-based MDF. However, the changing in pressing pressure is not significant on free-HCHO of MDF produced, where its value between 25.88 to 27.95 mg/100g board.

The explanation of the foregoing data may be ascribed to enhance the bond formation with increasing adhesive level and specific pressure, as well as due to the substrate constituents. Higher cellulose and hemicelluloses as well as lower extractives contents of sugar cane bagasse provide fibers easily adhered together during MDF formation, than palm frond fibers. Higher adhesive level and pressing pressure enhancing the affinity of palm frond fibers to response for adhering together during hot pressing.

For specifying the application of MDF produced from both fibers, as reported and based on ANSI and EN Standards, the MDFs produced from palm frond and bagasse fibers, especially with 12-14% UF and Specific pressure 30-35 bar have higher values than the requirements for general purposes. These boards fulfill the requirements for strength and thickness swelling properties reported in ANSI Standard. According to TS and HCHO these boards are possible for load-bearing applications and specified as E2 type boards.

The ANOVA analysis was carried out for mechanical properties and physical properties as shown, for example, in Tables 7-8. These data show a significant differences between the static bending of MDF samples made from bagasse fiber , fronds fiber, and mixed frond and frond leaves, at different addition of UF adhesive. The levels of glue addition are slightly significant on properties of MDF boards made from mixed and frond fiber than those made from bagasse, with higher specific pressure of hot press.

For economic potential of using palm frond fibers, it will preserve the additional cost required for storage of bagasse and transportation. Moreover, this waste is available in many provinces, and persuades to construct wood Mills, without concerning on the Mills of upper-Egypt.

Conclusions
Due to the availability of date palm at different provinces, and for trial to minimize the economic and environmental impacts from storage of sugar-cane bagasse (SCB), as a substrate for production MDF, in this article we evaluated the palm fibers components on the quality of produced MDF. The performance of palm-based MDF was also evaluated, in comparison with traditional prepared SCB-based MDF. The results of chemical constituents of date palm components (leaves and fronds), their average weights, as well as Fibers-UF interactions (via DSC and TGA studies) lead us to recommend the use of date palm frond (DPF) in production of MDF. DPF-based MDF especially with 12-14% UF and specific pressure 35 bar have acceptable properties as compared to those produced from sugar-cane bagasse. It is static bending (MOR and MOE), internal bond strength (IB) fulfilled the high requirements in mechanical and thickness swelling properties of ANSI and EN standards. Where, their values were 34.4 MPa, 3579.2 MPa, 1.15 MPa and 10.2 %, respectively.

Acknowledgment
The authors are grateful to Nag-Hamady Co. for Fiberboard-Quena, Egypt for providing the raw materials and equipment for the experimental study.

References

[1] A. Kargarfard, A.J. Latibari, The performance of corn and cotton stalks for medium density fiberboard production, Bioresources 6 (2011) 147-57.

[2] A.A. Adam, A.H. Basta, H. El-Saied, Performance of bagasse-based medium density fiberboard produced from different steam digestion retention times. Forest Product Journal 62-5 (2012) 400–405. https://doi.org/10.13073/0015-7473-62.5.400

[3] M. Akgul, A. Toslughu, Utilizing peanut husk in the manufacture of medium density fiberboards, Bioresource Technology, 99 (2008) 5590-5594. https://doi.org/10.1016/j.biortech.2007.10.041

[4] M. Akhtar, W.R. Kenealy, E.G. Horn, R.E. Swaney, J.E. Winandy, Method of making medium density fiberboard, US Patent No: US 2008/0264588 A1, 2008.

[5] ANSI 208.2., American National Standardization Institute. Medium Density Fiberboard, 1994.

[6] ASTM D1037, Standard test for evaluation the properties of wood-based and particle panel materials, American society and materials, Philadeplia, PA.: ASTM D1037-94, 1994.

[7] ASTM E1690, Determination of ethanol extractives in bagasse, in Annual book of ASTM Standards, Philadelphia, PA: American Society for Testing and Materials 11.05, E1690, 2003.

[8] N. Ayrilmis, J.E. Winandy, Effect of post heat treatment on surface characteristics and adhesive bonding performance of MDF, Materials and Manufacturing Processes 24 (2009) 594-599. https://doi.org/10.1080/10426910902748032

[9] A.H. Basta, H. El-Saied, New approach for utilization of cellulose derivatives metal complexes in preparation of durable and permanent colored papers, Carbohydrate. Polymers 74-2 (2008) 301-308. https://doi.org/10.1016/j.carbpol.2008.02.021

[10] A.H. Basta, V. Fierro, H. El-Saied, A. Celzard, Effect of Deashing Rice Straws on their derived Activated Carbons produced by Phosphoric Acid Activation, Biomass and Bioenergy 35 (2011) 1954-1959. https://doi.org/10.1016/j.biombioe.2011.01.043

[11] A.H. Basta, H. El-Saied, J.E. Winandy, R. Sabo, Preformed amide-containing biopolymer for improving the environmental performance of synthesized urea–formaldehyde in Agro-fibre Composites, Journal of Polymers and the Environment 19-2 (2011) 405–412. https://doi.org/10.1007/s10924-011-0286-4

[12] A.H. Basta, H. El-Saied, O. El-Hadi, C. El-Dewiny, Evaluation of rice straw-based hydrogels for purification of wastewater. Polymer Plastic Technology and Engineering 52-11 (2013) 1074-1080. https://doi.org/10.1080/03602559.2013.806548

[13] A.H. Basta, H. El-Saied, V.F. Lofty, Performance assessment of deashed and dewaxed rice straw on improving the quality of RS-based composites, RSC Advances 4-42 (2014) 21794-21801. https://doi.org/10.1039/c4ra00858h

[14] A.H. Basta, H. El-Saied, V.F. Lotfy, Performance of rice straw-based composites using environmentally friendly polyalcoholic polymers-based adhesive system, Pigment and Resin Technology 42-1 (2013) 24-33. https://doi.org/10.1108/03699421311288733

[15] A.H. Basta, H. El-Saied, E.M. Deffallah, Optimising the process for production of high performance bagasse-based composites from rice bran-UF adhesive system, Pigment and Resin Technology 43-4 (2014) 212-218. https://doi.org/10.1108/prt-08-2013-0077

[16] A.H. Basta, H. El-Saied, E.M. Deffallah, Effects of denaturisation of rice bran and route of synthesis of RB-modified UF adhesive system on eco-performance of agro-based composites, Pigment and Resin Technology 54-3 (2016) 172-183. https://doi.org/10.1108/prt-04-2015-0037

[17] E. Ciannamea, P.M. Stefani, R.A. Ruseckaite, Medium Density fiberboard from rice husks and soybean protein concentrate-based adhesive, Bioresource Technology 101 (2010) 818-25. https://doi.org/10.1016/j.biortech.2009.08.084

[18] A.W. Coats, J.P. Redfern, Kinetic parameters from thermogravimetric data, Nature 201 (1964) 68–72. https://doi.org/10.1038/201068a0

[19] Y. Copur, C. Gular, C. Tascioglu, A. Tozluoglu, Incorporation of hazelnut shell and husk in MDF production, Bioresource Technology 99 (2008) 7402-7406. https://doi.org/10.1016/j.biortech.2008.01.021

[20] D.E. El Nashar, S.L. Abd-El-Messieh, A.H. Basta, Newsprint paper waste as a fiber reinforcement in rubber composites, Journal of Applied Polymer, Science 91-5 (2004) 3410-3420. https://doi.org/10.1002/app.13726

[21] H. El-Saied, M.H. Fadl, A.H. Basta, Properties of bagasse hardboard made by in-situ formation of phenol-lignin formaldehyde resin, Polymers & polymer composites 4-7 (1996) 519-522.

[22] H. El-Saied, A.H. Basta, M.E. Hassanen, H. Korte, A. Helal, Behaviour of Rice-Byproducts and Optimizing the Conditions for Production of High Performance Natural Fiber Polymer Composites, Journal of Polymers and the Environment 20-3 (2012) 838-847. https://doi.org/10.1007/s10924-012-0439-0

[23] EN 120, The European Standard for "determination of free formaldehyde in wood-based panels Perforator method, 1992.

[24] H.R. Faraji, Investigation on properties of medium density fiberboard (MDF) produced from bagasse: M.Sc. Thesis, Tarbiat Modares University, Faculty of Natural Resources and Marine Science, 1998.

[25] N.A. Fathy, V.F. Lotfy, A.H. Basta, Comparative study on the performance of carbon nanotubes prepared from agro-and xerogels as carbon supports. Journal of Analytical and Applied Pyrolysis 128 (2017) 114-120. https://doi.org/10.1016/j.jaap.2017.10.019

[26] S. Halvarsson, H. Edlund, M. Norgren, Properties of MDF based on wheat straw and melamine modified urea- formaldehyde (UMF) resin, Industrial Crops and Products 28 (2008) 37-46. https://doi.org/10.1016/j.indcrop.2008.01.005

[27] Y. Hossein, Canola straw as a bio-waste resource for medium density fiberboard manufacture, Waste Management 29-10 (2009) 2644-2648. https://doi.org/10.1016/j.wasman.2009.06.018

[28] A. Istek, D. Aydemir, H. Eroglu, Surface properties of MDF coated with calcite/clay and effects of fire retardants on surface properties, Maderas Ciencia y tecnología 14-2 (2012) 135-144. https://doi.org/10.4067/s0718-221x2012000200001

[29] G. Jayme, P. Sarten, Über die quantitative bestimmung von pentosen mittels bromwasserstoff-sauer, Naturewise 28-52 (1940) 822–823. https://doi.org/10.1007/bf01489045

[30] S. Lee, T.F. Shupe, C.Y. Hse, Mechanical and physical properties of agro-based fiberboard. Holz-als-Roh-und-Werkstoff 64 (2006) 74-79. https://doi.org/10.1007/s00107-005-0062-z

[31] X. Li, Z. Cai, J.E. Winandy, A.H. Basta, Effect of Oxalic Acid and Steam Pretreatment on the Primary Properties of UF-bonded Rice Straw Particleboards, Industrial Crops and Products 33 (2011) 665- 669. https://doi.org/10.1016/j.indcrop.2011.01.004

[32] H. Mahmoudi, G. Hosseininia, H. Azadi, M. Fatemi, Enhancing date palm processing marketing and pest control through organic culture, J Organic Systems 3-2 (2008) 29-39.

[33] E. Roffael, C. Behn, D. Krug, A. Weber, C. Hartwig-Gerth, G. Gräfe, UF- und PMDI-Doppelbeleimung bei Faserplatten,. Holz-Zentralblatt 137 (2011) 1216–1217.

[34] V. Sivakumar, M. Asaithambi, P. Sivakumar, Physico-chemical and adsorption studies of activated carbon from agricultural wastes, Advances in Applied Science Research 3-1 (2012) 219-226.

[35] TAPPI Test Method T429, Determination of cellulose, In TAPPI Test Methods. Atlanta, GA: Technical Association of the Pulp and Paper Industry, 2012.

[36] TAPPI .Test Method T222, Determination of acid insoluble lignin, In TAPPI Test Methods. Technical Association of the Pulp and Paper Industry, 1998.

[37] V.K. Thakur, A.S. Singha, M.K. Thakur, Graft copolymerization of methyl acrylate onto cellulosic biofibers: Synthesis, characterization and applications. Journal of Polymers and the Environmen, 20-1 (2012) 164-174. https://doi.org/10.1007/s10924-011-0372-7

[38] V.K. Thakur, A.S. Singha, M.K. Thakur, Surface modification of natural polymers to impart low water absorbency. International Journal of Polymer Analysis and Characterization, 17 (2012) 133-143. https://doi.org/10.1080/1023666x.2012.640455

[39] V.K. Thakur, A.S. Singha, M.K. Thakur, In-air Graft copolymerization of Ethylacrylate onto Natural Cellulosic Polymers, International Journal of Polymer Analysis and Characterization, 17 (2012) 48-60. https://doi.org/10.1080/1023666x.2012.638470

[40] V.K. Thakur, M.K. Thakur, R.k. Gupta, Graft copolymers of natural cellulose for green composites, Carbohydrate Polymers, 104 (2014) 87-93. https://doi.org/10.1016/j.carbpol.2014.01.016

[41] H.R. Taghiyari, P. Nouri, Effect of nano-wollastonite on physical and mechanical properties of medium density fiberboard. Maderas. Ciencia y tecnología 17-4 (2015) 833-842. https://doi.org/10.4067/s0718-221x2015005000072

[42] H.R. Taghiyari, P. Nouri, Effect of wollastonite on physical and mechanical properties of medium density fiberboard (MDF) made from, Wood fibers and camel-thorn. Maderas. Ciencia y tecnología 18-1 (2016) 157-166. https://doi.org/10.4067/s0718-221x2016005000016

[43] L.F. Wise, M. Murphy, A.A. D'assiece, Chlorite holocellulose, its fractionation and bearing on summative wood analysis and on stiudes on the hemicellulosi. Paper Trade Journal 122-2 (1946) 35-43.

[44] X.P. Ye, J. Julson, M. Kuo, A. Womac, D. Myers, Properties of medium density fiberboard made from renewable biomass, Bioresource Technology 98 (2007) 1077-1084. https://doi.org/10.1016/j.biortech.2006.04.022

[45] H. Younesi-Kordkheili, A. Pizzi, Properties of plywood panels bonded with ionic liquid-modified lignin–phenol–formaldehyde resin, The Journal of Adhesion, http://dx.doi.org/10.1080/00218464.2016.1263945, 2017

Energy and Fertilizers

By-Products of Palm Trees and Their Applications
Materials Research Proceedings 11 (2019) 135-142

Materials Research Forum LLC
doi: https://doi.org/10.21741/9781644900178-8

Scenarios of Palm-Oil Biodiesel in the Mexican Transportation Sector

Jorge M. Islas S.[1,a*], Genice K. Grande A.[1,b] and Fabio L. Manzini Poli[1,c]

[1]Instituto de Energías Renovables, Universidad Nacional Autónoma de México, México

[a]jis@ier.unam.mx, [b]gkga@ier.unam.mx, [c]fmp@ier.unam.mx

Keywords: palm-oil biodiesel; scenarios; Mexican transportation sector; GHG mitigation; cost-benefit analysis

Abstract. This work analyses the environmental and economic feasibility of producing palm oil-based biodiesel in Mexico in order to substitute of diesel fuel consumption using B5 first and B10 to 2031 in the transportation sector. Two scenarios were created by projecting demand and costs for biodiesel as well as greenhouse gases emissions reduction and area requirements. In the economic section, the cost-benefit analysis of biodiesel and the mitigation costs of carbon dioxide were estimated. This work shows that Application of tax incentives could make biodiesel competitive against diesel.

Introduction

Biodiesel has been used in some countries as a substitute for diesel fuel in the transportation sector. In 2017 the production of biodiesel was increased by 82% compared to the year 2016 in the European Union. Countries with the highest production of rapeseed-based biodiesel are Germany 4,005 K t/year (151 PJ[1]), Spain 3,398 K t/year (128 PJ), and The Netherlands 2,505 K t/year (95 PJ) [1]. Production of biodiesel in United States mainly derived from soybean oil was 9,275 K t/year (351 PJ) in 2016, which represented an increase of 40% compared to the year 2015 [3].

Mexico has the problem of declining proven oil reserves and official sources estimated them in 8.5 years [4]. On the other hand, the use of energy generates a large amount of greenhouse gases (GHG), so in 2015 at the country level were emitted 442.3 million tons of CO_2 (Mt CO_2) of which 33% were generated by the road transportation sector; 7% corresponds to diesel vehicles [5].

In 2016, internal demand of diesel fuel in the Mexican transportation sector accounted nearly 26% in relation to the other fuels and grew at an average annual growth rate of 2% in the last 10 years [6].

The use of palm oil in Mexico as B5 and B10 can help reduce CO_2 emissions and reduce dependence on fossil fuels in the transportation sector. Given that the main raw material is vegetable oil, biodiesel is becoming a notable factor for promoting the regional development in Mexico.

In this work, we develop scenarios to use B5 and B10 in the Mexican transportation sector and we evaluate these scenarios in terms of a cost-benefit analysis, the amount of carbon dioxide CO_2 reduced and the area cultivated with oil palm. Likewise, CO_2 mitigation costs were estimated, and the impact of tax incentives on the economic feasibility of biodiesel was analyzed.

Methodology

To develop this work the following steps are made base on [7]:

---

[1] The calorific value of biodiesel considered in this article is 37.8 MJ/Kg [2]

A. The trend scenario corresponds to the scenario based on diesel while the alternative scenario was developed to use in a large scale the biodiesel in the Mexican transport sector.
B. Scenarios were built and simulated using LEAP (Long-range Energy Alternative Planning System [8]).
C. In this study the base year is 2005, due to the data were available for that year meanwhile the period of analysis was until 2031
D. The energy consumption was obtained in the trend scenario and the same energy consumption was considered for the alternative scenario.
E. For the alternative scenario oil palm-cultivated area requirements were obtained.
F. The CO_2 emissions were calculated for each of the analyzed scenarios.
G. The overall and mitigation costs of the alternative scenario were calculated.
H. Finally, the economic feasibility of alternative scenario is analyzed considering the implementation of tax incentives.

Scenarios construction
1. Trend scenario
The establishment of the trend scenario were conducted based on the followings two components:
1. Evolution of diesel vehicle fleet: The first step consists of estimating the evolution of vehicle fleet in the reference year, based on the existing stock, sales and vehicles that will be retired over the analysis period. This fleet was divided into the following categories: a) heavy-duty trucks, and b) passenger vehicles and private cars. The statistics on heavy-duty trucks reported at federal level are considered as a good approximation to depict the size of this fleet at national level. According to these, 97% of diesel-powered vehicle fleet was composed of heavy-duty trucks, while passenger vehicles accounted for slightly over 2%. The remainder was private cars. The second step consists of assigning a life cycle profile for each vehicle category so that the distribution of vehicles of different ages can be described in the reference year. According to our results the heavy-duty fleet is very old, since 78% of its vehicles are over 10 years old. In the 2006-2031 period, the growth of vehicle fleet is determined by sales and the survival of vehicles as they get older. Vehicle sales totaled 450 thousand for heavy-duty trucks [9], 12,500 for passenger vehicles [10], and 1,000 for private cars [11] in the reference year. With regard to the trend in vehicle sales, it is considered an average annual growth rate of 4% for both, heavy-duty and passenger vehicles, according to their historical growth [9]. Thus, it is expected that heavy-duty and passenger vehicles continue to grow at their historical growth rates; however, it is foreseen a further expansion of vehicle sales, owing to the replacement of some units of the existing vehicle stock, which is mostly composed of old units. In order to gradually replace existing vehicle stock, a survival profile describing the retirement of old vehicles is used. This profile represents the percent survival of vehicles as they get older as well as the percent share of vehicles that gradually will be retired from the existing vehicle stock in the country and always takes a percent share of 100% during the first year. This profile can be expressed by the following function:

$$F(t) = F(t-1) e^{tK} \qquad (1)$$

Where F is the fraction of surviving vehicles and t is the age in years of the vehicle.

$$K = \frac{\ln F(t) - \ln(F(t-1))}{t} \qquad (2)$$

K represents a decreasing rate of the existing vehicle stock in time t, and takes a negative value.

The survival profile of heavy-duty trucks and passenger vehicles was represented by these equations. In both cases K was obtained by averaging calculations for 4 years, resulting in K= -0.01 and K= -0.04 for heavy-duty trucks and passenger vehicles, respectively. For private cars, and due to the fact that related information was not available, the value of K was assumed -0.0236, which corresponds to the one reported for gasoline vehicles in Mexico [12].

Once the survival profiles and existing stocks in the reference year as well as the annual growth in future sales have been obtained for each vehicle category, the trend in vehicle fleet is simulated using LEAP for the analyzed period. Thus, for each year of analysis, annual sales are summed to the existing vehicle stock in the reference year, while the number of vehicles that will be retired from this stock is subtracted according to the survival profile of each vehicle category.

According to our results diesel vehicle fleet would reach 20.45 million in the year 2031, of which 16.5 million corresponds to heavy-duty trucks, 480 thousand to passenger vehicles, and 3.3 million to private cars, respectively (see Figure 1).

2. Estimation of diesel demand: In order to calculate the total annual diesel fuel demand in the transportation sector it is required to estimate for each vehicle category the fuel consumption of existing vehicle fleet in the year t. This is obtained from an estimate of the number of existing vehicles in the year t, their average annual mileage (in kilometers), and the average annual diesel consumption.

Fuel consumption of heavy-duty trucks and passenger vehicles was calculated based on data reported for these vehicle categories in the USA [13]. The annual mileage (in kilometers) was calculated using the average value reported in the 2002 National Inventory of Emissions. Based on these data, energy consumption for each vehicle category was calculated in LEAP by multiplying the existing vehicle stock in the year t, the annual mileage (in kilometers), and the fuel consumption. In trend scenario, the vehicle fleet would consume 1,543 PJ by the year 2031.

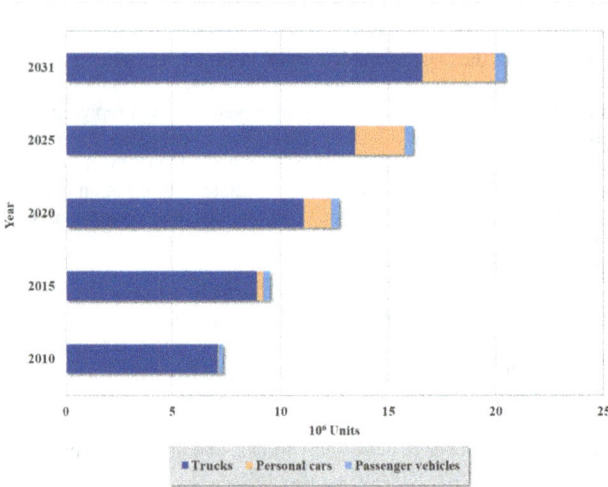

Fig. 1. Evolution of diesel-powered vehicle fleet in the Mexican transportation sector [7].

2. Alternative scenario

The alternative scenario considers the same evolution in vehicle fleet at national level that the Trend Scenario and aims at identifying the main parameters related to the substitution of diesel fuel in Mexico. The main assumptions for the construction of this scenario are:

- It considers the use of B5 and B10, which implies the massive use of palm oil-based biodiesel.
- The plantations and infrastructure for production and distribution of biodiesel have been developed in the country. This considers a potential for growing oil palm, taking account of requirements such as water, temperature, soil and fertility have an optimal gross potential of 2.5 million hectares (Ha) in Mexico [14] and a yield of 3,239 liters (L) per hectare or 20 tons of fresh fruit bunches (FFB) per Ha [7].
- The calorific value of biodiesel is lower 13% in mass terms than that of diesel fuel [15].
- It considers that the emissions of carbon dioxide (CO_2) for biodiesel (B100) are neutral.
- It considers that biodiesel is produced in plants with a capacity of 38 million of L per year, which requires investment costs of $12.5 million [7].
- The structure of operating costs used in this article, in terms of USD/L of biodiesel is shown in Figure 2. As can be seen in this figure, most of the costs per liter of biodiesel are occupied by raw material (79%) and the rest costs such as services and fuels, operation and maintenance and other costs (21%). It is important to note that there is income from the sale of glycerol as a co-product that can represent a benefit of 17% of the unit cost per liter of palm oil biodiesel, which has a cost of 0.380 USD/L.
- This study considers a discount rate of 10%.

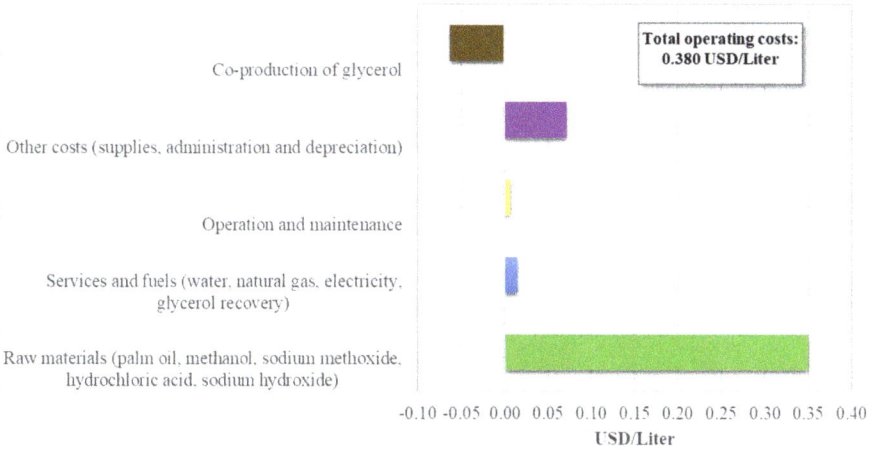

Fig. 2. Structure of operating costs for Palm-Oil biodiesel production (USD/L) [7].

Results
The biodiesel replaces 9% of diesel fuel consumption of the transport sector in the alternative scenario in 2031. This represents 8% of total energy required in the period analysis (see Figure 3).

The small biodiesel plants need a production capacity of 133.7 million of L/year by 2031 that means 113 biodiesel plants would be required. The cumulative investment costs of these plants would be approximately $7882 million.

To meet this scenario, it is required approximately 1 million hectares of cultivated area to satisfy the palm-oil biodiesel demand. This cultivated area is far below of good resource potential equivalent to 2.5 million of Ha. These results indicate that the proposed alternative biodiesel scenario would be enough to meet the palm-oil biodiesel demand of Mexican road transport sector.

---

[2] In this article the monetary unit is US dollar of year 2007.

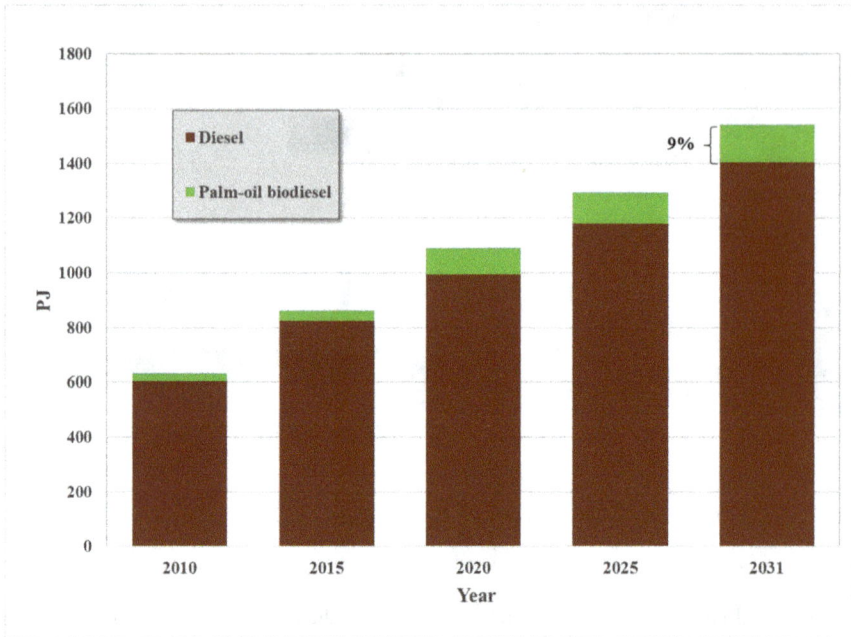

Fig. 3. Palm-oil biodiesel and diesel demand in alternative scenario [7].

The CO_2 emissions would be reduced by about 10 million tons by 2031. This reduction would also account for 9% in relation to trend scenario. The cumulative reduction could arrive to 148 million tons of CO_2 emissions (see Figure 4), which represents a total reduction of 8%.

The results show that biodiesel costs, when compared with the costs of diesel fuel, would represent overall costs of 2.3 billion dollars (BUSD) (4.7 BUSD of avoided costs of diesel fuel vs 7 BUSD of the costs of using biodiesel). Finally, mitigation costs would total 16 USD/ton of CO_2. Nevertheless, we found that if diesel price and biodiesel prices, this last one exempting the Special Tax on Production and Services (IEPS), are compared, the massive use of biodiesel would lead to a net benefit of 1.24 BUSD.

Conclusions

The potential to produce oil palm in México, is more than enough to cover the needs of B5 and B10 of the alternative scenario.

CO_2 emissions would reach cumulative reductions of 148 million tones that means 8% of the trend scenario.

Finally, the cost–benefit analysis points out that the substitution of diesel fuel for palm oil-based biodiesel is feasible when a tax-exemption policy (e.g. exemption of IEPS) is implemented.

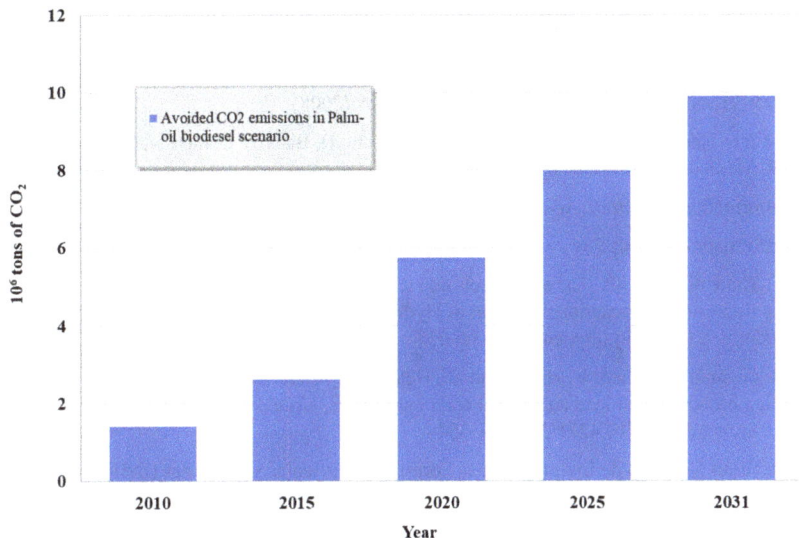

Fig. 4. Avoided CO2 emissions under palm-oil biodiesel scenario [7].

Acknowledgments
The authors acknowledge Project CEMIE-BIO "Clúster Biocombustibles Sólidos - Línea 4 Sustentabilidad y Políticas Públicas de Biocombustibles Sólidos, Fondo de Sustentabilidad Energética SENER-CONACYT, for the partial financing of this work.

References

[1] European Biodiesel Board (EBB), Statistics, The EU biodiesel industry, Available at: http://www.ebb-eu.org/stats.php, 2018.

[2] D. Pimentel, Global economic and environmental aspects of Biofuels. Ed. Taylor & Francis Group, 2012, pp. 428.

[3] National Biodiesel Board (NBB), Available at: http://www.biodiesel.org, 2018.

[4] Petróleos Mexicanos (PEMEX), Reservas 1P al primero de enero de 2018, PEMEX, Mexico.

[5] International Energy Agency (IEA), CO2 emissions from fuel combustion. Paris, France, OECD/IEA (2017). https://doi.org/10.1787/804a05bf-en

[6] Secretaría de Energía (SENER), Balance Nacional de Energía 2016, SENER, Mexico (2017).

[7] I. Lozada, J. Islas, G. Grande, Environmental and economic feasibility of palm oil biodiesel in the Mexican transportation sector, Renew Sust Energ Rev 14 (2010) 486–492. https://doi.org/10.1016/j.rser.2009.06.034

[8] C.G. Heaps Long-range energy alternatives planning (LEAP) system. [Software version 2008.0.0.33], Stockholm Environment Institute, USA (2008).

[9] Instituto Nacional de Estadística y Geografía (INEGI), Balanza Comercial de México. INEGI, Mexico (2015).

[10] Information on: http://www.anpact.com.mx.

[11] Information on: http://www.volkswagen.com.mx.

[12] F. Manzini, Inserting renewable fuels and technologies for transport in Mexico City Metropolitan Area. International Journal of Hydrogen Energy 31 (2007), pp. 335-327. https://doi.org/10.1016/j.ijhydene.2005.06.024

[13] R. Giannelli, E. Nam, Y. Helmer, et al., Heavy-duty diesel vehicle fuel consumption modeling based on road load and power train parameters, United States. Paper offer #:05cv-3 (2005). https://doi.org/10.4271/2005-01-3549

[14] Instituto Nacional de Investigaciones Forestales, Agrícolas y Pecuarias (INIFAP). Tecnología para la producción de palma de aceite Elaeis guineensis Jacq. en México, 2a ed. INIFAP, México (2006)149. https://doi.org/10.24850/j-tyca-2017-03-04

[15] M. Canakci, Combustion characteristics of a turbocharged DI compression ignition engine fueled with petroleum diesel fuels and biodiesel, Bioresource Technology 98 (2007) 1175-1167. https://doi.org/10.1016/j.biortech.2006.05.024

By-Products of Palm Trees and Their Applications
Materials Research Proceedings **11** (2019) 143-149

Materials Research Forum LLC
doi: https://doi.org/10.21741/9781644900178-9

A Study of the Potentiality of use of Siwei Palm Midribs in Charcoal Production

Maysa Muhammad[1,a*], Hamed Elmously[1,b]

[1]Faculty of Engineering, Ain Shams University, 1 Elsarayat St., Abbaseya, 11517, Cairo, Egypt

[a]eng.maysa.asu@gmail.com, [b]hamed.elmously@gmail.com

Keywords: charcoal production, biochar production, palm midrib charcoal, production of charcoal from agriculture residues

Abstract. The objective of this study utilizes the residues of Siwei palm midrib to produce charcoal with satisfactory environmental, medical, and industrial applications. Choosing the Siwei palm midrib residues is based on its distribution all over Egypt and availability for Egyptian farmers. The study objective was achieved by passing with some steps. The first step prepared the samples, where used the Siwei palm midrib samples and then divided the Siwei palm midrib into five parts (top, middle, base, knee, and end), according to the dimension of the inner reactor. The second step is to design and manufactures a pyrolysis reactor (test rig) to produce charcoal. The third step is carbonization cycle process for the samples of Siwei palm midrib five parts with quantity for all part, where the carbonization cycle process steps according to food and agriculture organization (FAO) standard. The four-step is experimental analysis for ten samples of Siwei palm midrib five parts (row material before carbonization) and Siwei palm midrib five parts (after carbonization) in labs according to American society for testing and materials (ASTM) standards. The experimental analysis divided into proximate analysis such as (moisture content, ash content, volatility matter content, and fixed carbon content), ultimate analysis such as sulfur, and calorific value(also known as heating value or a specific value). Finally, after comparing the results of the experimental analysis for samples Siwei palm midrib parts (after carbonization) to FAO standard values. The potentiality of production of charcoal from Siwei palm midribs with satisfactory properties has been proven. The procedure charcoal is suitable for environmental, medical, and industrial applications. According to FAO, the best samples are the top part of palm midrib in Siwei, followed by the base, middle, knee, and end. The whole Siwei palm midrib could be utilized realizing the calorific value 88% of the FAO standard. The designed reactor in this work could serve as a model for the production of charcoal from palm midribs in the village conditions.

Introduction

The charcoal is the black carbon and ash residues, which come from animal or vegetation substances by removing water and volatile matter during slow heating in the absence of oxygen by pyrolysis process. Charcoal marketing shapes, the first lump charcoal is low ash, high calorific value, and ability to be used in many applications, the other briquette charcoal is high ash, medium calorific value, and ability to use in low energy applications. The charcoal applications are environmental, medical and industrial. The environmental applications are using the charcoal in soil amendment is considerably required, because it increases the carbon concentration in soil and reduces the emissions of green carbon gases. The medical applications with activated carbon mean that the carbon structure of the charcoal has a pore in low volume to do absorption of chemical substances. It acts as filters and has excellent health and medical benefits. The industrial applications have required the sulfur at low levels as much as possible to avoid environmental effects, the ash content at high to realize the most significant energy consumption, the stable pore structure, and chemical compatibility. Almost smokeless, because of its low ash content and chemical stability.

Experimental Procedures

Siwei Palm Midribs Specimens were obtained from Al-Qayat Village, Menia governorate, Egypt. (Fig. 1) illustrates the different parts of palm midrib including: Palm end left on the palm after pruning, the knee (the bent part of the palm midrib), difficult to manufacture, and the three parts of the midrib: base, middle and top.

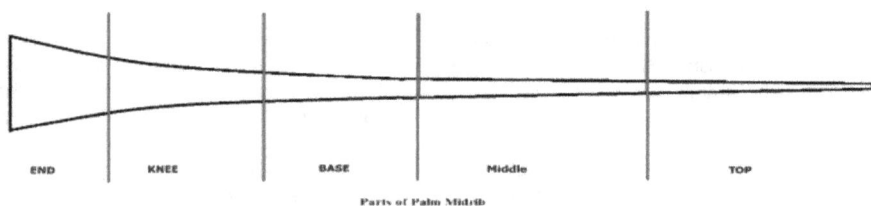

END   KNEE   BASE   Middle   TOP

Parts of Palm Midrib

Fig. 1 Parts of Siwei palm midrib

The following test samples (Table1) have been sourced from Al-Qayat village, Menia Governorate, Egypt.

Table 1 Quantity of items

Item	Quantity (kg)*
End	3
Knee	2
Base	2
Middle	2
Top	1.5

*Air-dry weight

Test Rig (Pyrolysis Reactor) was considered the step to produce the charcoal from Siwei palm midrib parts. The pyrolysis reactor is consisting of three main assemblies, pyrolysis reactor assembly, control system, and condensing unit. The (fig.2) illustration the 3D assembly for pyrolysis reactor. The pyrolysis reactor assembly is a double wall chamber made of steel sheets with a thickness of 6mm, and its capacity is about 120L, where take into consideration the space between species. The outer dimension of the reactor is 70cmx70cmx95cm, and the inner dimension is 40cmx40cmx65cm, where the isolation layers exist made of ceramic fiber to isolate the inner kiln. The inner kiln is supported with three heaters in three different positions, at all side one and in bottom one. The reactor has a door and fixes the thermocouple with weldment in the door to measure the carbonization process (inner reactor). The pyrolysis reactor assembly is connected to a central pipe to condensing unit. The central pipe in the chamber takes out the evolved gases and vapers directly through a condenser unit.

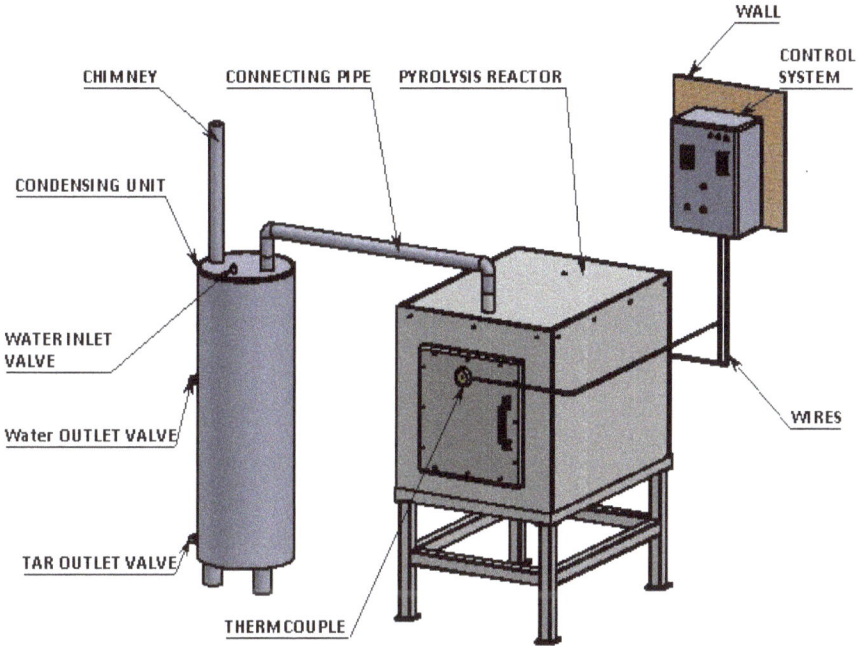

Fig.2 A Sketch of the pyrolysis reactor used in the research

Carbonization Cycle Method by pyrolysis process, and slow heating rate around 5-7°C/min. Until reaching 400-500°C. The method of heat with pyrolysis process in the study is indirect heating. The process includes stages of carbonization cycle according to food and agriculture organization (FAO) standards. The process includes the following stages: At temperatures between 100–120 °C, drying of the input material and moisture goes out. At around 275 °C gases like N2, CO, and CO2, go out. Also, methanol is distilled. Around between temperatures of 280 – 350 °C, exothermic chemical reactions. At more than 350 °C biochar remains, H2 reacts with CO and goes out in the form of tar. The (Table.2) and (Fig.3), illustrate a sample of the carbonization cycle for the top part of the Siwei palm midrib.

Table 2 Carbonization cycle of Siwei palm midrib (top part)

Time (min.)	Temperature (°C)	Remarks
0	33	
6	38	
12	66	
18	123	Starting Smoke
24	180	
30	257	
36	315	
42	363	
48	389	
54	423	End Smoke
60	447	
66	463	
72	490	
74	501	

Top Part Pyrolysis Temp. - Time Graph

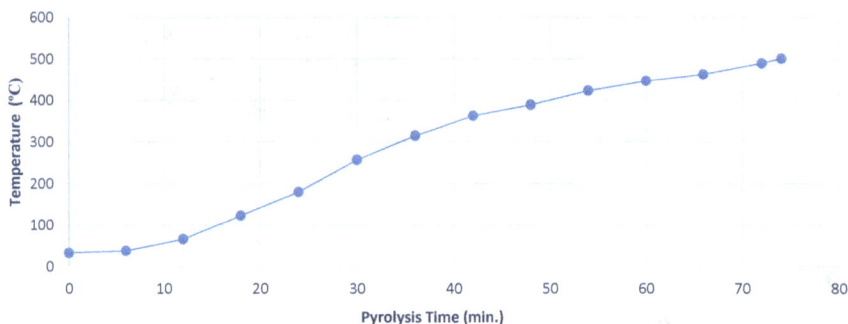

Experimental Analysis Method is standard chemical laboratory analysis is done up samples of the Siwei palm midrib parts (raw material), and the Siwei palm midrib parts (after carbonization cycle). The experimental analysis is calculated according to American society for tooling and materials (ASTM) standard. This experimental analysis determines the charcoal quality and suitability to be used in environmental, medical, and industrial applications. The tests were carried out, proximity analysis, ultimate analysis, and specific energy analysis. The current work of the experimental analysis of the palm midrib parts is done at agriculture research center (ARC). The proximity analysis is divided into the analysis of moisture content, the analysis of volatile matter content, the analysis of ash content, and the analysis of fixed carbon. The ultimate analysis is an analysis of sulfur content. The specific energy analysis is an analysis of the calorific value.

Results and Discussions

The results of the experimental analysis are according to ASTM standard for Siwei palm midrib parts after carbonization cycle, and Siwei palm midrib parts (raw material) are a shown in (table. 3). These results are significant to the limited quality of all part of Siwei palm midrib after carbonization, and its application is suitable for each part. As for the Siwei palm midrib parts (raw material), results are compared to the Siwei palm midrib parts after carbonization to illustrate the changes, which occur for chemical and physical composition of Siwei palm midrib under pyrolysis process.

Table 3 Results of analysis for Siwei palm midrib parts after carbonization

Specimen No.	Specimen Name	Calorific Value kJ/kg	Fixed Carbon %	Volatile Matter %	Ash Content %	Sulfur Content %	Moisture Content %
1	Siwei End	23,300	40.74	22.90	34.29	0.23	3.09
2	Siwei Knee	24,800	38.88	28.25	31.65	0.12	1.09
3	Siwei Base	27,000	48.84	29.42	20.49	0.11	2.29
4	Siwei Middle	25,700	56.61	27.42	15.30	0.12	1.87
5	Siwei Top	28,600	57.36	26.89	14.94	0.14	1.82
6	**Average of palm midrib**	**25,900**	**48.84**	**26.98**	**23.33**	**0.15**	**2.19**

By comparison the calorific value, fixed carbon, volatile matter, ash content, sulfur content, and moisture content experimental analysis results for Siwei palm midrib parts after carbonization concerning FAO lower and higher limits. The Siwei palm midrib parts after carbonization are accepted to produce charcoal. These differences in results of midrib parts are referred to fiber ratio of the chemical composition of midrib, where the top, middle, and base parts have fiber ratio large then knee and end. The percentage of calorific value concerning FAO lower limits. The average Siwei palm midrib charcoal is 88.39%, 97.74% for the top part, 87.76% for the middle part, 92.07% for base part, 84.84% for knee part, and 79.51 for end part, where FAO higher limits are not verified. The percentage of fixed carbon concerning FAO lower limits. The average Siwei palm midrib charcoal is 97.68%, 114.72% for the top part, 113.22% for the middle part, 97.68% for base part, 77.76% for knee part, and for 81.48% for end part, where FAO higher limits are not verified. The top, base, and middle parts are more suitable parts for the industrial application. (Fig.4) illustrates the calorific value and fixed carbon of Siwei palm midrib parts after carbonization and the lines of FAO standard with upper and lower limits.

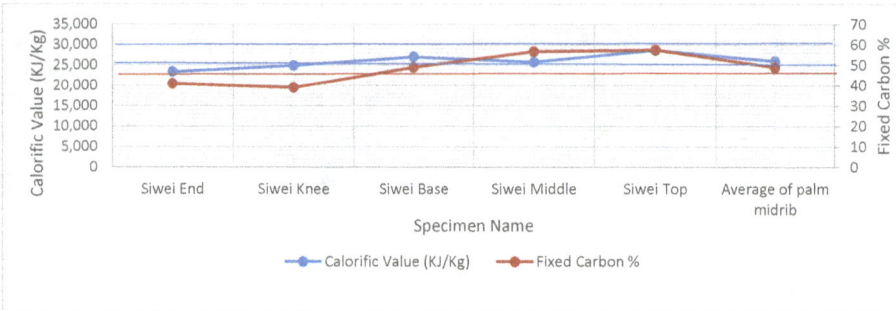

Fig. 4 Calorific value and fixed carbon of Siwei palm midrib parts (After carbonization)

The percentage of the volatile matter concerning FAO lower limits for all parts are verified. AS for FAO higher limits, the average Siwei palm midrib charcoal is 67.45%, 67.23% for the top part, 68.55 for the middle part, 73.55% for base part, 70.63% for knee part, and 57.25% for end part. The middle and top parts are more parts suitable for the environmental application. The percentage of ash content concerning FAO lower and higher limits are not verified. However, can edit these results by reducing the carbonization temperature to 450°C. (Fig.5) illustrates the volatile matter and the ash content for Siwei palm midrib parts after carbonization and the lines of FAO standard with upper and lower limits.

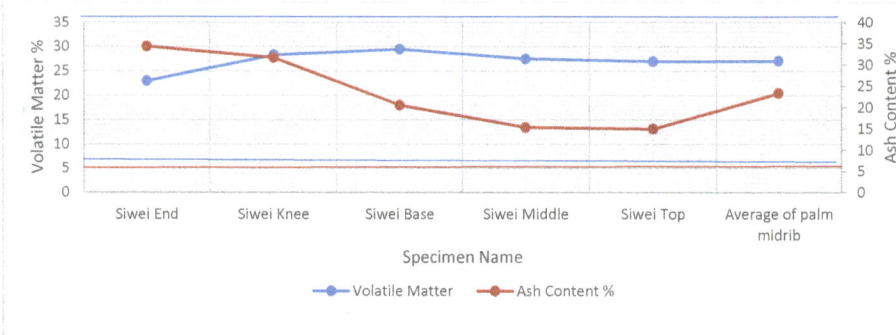

Fig. 5 Volatile matter and ash content of Siwei palm midrib parts (After carbonization)

Fig. 6 is shown the sulfur content, and moisture content concerning FAO lower and higher limits are verified, where the impact on the environment is positive.

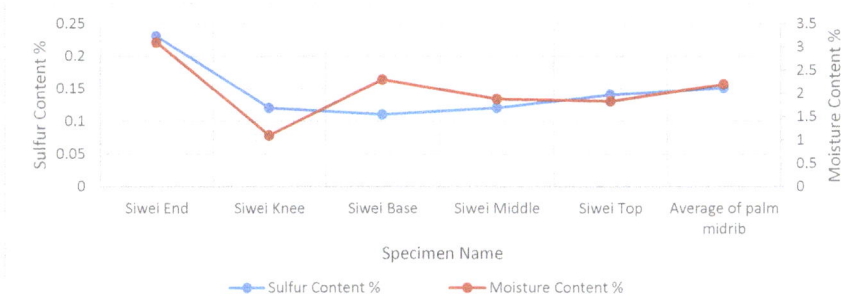

Fig. 6 Sulfur content and moisture content of Siwei palm midrib parts (After carbonization)

Summary

From the current study, the following conclusions, the potentiality of production of charcoal from Siwei palm midribs with satisfactory properties has been a proven. According to FAO, the best samples are the top part of palm midrib in Siwei, followed by the base, middle, knee, and end. The whole Siwei palm midrib could be utilized realizing the calorific value 88% of the FAO standard. The designed reactor in this thesis could serve as a model for the production of charcoal from palm midribs in the village conditions.

References

[1] John Twidell, Tony Weir, Renewable energy resources, Taylor and Francis, 2006.

[2] Global Status Report, Renewable Energy Policy Network for the 21st Century, Renewables, 2017.

[3] Atanu Mukherjee and Rattan Lal, Biochar impacts on soil physical properties and greenhouse gas emissions, Agronomy, 2013. https://doi.org/10.3390/agronomy3020313

[4] Ramadan A. Nasser, Evaluation of the use of midrib from common Date Palm cultivars grown in Saudi Arabia for Energy Production, Saudi Arabia, 2014. https://doi.org/10.15376/biores.9.3.4343-4357

[5] Hassan A. Gomaa, Peter Steele, and Yousef A. Hamdy, Charcoal in Egypt, FAO Project, 2011.

[6] Hassan K.M. Bakheit and Magdi Latif, Date palm in Egypt, Egypt, 2015.

[7] Julije Domac and Miguel Trossero, Industrial charcoal production, FAO, 2008.

[8] William McDonough, The Hannover principles, design for sustainability, 1992.

[9] Sean Case, Biochar amendment and greenhouse gas emissions from agricultural soils, 2013.

[10]Public Par, Methods for producing biochar and advanced biofuels, 2017.

[11]Robert Prins, Wayne Teel, John Marier, Geoff Austin, Tim Clark, and Brandon Dick, Design, construction, and analysis of a farm scale biochar production system, Washington, 2011.

[12]Stefan Czernik, Fundamentals of charcoal production, National bioenergy center, 2010.

Palm Secondary Products as a Source of Organic Material for Compost Production: Applied Examples from Egypt

Mohamed Mahmoud Mohamed Ahmed

Soil, Water and Environment Research Institute, Agriculture Research Center, Giza, Egypt

m311332@hotmail.com

Keywords: pruning products, composting, Date Palm Residues, Doum Palm residues

Abstract. The increasing amounts of fruit pruning wastes in local societies of Egypt cause environmental problems closely related to human health. Its utilization as renewable materials has received a great attention in our present days and encouraged the use of it as an organic fertilizer. Composting these wastes not only reduces their weight and volume but also produces high-value-added products (compost). Manure is the most common organic and natural fertilizer form in our Egyptian rural Village. Due to the high cost of chemical fertilizers and the increase of organic fertilizers demand, it is necessary to use the local crop residues, "Palm Secondary Products" (PSP) as a basic (raw) material which contains 92.99% organic matter. Compost manufacturing provides successfully produce an organic fertilizer from available waste in each region which serves as both fertilizers and soil conditioners. In this study, we present Egyptian examples of the recycling of date palm pruning products mixed with other organic wastes in small scale (Faris rural village, Kom-Ombo, Aswan Governorate and Mandisha village, Baheria oases, Giza Governorate) and in large scale (Al-Kharga, New Valley Governorate).

Introduction

The total world number of date palms is about 120 million trees, distributed in 30 countries and producing nearly 7.5 million tons of fruit per year [1]. Arab countries account for 70% of the world's date palms number and are responsible for 67% of the global production of date palm [2]. The total number of palm tree planted in Egypt is 16 million including 12 million fruiting tree [3]. Mohamed [4] reported that the significant annual increase of fruit dates was about 298.9 thousand palm trees, equivalent to 2.75% of the average number of fruitful dates during the period (1997 - 2012). Due to its adaptation to various stress condition, its plantation is nowadays spread out all over Egypt including the new reclaimed land in the desert and in saline-affected areas. The utilization of by-products of date palm as raw material source for industrial activities gave a promising issues. Some studies have reported that Egypt alone generates more than 300,000 tons of date palm biomass each year [5].

Although date palm residues (DPR) consist of hardly decomposed elements (Cellulose, hemicelluloses, lignin and other compounds) they could be composted with microbiological process instead of burning in farms and causing serious threat to environment [6]. Many researchers reported about compost production from date palm by products [7, 8, 9, 10, and 11].

It should be noted that Egyptian agriculture is mainly dependent on chemical fertilizers (nitrogen, phosphates and potassium) and organic fertilizers. So, recycling palm residues could reduce chemical fertilizers as well as the impact of drought and desertification and pesticides. Moreover, social, economic and environmental benefits could be obtained from the Recycling palm residues including increasing agricultures production in quantity and quality.

By-Products of Palm Trees and Their Applications
Materials Research Proceedings **11** (2019) 150-158

Materials Research Forum LLC
doi: https://doi.org/10.21741/9781644900178-10

Materials and Methods
There are 3 experiments of DPR composting; two of them were conducted in small scale in compost units in different Egyptian village (Faris village Kom-Ombo – Aswan and Mandisha village, Baheria Oasis – Giza Governorate) and the other experiment was made in large scale in sustainable integrated system in (Al-Kharga, New Valley Governorate).

Experiment 1. Faris village is an Egyptian rural village, which located in the west side of the Nile in Kom Ombo, Aswan Governorate. It produces annually about 5000 tons of renewable biomass from fruit trees as date palm (*Phoenix daetylifera*), Mango (*Anacardiaceae*) and doum palm (*Hyphaene thebica*). Palm secondary products (PSP) represent about 80% of this total [11]. The first stage was conducted in site which residues were collected and transported to this site which called (fixed shredding system). Meanwhile, in pilot stage shredding and composting process was made where the residues accumulated (mobile shredding system). The amount of compost produced at pilot stage is equivalent to 70% of the total waste of palm trees. This is in addition to 20% Mango residues and 10% Doum palm residues. Filter mud cake by product produced from sugar can industry in this region was used as organic activator in composting process. About 45 tons and 33 tons of compost were produced in the first and pilot stage respectively. The produced compost was distributed on the local farmers to use in the cultivation of onion crop.

Experiment 2. Baheria Oasis is a depression and lies in the Western Desert of Egypt. Located in Giza Governorate, the main economic sectors are agriculture. The number of palm trees in the oases is estimated at 1.3 million palms; the wealth of palm oases is about 10% of Egypt's total palm wealth (according to estimates that the number of palm trees in Egypt is approximately 13 million palm trees) El-Mously [12]. Large quantities of PSP are available; currently estimated at 69.5 thousand tons / year (dry weight); these quantities are not only abundant but also renewable diversity and pricing, and the possibility of using them as inputs to the manufacturing of many products (as confirmed in this study). A pilot experiment of compost production was conducted in the village of Mandisha. The main additive material in composting process was poultry manure wastes as a common by product of Poultry Industry used as compost activator material. About 3 ton compost was produced from 8 ton DPR+8 m^3 poultry manure.

Experiment 3. The New Valley Governorate is considered the largest governorate in Egypt with a total area represents about 44% of Egypt's area. Date palms (more than 1.5 million palm trees) represent the economic axis of the governorate. A study of the development of the number of fruitful dates in the New Valley Governorate reveal that these numbers has increased significantly by 1.28% during the period from 1997 to 2012 [4]. The date palm pruning produces large amounts of PSP, leading to serious environmental problems. The governorate has approximately 92503 tons per year according to the report of the Directorate of Agriculture in the Governorate of [13]. Experimental pilot attempts produce about 170 ton compost. The main additive materials in composting process was Farmyard manure (FYM) as common by product of cattle Husbandry as organic activator.

Analysis of Fiber Derivatives
Fiber Derivatives of date palm, Mango and Doum palm residues were determined according to TAPPI standard method. The samples were first placed into soxhlet extraction for 6 hours according to method T 264 cm-07 to remove plant extractives. The determination of cellulose, hemicellulose and lignin content were assessed by using the following respective standard method: Kurscher-Hoffner approach [14], chlorite [15], T 222 om-06 and T 211 om-07. As recommended by various pertinent standards, all experiments were conducted in triplicates.

Characterization of compost

Compost resulting from 3 experiments was analyzed. Compost samples were dried at 70 °C to constant weight ground. Values of pH and EC were determined as described by Jackson [16]. The organic matter (OM) content of compost was analyzed by weight loss on ignition at 43 °C for 24h and total organic carbon (TOC) was calculated from (OM) to the following equations [17]:

$$OM = [(W_{105} - W_{430}) / W_{105}] \times 100 \qquad (1)$$

Where W105 = oven dry weight of mass at105°C; W430 = furnace dry weight of mass at 430°C

$$TOC = 0.51 \times OM + 0.48 \qquad (2)$$

Compost samples were digested using a mixture of H_2O_2 and H_2SO4. Total nitrogen was determined by using the micro-kjeldahl procedure [16]. Total phosphorus and potassium were determined by Page et al. [18]. Moisture content throughout this study was measured by drying at 105°C for approximately 24 h or at constant weight. Bulk density (B_d) was measured by obtaining the dry weight of a known volume of the sample. Bulk density was calculated by the following formula [19]:

$$B_d = M_s/V_t \qquad (3)$$

Where M_s is mass of oven dry compost (g), and V_t is total volume of compost (cm^3).

Results and Discussion

DPR materials are rich in some nutrients and can be returned to the field as compost. The economic losses and the financial waste resulting from the non-use or use of the by-products of palm trees in the oases; most of which are disposed of either by open burning in the fields, which contributes to the increase of environmental pollution. Storage in the fields prevents land to cultivation, leading to many environmental problems caused by many insects and rodents, which helps to reproduce and the transmission of diseases and epidemics ... etc.

Table 1: Quantity of PSP in dry weight and its nutrient content in all sites under the study.

Oasis	Residues quantity Thousand (ton)	Nutrients content %)(		
		Nitrogen N	Phosphorous P	Potassium K
Faris*	5.000	0.84		
Baheria Oases**	69.532		0. 22	0.80
Al-Kharga***	29.075			

* El-Mously [20]; ** El-Mously [12]; ***Official Records at the Directorate of Agriculture, Markaz El Waadi El Gadid Governorate [13].

Based on nutrient content in palm residues (Table 1), these residues are rich in nitrogen and potassium. Consequently, each ton from PSP content has approximately about 8.40 kg of Nitrogen and 8.0 kg of potassium. The value of nutrients removed is estimated as the cost of fertilizers that will be needed to replace these nutrients. In view of the available large quantities of PSP in the oases under the study, these quantities are not only abundant but also renewable Diversity and pricing, and the possibility of using it as inputs to the manufacturing of many products (as confirmed in this study).

Experiment 1. Composting of PSP in small scale (Faris rual village Kom-Ombo – Aswan):

Data in table 2 show the percentage content of Fiber Derivatives of date palm, Mango and Doum palm residues is which show clearly the Plant-derived cellulose materials which play a critical role as organic wastes in composting to produce a beneficial amendment for topsoil. Hubbe et al. [21] reported that Cellulose has been described as a main source of energy to drive the biological transformations and the consequent temperature rise and chemical changes that are associated with composting. Lignin can be viewed as a main starting material for the formation of humus.

Table 2: Analysis of Fiber Derivatives of palm, Mango and Doum residues

Residues type	Lignin (%)	Hemicellulose (%)	Cellulose (%)
Date palm residues	35,74	4.75	12.27
Mango residues	40.50	0.57	14.57
Doum palm residues	35.19	13.82	10.81

The analysis of produced compost with its highly organic matter (40.69%), which improves soil fertility, was shown in table 3. Compost acts as sponge to help retain water in the soil that would otherwise drain down below the reach of plant roots, protecting the plant against drought. Results of the pilot fields on onion cultivation showed that the use of Faris compost increased the average onion yield by 10.22% compared to the use of they own organic manure (fig. 1). This increase represents the direct effect of compost application. Indirect impacts were known that organic fertilization helps keep the soil moist for a long time which reduces irrigation periods and thus reduces irrigation costs of energy and labor and helps to rationalize the consumption of irrigation water.

Table 3: Physiochemical analysis of Faris compost

Parameter	Faris compost
pH (1:10)	7.75
EC (dSm^{-1}) (1:10)	3.87
Organic C (%)	23.66
Organic matter (%)	40.69
Nitrogen (%)	0.92
C/N ratio	14.22:1
Phosphorous (%)	0.48
Potassium (%)	0.67
Bulk density (g cm^{-3})	715
Moisture (%)	28

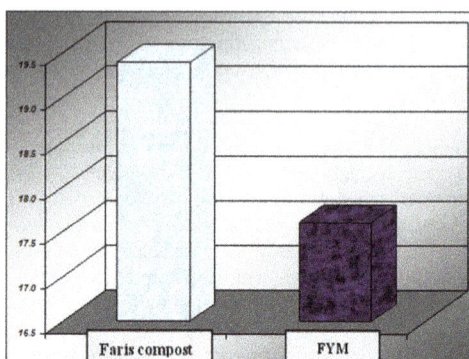

Fig. 1: A graph showing the productivity of onion crop by using faris compost (19.4ton/fed) compared to the use of Farmyard manure (17.6 ton/fed).

Experiment 2. Composting of PSP in small scale (Mandisha village, Baheria Oases – Giza Governorate:

A part of the experiment demonstrates a study on compost production in large scale in the Baheria oases reveal that there are a great possibilities to provide a compost facility with 100 ton PSP each day with a total of 30,000 ton of PSP annually to produce 60-70 ton compost daily. So, it's suggested to establishing compost facility utilized half of the PSP in Baheria oasis and the remained quantity could be used in another industry activity for Medium density fiberboard (MDF) production. Some physical and chemical analyses of the produced compost were showed in table 4.

Table 4: Physiochemical analysis of Baheria oases compost.

Parameter	Baheria oasis compost
pH (1:10)	7.84
EC (dS m^{-1}) (1:10)	4.61
Organic C (%)	26.54
Organic matter (%)	45.66
Nitrogen (%)	1.12
C/N ratio	21.75:1
Phosphorous (%)	0.61
Potassium (%)	0,91
Bulk density (g cm^{-3})	678
Moisture (%)	27

Experiment 3. Composting of PSP in large scale (Al-Kharga) New Valley - New Valley Governorate:

Experimental pilot attempts produce about 170 ton compost with high quality value sold by 36524 LE [22].The main additive materials in composting process was Farmyard manure (FYM) as common by product of cattle Husbandry as organic activator. Analysis of palm compost in the Experimental pilot attempts of the New valley are demonstrated in table 5.

Table 5: Physiochemical analysis of New valley compost.

Parameter	New valley compost
pH (1:10)	8.38
EC (dS m^{-1}) (1:10)	3.45
Organic C (%)	22.97
Organic matter (%)	46.38
Nitrogen (%)	0.95
C/N ratio	17.17:1
Phosphorous (%)	0.74
Potassium (%)	0.94
Bulk density (g cm^{-3})	600
Moisture (%)	27

Adoption of some industry activities targeted to provide PSP raw materials in the Egyptian oases (such as compost industries) achieves two goals:

First: Economic loss and financial waste resulting from the non-use or utilization of the by-products of palm trees in the oases:

The chemical fertilizer that equivalent to nutrient content in the PSP and its values are estimated according to nutrient analysis and demonstrated in table 6. Data clearly expressed that the total losses were 1398, 19463 and 8130 LE due to burning PSP in Faris village, Baheria oasis and Al-Kharga, respectively.

Table 6: Quantity of mineral fertilizer equivalent to nutrient content of PSP in dry weight and its value in each site under the study.

Oasis	PSP quantity Thousand (ton)	Fertilizers (kg)						Total Values LE
		Urea		Super phosphate		Potassium sulphate		
		Q Kg	V LE	Q	V	Q	V	
Faris*	5.000	9o	361	16	33	100	1004	1398
Baheria Oases**	69.532	1256	5024	226	452	1397	13970	19463
Al-Kharga***	29.075	525	2101	95	189	584	5840	8130

* El-Mously [19]; ** El-Mously [12]; ***Official Records at the Directorate of Agriculture, Markaz El Waadi El Gadid Governorate [13]; Q Kg = Quantity of fertilizer; V LE =Value Egypt Pound; Urea= N X 0.46, Tri-calcium; phosphate 37.5% = P X 0.375; Potassium sulphate 48%= N X 0.48; Fertilizers price: Urea = 4000 EL/ ton; Super Phosphate = 2000 EL / ton; Potassium sulphate = 10000 EL / ton.

Second: Environmental loss: Open burning in the fields contributes to the increase of environmental pollution and CO^2 emissions:

Palm trees have a big potential in absorbing CO_2 from the atmosphere. It was demonstrated that one million mature date palm trees can absorb 2.0 million tons of CO_2. Based on photosynthesis calculations, It is common that burning one ton of carbon produces 3.66 tons of

Materials Research Forum LLC
doi: https://doi.org/10.21741/9781644900178-10

CO_2 as per the following reaction: $C + O_2 = CO_2$ [23]. It's known that carbon constitutes 50% of the dry wood, so the carbon part represents half of the demonstrated PSP.

Data in Table 7 show the huge amount of CO_2 that will produce annually from burning date palm residues in open field which contributes to regional and global climate change by producing CO_2, methane, and cause loss of human life in our societies. Data in table 7 shows that CO_2 emissions increased with increasing the burning residues, the emission of CO_2 was differing from a small quantity (9150 ton) to a large quantity in the new valley (127243 ton). Data also shows that carbon quantity = 50% from dry wood materials (residues) and Burning 1 ton residues= 3.66 ton CO_2.

Composting the agricultural residual lowered emission and sequestrate carbon in soils. Luske [24] reported that composting facility in Egypt show that the composting scenario causes significant lower emissions than the baseline scenario (organic waste is not recycled and chemical fertilizer is used on the farm). It must be recommended that farmers in our land should receiving information about modern methods of agricultural waste management and cleaning fields for planting.

Table 7: The expected quantities of CO_2 emission (thousand ton) in case of burning PSP in each site.

Oasis	PSP Thousand (ton)	Carbon quantity Thousand (ton)	CO_2 emissions Thousand (ton)
Faris*	5.000	2.500	9.15
Baheria Oasis**	69.532	34.766	127.243
Al-Kharga***	29.075	14.537	53.207

* El-Mously [19]; ** El-Mously [12]; ***Official Records at the Directorate of Agriculture, Markaz El Waadi El Gadid Governorate [13].

Conclusion

According to the experience, it could be concluded that pruning products are considered as recyclable materials and useful resource. The successful composting of PSP on a small scale in rural communities of Egypt will depend on farmers' awareness of the importance of exploiting the various date palm residues in compost production which will encourage the farmer of each local community over time to participate in producing his own compost.

Acknowledgement
The author wishes to express his sincere gratitude and appreciation to Prof. Dr. H.I. El-Mosely and all members the Egyptian Society for Endogenous Development of Local Communities (EGYCOM).

References
[1] FAO, Food and Agriculture Organization statistical database (FAOSTAT), Retrieved from http://faostat3.fao.org/, 2013.

[2] L.I. El-Juhany, Surveying of Lignocellulosic Agricultural Residues in Some Major Cities of Saudi Arabia; Research Bulletin No. 1-Agricultural Research Center, College of Agriculture,

King Saud University: Riyadh, Saudi Arabia, 2001. https://doi.org/10.18006/2017.5(spl-1-safsaw).s136.s147

[3] FAOSTAT, Crop Production 2008, Statistics Division, Food and Agriculture Organization of the United Nations, 2009.

[4] M.M.A.H. Mohamed, An economic head to promote the production and marketing of dates in the New Valley Governorate: Ph.D Thesis, Assiut University - Faculty of Agriculture - Agricultural Economics, 2014.

[5] E.S. Setyawan, Charcoal Briquette Production in the Middle East: Perspectives, BioEnergy Consult, Powering Clean Energy Future, https://www.bioenergyconsult.com/tag/date-palm-biomass/, 2018.

[6] S.Y. Wong, S.S. Lin, Composts as soil supplement enhances plant growth and fruit quality of straw berry. Plant Nutrition, 25 (2002) 2243-2259. https://doi.org/10.1081/pln-120014073

[7] M.M.M. Ahmed, Composting of date palm residues to produce high value organic fertilizer, Conference titled "The role of scientific research in the development of small industries and the environment in the New Valley", Al-Kharja. New Valley governorate, 2005.

[8] M.S.A. Safwat, Organic Farming of Date Palm and Recycling of Their Wastes, African Crop Science Conference Proceedings, 8 (2007) 2109-2111.

[9] O.A.M. Mahmoud, Composting of date palm wastes and its effect on soil productivity and some soil properties: M.Sc. Thesis, Botany Department, Fac. Of Agric, Assiut Univ., Egypt, 2010.

[10] M.W. Sadik, A.O. Al Ashhab, M.K. Zahran, F.M. Alsaqan, Composting mulch of date palm trees through microbial activator in Saudi Arabia, International Journal of Biochemistry and Biotechnology 1 (2012) 156-161.

[11] M.M.M. Ahmed, Recycling of pruning products of date palm, doum palm and mango as a source of organic fertilizer, The 2nd International conference for Date Palm (ICDP), Al-Qassim Uni. Saudi Arabia, 2016.

[12] H. El-Mously, The final report "The improvement of environmental conditions and palm care and the economic use of its secondary products In the Baheria Oases", 2016.

[13] Official Records at the Directorate of Agriculture, Markaz El Waadi El Gadid Governorate, 2013.

[14] N. Cordeiro, M.N. Belgacem, I.C. Torres, J.C.V.P. Moura, Chemical Composition and Pulping of Banana Pseudo-Stems. Industrial Crops and Products, 19 (2004) 147-154. https://doi.org/10.1016/j.indcrop.2003.09.001

[15] J.S. Han, J.S. Rowell, Chemical Composition of Fibers, In R. M. Rowell, R. A. Young, J. K. Rowell, (Ed.), Paper and Composites from Agro-Based Resources. United States: CRC Press, 1997, pp. 83- 134. https://doi.org/10.1016/s0144-8617(99)00096-x

[16] M. L. Jackson, Soil Chemical Analysis, Prentice-Hall of India Private Limited, New Delhi, 1973, pp. 498.

By-Products of Palm Trees and Their Applications   Materials Research Forum LLC
Materials Research Proceedings 11 (2019) 150-158   doi: https://doi.org/10.21741/9781644900178-10

[17] A.F. Navarro, J. Cegarra, A. Roig, D. Garcia, Relationships between organic matter and carbon contents of organic waste, Bioresource Technology, 44 (1993) 203-207. https://doi.org/10.1016/0960-8524(93)90153-3

[18] A.L. Page, R.H. Miller, D.R. keeney, Methods of Soil Analysis Π: Chemical and Microbiological Properties, American Society of Agronomy Inc. Bull., Madison, Wisconsin, USA, 1982.

[19] S.D. BAO, Soil Physical and Chemical Analysis, Chinese Agricultural Press: Beijing, 2000, pp. 67-127.

[20] H. El-Mously, The final report "Project of compost production of palm, mango and Doum residues". 2012.

[21] M. A. Hubbe, M. Nazhad, C. Sánchez, Composting of lignocellulosics. BioResources, 5 (2010) 2808-2854.

[22] Official Records at the Directorate of Agriculture, Markaz El Waadi El Gadid Governorate, 2011.

[23] A. O. M. S. Sharifa, H.M. Talebb, The Date Palm and its Role in Reducing Soil Salinity and Global Warming, Blessed Tree, 2011.

[24] B. Luske, Composting as an emission reduction methodology and carbon footprint calculations of products, Reduced GHG emissions due to compost production and compost use in Egypt, Comparing two scenarios, Louis Bolk Instituut, 2010, pp. 30.

By-Products of Palm Trees and Their Applications
Materials Research Proceedings 11 (2019) 159-168

Materials Research Forum LLC
doi: https://doi.org/10.21741/9781644900178-11

Production of Biochar from Date Palm Fronds and its Effects on Soil Properties

Mohamed A. Badawi

Soils, Water and Environment Research Institute, ARC, Egypt and General Manager, Emirates Biofertilizers Factory, UAE

dr_badawi22@hotmail.com

Keywords: biochar, date palm wastes, soil conditioning

Abstract. The UAE has the largest number of date palm trees in the Arab world, there are about 42 million date palm trees. Each tree generates about 15 kilograms (kg) of waste biomass annually, totaling 600 million kg of green waste. Converting date palm waste into biochar can reduce carbon dioxide (CO_2) and methane (CH_4) emissions generated by the natural decomposition or through burning of the waste. Biomass produced from date palm trees can't be composted easily in normal composting process due to its high content of lignocellulose compound, while the biochar production can be the option to generate both energy and soil conditioner for the improvement of sandy soil under the gulf countries severe climate. Biochar is one of the most stable biologically produced carbon sources that can be added to soil. It processes agricultural waste into a soil enhancer that improves soil fertility, saves water, helps to mitigate greenhouse gas (GHG) emissions and fight global warming. The United Arab Emirates has sandy soil with very low water and nutrient holding capacities, using biochar improved its soil WHC, and biological activities. In this paper we did several trials to evaluate the produced biochar from date palm tree green wastes as a soil conditioner in sandy soil. Research has been undertaken in a pilot plant of 200-liter capacity. The produced biochar (25% w/w) of raw materials was used as a soil conditioner for sandy soil. The soil physical, chemical and biological properties were tested in pot experiment with different mixing ratios and the results showed better improvements in its properties. The aim of this study was to evaluate the effect of biochar and organic soil amendments on soil physicochemical and microbial load, carbon sequestration potential.

Introduction

The topic of utilizing agricultural or forestry waste as co-products such as biochar has received great interest from the scientific community.

Around the world, there are growing initiatives on finding strategies to encourage collection of agricultural waste as new products in a logical and economical way to be used in other applications rather than disposal in landfill.

Date palm trees in the UAE generate around 600000 tons of date fronds which is an abundantly available agricultural waste and small percentage is economicaly used and recycled as well as in many date producing countries. Date trees are cultivated in arid and semi-arid regions and can thrive in long and hot summers, low rainfall and very low relative humidity [1]. About 105 million trees are available around the world covering over a million hectares. UAE is one of the largest producers of date fruits with more than 42 million date palm trees and an annual production rate of 770,000 tons of date fruits [2].

The United Arab Emirates (UAE) has sandy soil with very low water and nutrient holding capacities.

In these conditions, date palm is considered one of the most resilient crops in the region.

Over the years, with rising temperatures and scarce precipitation, there have been calls for new ways to conserve water, improve soil properties and prevent nutrient loss to achieve future food and nutrition security.

Biochar is a solid product produced from thermal conversion of unstable carbon-enriched materials into stable carbon-enrichedcharred materials that can be incorporated into the soils as a mean for agronomic or environmental management.

Biochar can be produced out of a long list of feedstock. Some of this raw materials can be agricultural waste (wheat straw, nuts and coconuts shells, waste wood, etc.) biomass energy crops (corn, cereals, wood pellets, palm oil, oilseed rape), bioenergy residues, compost (green waste), animal manure (camel, chicken...), sewage sludge, etc [3].

The composition of biochar (content in carbon, nitrogen, potassium, calcium, etc) is directly related on the feedstock used and the duration and temperature of pyrolysis.

At higher temperatures, biochar showed high particle density, high porosity but the bulk density showed small differences between varying temperatures as shown in Table (1). However, [4] stated that increasing pyrolysis temperatures produced high bulk density and an increase of porosity for rice husk and empty fruit bunch biochars.

During the pyrolysis process, biomass porosity increased with increasing pyrolysis temperature.

This can be attributed to volatilization processes and the loss of organic compounds, which creates more voids [5]. And found that particle density of wood derived biochar increased from 1344 kg/m3 to 1742 kg/m3 with an increase of pyrolysis temperature from 300 $^{\circ}$C to 700 $^{\circ}$C. [6] also reported that the solid density of biochar prepared from sewage sludge increased with increasing pyrolysis temperature while the bulk density decreased, as a result; the porosity increased.

In 2012, the International Biochar Initiative defines biochar as 'solid material obtained from thermochemical conversion of biomass in an oxygen limited environment [7]. Pyrolysis is an industrialized thermochemical conversion process of biomass into biochar, syngas and bio-oil .

The yield of end products depends upon the temperature elevated in the pyrolysis.

Biochar has been produced with a range of pH values between 4 and 12, dependent upon the feedstock and pyrolysis temperature[8].

Generally, low pyrolysis temperatures (< 400° C) yield acidic biochar, while increasing pyrolysis temperatures produce alkaline biochar.

Once incorporated to the soil, surface oxidation occurs due to reactions of water, O_2 and various soil agents [8,9]. The cation exchange capacity (CEC) of fresh biochar is typically very low, but increases more with time as the biochar ages in the presence of O_2 and water [8,10,11].

Biochar is used as a soil amendment to improve soil nutrient status, C storage and/or filtration of percolating soil water [12]. Biochar from pyrolysis and charcoal produced through natural burning share key characteristics including long residence time in soils and a soil conditioning effect [13]. Research has claimed that application of biochar can increase soil organic carbon (SOC), improve the supply of nutrients to plants and therefor enhance plant growth and soil's physical, chemical, and biological properties [13,14].

Biochar can alter soil physical properties such as structure, pore size distribution and density, with implications for soil aeration, water holding capacity, plant growth, and soil workability. Consequently this may improve soil water and nutrient retention [15].

Biochar may increase the overall net soil surface area [16]; reduce soil bulk density which is generally desirable for most plant growth [17].

Biochar has a higher surface area and greater porosity relative to other types of soil organic matter, and can therefore improve soil texture and aggregation, which improves water retention in soil. Improved water holding capacity with biochar additions is most commonly observed in coarse-textured or sandy soils [13,18].

Biochar has a higher sorption affinity for a range of organic and inorganic compounds, and higher nutrient retention ability compared to other forms of soil organic matter [20,21] and [22-24].

Previous analysis have shown that it is feasible to prepare biochar with relatively high BET surface areas from date palm fronds, which is favorable for microbial communities to grow and therefore enhancing fertility of the soils,. Biochar enhances soils. By converting agricultural waste into a powerful soil enhancer that holds carbon and makes soils more fertile, we can boost food security, discourage deforestation and preserve cropland diversity [25].

Biochar addition may increase soil microbial biomass (population size), and affect microbial community structure (species present) and enzyme activities.

Australian researchers observed an increase in microbial biomass in the presence of poultry litter biochar in a hard-setting soil growing radishes. While increased microbial biomass has been observed, it has often been accompanied by a reduction in microbial activity, most probably due to sorption of labile organics, nutrients, and enzymes on the biochar.

Biochar application has been shown to increase the rate of mycorrhizal fungal colonization in roots, although it depends on the biochar, soil type and plant species.

There is growing evidence that biochar addition can reduce disease severity for several crop species [26-29].

Materials and methods

Biochar preparation

A-Bench scale biochar preparation: Thermal decomposition of the biomass was carried out using a bench-scale slow pyrolysis process in an oxygen deprived condition to convert the date fronds, biomass into stable biochar. For each experiment, about 100 g of dry biomass sample were placed in a ceramic dish and purged with nitrogen gas to provide a low-oxygen environment.

The container was covered with aluminium foil with two small vents allowing only the evolved volatiles to escape.

The biomass was pyrolysed inside a muffle furnace. In this study, the evolved volatile was not collected or quantified.

The pyrolysis temperature was increased to four terminal temperature levels (350, 450, 550 and 650 °C) and kept constant for a residence time of 2 h.

The biochar was cooled in the furnace to room temperature and then placed inside a desiccator for 15 min.

The mass yield was obtained from the final weight of the biochar. The produced biochars are named as BC 1, BC-2, BC 3, and BC4 , where BC and numbers denoting biochar and pyrolysis temperatures, respectively.

Sandy soil was collected from Emirates bio farm, analysis is in table 1). Date palm tree waste leaves, was collected from al Ain City, UAE. The waste was dried in air under sunshine and then chopped to small pieces. Pyrolysis of the processed date palm waste was carried out in a closed stainless steel container, at 350 °C were maintained for 4 h under a limited supply of air.

Feedstock samples were pyrolyzed to the desired temperature at the rate of 5°C min−1.

The biochars produced were left to cool inside the furnace overnight.

Characterization is an essential step to evaluate the physicochemical properties of biomass prior to the pyrolysis process.

The biomass characterization was carried out by using proximate and ultimate analysis on dry matter. Proximate analysis included the measurements of moisture content, ash content, volatile matter and fixed carbon.

The moisture content was determined by measuring the loss in mass at 105 °C for 24 h. The volatile matter content was measured by placing dry biomass inside a muffle furnace at 550 °C for 6 h to measure mass loss due to volatilization of volatile components Black et al. [30].

Bulk density was determined according to Mudoga et al., (2008) method by filling a 10- ml tube with dry adsorbent.

The tubes were capped, tamped to reach a constant volume by tapping on a table, and weighed. Whereases, particle density was determined by using the method of volume displacement according to Khanmohammadi et al., (2015) by using water instead of kerosene. The pH of the biomass was measured by adding 1 g of powdered biomass to 20 mL of deionized water (1:20) and heated to 90 oC with continuous stirring for 20 min and was subsequently determined by using a pH meter. Soils: Soil physical and chemical analysis were done according to [30].

Electrical conductivity (EC) and pH were determined potentiometrically on a 1:2 soil mass-to-water volume paste, Effect of biochar on sandy soil properties was carried out in 5 kg pots and three replicates from each. Treatments include control, 1ton per hectare, two tons per hectare, three tons per hectare and 4 tons per hectare was evaluated for 180 days.

Irrigation was performed on schedule to maintain the moisture content in the soil. Pots were incubated at 28^0C in the green house, Periodical samples were taken after one week, 30 days, 60 days, 90 days and at the end of experiment at 180 days.

Soil samples were preapared and tested according to protocol followed by [30]

Table 1 criterial of materials used.

Materials	Sandy soil	Biochar
Water Holding Capacity, %	32.0	150.0
Cation Exchange Capacity	7.6	48.2
Organic matter	0.22	75.0
Organic carbon	0.12	43.6
Total nitrogen	0.12	1.12
C/N	---	38.9
pH	8.9	6.8
EC Ms/d	5.1	6.4

Total bacterial counts CFU, was measured using nutrient agar media, according to [31-32], while potato dextrose media was used to measure the total fungi in respective order, according to [32].

Results and discussion

Table (2) depicts the effects of pyrolysis temperature on mass yield, Biochar bulk density and pH of the biochar were also studied.

The biochar was prepared by using a slow pyrolysis process under different temperatures (350, 450, 550 and 650 °C) for 2 h. The results of the analysis indicated that the biochar bulk density were affected by the pyrolysis temperature, and it decreased with increasing pyrolysis temperature e.g. from 0.53, 0.49, 0.48, and 0.47 g/l,. The biochar mass yield was inversely proportional to the pyrolysis temperature.

The mass yields were 41.0 %, 31.8 %, 26.9 % and 22.6 % at temperatures (350, 450, 550 and 650 °C), respectively The pH of the biochar increased with the pyrolysis temperature, while the bulk density decreased. Therefore, the biochar prepared from date seed had a highly porous structure and thus it is expected to improve soil physical, chemical and biological properties [32-34].

Table 2 Effect of temperature variations on biochar quality.

Treatment	Temperature (°C)	Biochar yield (%)	Bulk density g/l	pH
BC 1	350	41.0	0.532	6.8
BC 2	450	31.8	0.498	7.2
BC 3	550	26.9	0.487	8.0
BC 4	650	22.6	0.471	8.9

The pH values of the biochar is shown in table 2. The values increased with the pyrolysis temperature ranged from 6.8 to 8.9. These results were similar to other literature values that indicate most of derived biochar are alkaline[33,37]. [37] found that the pH values of biochars derived from wheat straw, wood, spruce and needle mixture ranged from 6.9 to 9.2, which indicates a certain liming effect that may be achieved after biochar application.

The findings of [33] indicated that rice husk and empty fruit bunch derived biochars produced at 300°C have acidic nature but that those produced at 650°C were alkaline. [38] stated that biochar produced from canola straw, corn straw, peanut and soybean straws at different pyrolysis temperatures had shown the alkalinity nature and pH of biochars increased with the pyrolysis temperature.

Table 3 Periodical changes of organic carbon content in sandy soil amended with biochar through 180 days.

Treatment	Zero time	60 Days	120 Days	180 days
Control	0.14	0.13	0.13	0.09
Sandy soil + 1% biochar	0.20	0.18	0.17	0.17
Sandy soil + 2% biochar	0.27	0.25	0,24	0.24
Sandy soil + 3% biochar	0.35	0.34	0.33	0.31
Sandy soil + 4% biochar	0.42	0.40	0.39	0.38

In table (3). It is very clear that treatments recived biochar improved its organic matter content compared to control and increased with increasing biochar content from 1% to 4% reaching 0.42% organic carbon compared to 0.14 in the control. During the incubation time it showed big loss in the control treatment while organic carbon loss from treated soil was minimum. And this is mainly because carbon is very low release carbon, and this won't lead to carbon sequestration.

Table 4 Periodical changes of Water holding capacity in sandy soil amended with biochar through 180 days.

Treatment	Zero time	60 Days	120 Days	180 days
Control	17.6	16.0	16.5	16.1
Sandy soil + 1% biochar	21.2	21.0	20.7	20.5
Sandy soil + 2% biochar	26.5	26.7	26.0	25.3
Sandy soil + 3% biochar	28.1	27.8	26.9	26.8
Sandy soil + 4% biochar	33.7	33.0	31.9	31.7

Table (4) depicted the water holding capacity of soil treated with biochar in comparison to control.

All treatments received biochar showed water holding capacity improvement compared to control and increasing biochar content increased water holding content values [37-39].

Table 5 Periodical changes of Total Plate Counts, TPC content in sandy soil amended with biochar through 180 days, $CFU/10^6$.

Treatment	7 Days	60 Days	120 Days	180 days
Control	37.0	34.0	36.0	32.0
Sandy soil + 1% biochar	68.0	67.2	88.1	74.3
Sandy soil + 2% biochar	84.0	88.0	92.3	89.0
Sandy soil + 3% biochar	87.0	96.5	98.1	99.4
Sandy soil + 4% biochar	98.7	103.0	107.0	107.8

Table 6 Periodical changes of Total fungi colonies content in sandy soil amended with biochar through 180 days, $CFU/10^4$.

Treatment	7 Days	60 Days	120 Days	180 days
Control	7.20	7.29	7.31	7.28
Sandy soil + 1% biochar	12.10	12.90	13.20	13.90
Sandy soil + 2% biochar	14.30	16.10	17.20	17.00
Sandy soil + 3% biochar	16.00	17.40	18.60	19.00
Sandy soil + 4% biochar	18.30	19.20	19.40	22.00

Table 5 and 6 describe the total bacterial population and total fungi population measured by plate count technique in the soil amended with different rates of biochar from 1% till 4%. Numbers of CFU increased with increasing dose of biochar applied in all treatments. All treatments showed high microbial counts over control.

Bacterial population was in the range of 32-37 milions bacteria in the control treatment while increased sharply in all treatments reaching 107 millions micrope after 180 days.

Fungi population showed less numbers, e.g. control treatments showed 7×10^4, while for treatments received biochar, the fungi numbers grown in the range of 12-22 millions $\times 10^4$. The significant differences in bacteria, and fungi population were observed between biochar and control [40-42].

Conclusion

BIOCHAR is an organic soil amendment which is rich in organic carbon. A significant percent of the organic carbon is recalcitrant in nature. Recalcitrant organic carbon does not decompose easily in soil & keeps performing long after soil application. This ensures long term soil fertility unlike synthetic chemical fertilizers & other organic soil conditioners that only provide short term benefits. Chemical fertilizers & other organic soil conditioners have to be replenished regularly to ensure effectiveness. Biochar addition may increase soil microbial biomass (population size), and affect microbial community structure (species present) and enzyme activities.

Refrences

[1] T. Ahmad, M, Danish, M. Rafatullah, A. Ghazali, O. Sulaiman, R. Hashim, M.N.M. Ibrahim, The use of date palm as a potential adsorbent for wastewater treatment: a review, Environ Sci Pollut Res Int, 19(5), 1464-1484. DOI: 10.1007/s11356-011-0709-8 (2012). https://doi.org/10.1007/s11356-011-0709-8

[2] FAO (Food and Agriculture Organization of the United Nations), Statistical Databases (2012).

[3] S.P. Sohi, E. Krull, E. Lopez-Capel, R. Bol., A review of biochar and its use and function in soil, Advances in Agronomy, San Diego, Elsevier Academic Press Inc. 105 (2010) 47-82. https://doi.org/10.1016/s0065-2113(10)05002-9

[4] N. Claoston, A.W. Samsuri, M.H. Ahmad Husni, M.S. Mohd Amran, Effects of pyrolysis temperature on the physicochemical properties of empty fruit bunch and rice husk biochars. Waste Manag Res, 32(4), 331-339, 2014. https://doi.org/10.1177/0734242x14525822

[5] C.E. Brewer, K. Schmidt-Rohr, J.A. Satrio, R.C. Brown, Characterization of biochar from fast pyrolysis and gasification systems. Environ. Prog. Sustain. Energy 28 (2014) 386–396. https://doi.org/10.1002/ep.10378

[6] Z. Khanmohammadi, M. Afyuni, M.R. Mosaddeghi, Effect of pyrolysis temperature on chemical and physical properties of sewage sludge biochar. Waste Management & Research 33-3 (2015) 275-283. https://doi.org/10.1177/0734242x14565210

[7] D.R. Kasten, J. Heiskanen, K. Englund, A., Tervahauta, Pelleted biochar: Chemical and physical properties show potential use as a substrate in container nurseries. Biomass and Bioenergy, 35-5 (2012) 2018-2027. https://doi.org/10.1016/j.biombioe.2011.01.053

[8] J. Lehmann, A Handful of Carbon. Nature 447-7141 (2007) 143-144. https://doi.org/10.1038/447143a

[9] C.H. Cheng, J. Lehmann, J.E. Thies, S.D. Burton, M.H. Engelhard, Oxidation of black carbon by biotic and abiotic processes, Organic Geochemistry 37 (2006) 1477-1488. https://doi.org/10.1016/j.orggeochem.2006.06.022

[10] C.H. Cheng, J. Lehmann, M. H. Engelhard, Natural oxidation of black carbon in soils: Changes in molecular form and surface charge along a climosequence. Geochim.Cosmochim. Acta 72 (2008) 1598-1610. https://doi.org/10.1016/j.gca.2008.01.010

[11] B. Liang, J. Lehmann, D. Solomon, J. Kinyangi, J. Grossman, B. O'Neill, J.O. Skjemstad, J. Thies, F.J. Luizao, J. Petersen. E.G. Neves, Black carbon increases cation exchange capacity in soils. Soil Science Society America Journal, 70 (2006) 1719–1730. https://doi.org/10.2136/sssaj2005.0383

[12] J. Lehmann, S. Joseph, Biochar for Environmental Management: Science and Technology. Earthscan, London & Sterling, VA. (2009) 416.

[13] B. Glaser, J. Lehmann, W. Zech, Ameliorating physical and chemical properties of highly weathered soils in the tropics with charcoal - a review Biology and Fertility of Soils 35 (2002) 219-230. https://doi.org/10.1007/s00374-002-0466-4

[14] M.A. Rondon, J. Lehmann, J. Ramirez, M. Hurtado, Biological nitrogen fixation by common beans (Phaseolus vulgaris L.) increases with bio-char additions, Biology and fertility of soils 43 (2007) 699-708. https://doi.org/10.1007/s00374-006-0152-z

[15] A. Downie, A. Crosky P. Munroe, Physical properties of biochar. In Biochar for environmental management: science and technology Eds. J. Lehmann and S. Joseph. Earthscan, London; Sterling, VA (2009) 13-32.

[16] K.Y. Chan, L. Van Zwieten, I. Meszaros, A. Downie, S. Joseph, Agronomic values of greenwaste biochar as a soil amendment, Australian Journal of Soil Research, 45-8 (2007) 629-634. https://doi.org/10.1071/sr07109

[17] K. Chan , Z. Xu, Biochar: Nutrient Properties and Their Enhancement. In Biochar for Environmental Management: Science and Technology (Eds J Lehmann and S Joseph), Earthscan: London, UK (2009)53-66.

[18] J.W. Gaskin, K.C. Das, A.S. Tassistro, L. Sonon, K. Harris, B. Hawkins, Characterization of char for agricultural use in the soils of the southeastern United States, in: W.I. Woods (Ed.), Amazonian Dark Earths: Wim Sombroek's Visionl, Springer Science, Business Media, Heidelberg, Germany, (2009) 433–443. https://doi.org/10.1007/978-1-4020-9031-8_25

[19] N.C. Brady, R.R. Weil, he nature and properties of soils, Macmillan: New York, 2014.

[20] T.D. Bucheli, O. Gustafsson, Quantification of the soot–water distribution coefficient of PAHs provides mechanistic basis for enhanced sorption observations Environ. Sci.Technol 34 (2000) 5144–5151. https://doi.org/10.1021/es000092s

[21] T.D. Bucheli, O. Gustafsson, Soot sorption of non-ortho and ortho substituted PCBs, Chemosphere (2003) 53:515–522. https://doi.org/10.1016/s0045-6535(03)00508-3

[22] R.M. Allen-King, P. Grathwohl, , W.P. Ball, New modeling paradigms for the sorption of hydrophobic organic chemicals to heterogeneous carbonaceous matter in soils, sediments, and

rocks. Advances in Water Resources 25 (2002) 985–1016. https://doi.org/10.1016/s0309-1708(02)00045-3

[23] S. Kleineidam, C. Schuth, P.Grathwohl, Solubilitynormalized combined adsorption-partitioning sorption isotherms for organic pollutants. Environ. Sci. Technol. 21 (2002) 4689–4697. https://doi.org/10.1021/es010293b

[24] B.T. Nguyen, J. Lehmann, J. Kinyangi, R. Smernik, S.J. Riha, M.H. Engelhard, Long-term black carbon dynamics in cultivated soil. Biogeochemistry 89 (2008) 295- 308. https://doi.org/10.1007/s10533-008-9220-9

[25] T.A. Muhamed, B. Lopez, C.G. Lina J.E. Shmidt, Economic analysis of biochar production from date palm fronds. I Energy, MASDAR institute of science and technology, PO Box 54224, AD, UAE, 2018.

[26] J. Lehmann, M.C. Rillig, J. Thies, C.A. Masiello, W.C. Hockaday, D. Crowley, Biochar effects on soil biota—a review, Soil Biol. Biochem. 43 (2011) 1812–1836. https://doi.org/10.1016/j.soilbio.2011.04.022

[27] J. Lehmann, J. Gaunt, M. Rondon, Bio-char sequestration in terrestrial ecosystems-a review. Mitigation and Adaptation Strategies for Global Change 11 (2006) 403-427. https://doi.org/10.1007/s11027-005-9006-5

[28] M. Zainab, A. El Hanandeh, Q. J. Yu, Date Palm (Phoenix Dactylifera L.) Seed Characterization for Biochar Preparation, Journal of Analytical and Applied Pyrolysis 115 (2015) 392–400. https://doi.org/10.32738/ceppm.201509.0015

[29] H.L. Mudoga, H. Yucel, & N.S.Kincal, Decolorization of sugar syrups using commercial and sugar beet pulp based activated carbons. Bioresource Technology, 99-9 (2008) 3528-3533. https://doi.org/10.1016/j.biortech.2007.07.058

[30] C.A. Black, O.D. Evans ; L.E.Ensminger, J.L. White. F.E. Clark, R.C. Dinaver, Methods of Soil Analysis part II, Chemical and Microbiological properties, 2nd, Soil Sci., socities of Am. Inc., publications Madison Wisconsin, USA (1982)1573.

[31] A.O. Rolf, L.R. Bakken, Viability of soil bacteria: optimization of plate counting technique and comparison between total counts and plate counts within different size groups. Micro. Ecol. 13 (1987) 59-74. https://doi.org/10.1007/bf02014963

[32] Difco manual, Dehydrated culture media and reagents for microbiology, laboratories incorporated, Detroite, Mitchigan, 48232, USA, (1985) 621.

[33] N. Claoston, A.W. Samsuri, M.H. Ahmad Husni, M.S. Mohd Amran, Effects of pyrolysis temperature on the physicochemical properties of empty fruit bunch and rice husk biochars, Waste Manag Res, 32-4 (2014) 331-339. https://doi.org/10.1177/0734242x14525822

[34] J. Lehmann, M.C. Rillig, J. Thies, C.A. Masiello, W.C. Hockaday, D. Crowley, Biochar effects on soil biota – A review. Soil Biology and Biochemistry, 43-9 (2011) 1812-1836. Retrieved October 11, 2013. https://doi.org/10.1016/j.soilbio.2011.04.022

[35] M.K. Hossain, V. Strezov, K.Y. Chan, P.F. Nelson, Agronomic properties of wastewater sludge biochar and bioavailability of metals in production of cherry tomato (Lycopersicon

esculentum), Chemosphere, 789 (2010) 1167-1171.
https://doi.org/10.1016/j.chemosphere.2010.01.009

[36] M. Inyang, B. Gao, P. Pullammanappallil, W. Ding, A.R. Zimmerman, Biochar from anaerobically digested sugarcane bagasse. Bioresour Technol, 101-22 (2010) 8868-8872. https://doi.org/10.1016/j.biortech.2010.06.088

[37] S. Kloss, F. Zehetner, A. Dellantonio, R. Hamid, F. Ottner,V. Liedtke, G. Soja, Characterization of slow pyrolysis biochars: effects of feedstocks and pyrolysis temperature on biochar properties, J Environ Qual, doi: 10.2134/jeq2011.0070, 41-4 (2012) 990-1000. https://doi.org/10.2134/jeq2011.0070

[38] J. Yuan, R. Xu, & H. Zhang, The forms of alkalis in the biochar produced from crop residues at different temperatures. Bioresource Technology, doi:10.1016/j.biortech.2010.11.018, 102-3 (2010) 3488-3497. https://doi.org/10.1016/j.biortech.2010.11.018

[39] D. M. Glazonova. P.A. kuryntseva, S.Y. Selvenovskaya P.Y. Galytskaya, Assessing the potential of using biochar as a soil conditioner. IOP conf. series: earth and Environmental sci., (2018) 107 (012059). https://doi.org/10.1088/1755-1315/107/1/012059

[40] R.A. Adel , A. Usmana, M.V. Abduljabbarc, Y.S. Oke,M. Ahmada, M. Ahmada, J.Elfakia, S.S. Abdulazeema, M. I. Al-Wabela, Biochar production from date palm waste: Charring temperature induced changes in composition and surface chemistry, 2015.

[41] H.A. Qasim, A.l. M. Rahman, M. H. Salman, Z. H. Aly, A.K. Abdul Satar, A. Sh., J. Abdul Ameer, Characterization of Biochar Produced from IRAQI Palm Fronds by Thermal Pyrolysis Al-Khwarizmi Engineering Journal 11-2 (2015) 92-102.

[42] E.B. Katy, K. R. Brye, Mary C. Savin, D.M. Longer, Biochar Source and Application Rate Effects on Soil Water Retention determined Using Wetting Curves, Open Journal of Soil Science, 5 (2015) 1-10. https://doi.org/10.4236/ojss.2015.51001

By-Products of Palm Trees and Their Applications
Materials Research Proceedings **11** (2019) 169-185

Materials Research Forum LLC
doi: https://doi.org/10.21741/9781644900178-12

Application of Date Palm Trees Mulch as a Bedding Material for Dry Heifers, Part 2 –Preparing the Bedding Materials

Elashhab A.O.[1,a*], M.W. Sadik[2,b], M.K. Zahran[3,c]

[1]Agricultural Engineering Research Institute (AERI), ARC, Giza, Egypt

[2]Cairo University, Faculty of Agriculture, Department of Microbiology, Giza, Egypt

[3]Al Hofuf Stars Trading Est., P.O.220540, Riyadh, Saudi Arabia

[a]elashhabahmed@yahoo.com, [b]mwas20032004@yahoo.com,
[c]mahmoud.zahran60@gmail.com

Keywords date palm trees mulch, bedding material, cow manure, dry heifers, mechanical preparing bedding, waste recycling, compost

Abstract. Date palm trees mulch can be safely and effectively used as a bedding material for cow feedlots. Feedlot managers will need to adjust bedding rates according to facilities, environment, and cow comfort. Feedlot managers interested in using date palm trees mulch as bedding will recognize that absorbency of date palm trees mulch is lower than that of sand. However, utilizing of date palm trees mulch eliminates costs of harvesting sand. A total power consumption for horizontal grinder machine, grapple to loading the grinding machine and loader for handling mulch material. Were 468 kW/h, specific energy was 46.8 kW/ton and 7.49 kW/m3. Total Power consumption to preparing the barn to use mulch bedding material for m3 were 7.84 kW/m3 every day and 78.4 kW/m3 after finish experimental time 10 days. Total cost for using date palm trees mulch as a bedding every 10 days were 3180.0 SR/m. Total cost for traditional manure management every 10 days were 3180.0 SR every 10 days for all operation

Introduction
Bedding for livestock animals must be comfortable, clean, and absorbent. There are several materials, both organic and inert, that may be used for bedding, and most may be used for all types of livestock. When organic materials are used, ammonia volatilization is reduced, improving the air in the housing facility. Bedding, as with other aspects of livestock management, can be manageable through proper care and attention [18]. The removal of accumulating manure reduces odors, a control fly larvae, and minimizes the potential for surface and groundwater contamination. Maintaining a firm, dry feedlot surface is an important factor in good animal health and a healthy environment. While this is labor intensive for feedlots, it does indicate that pen cleaning as frequently as feasible for your specific operation is good management [9]. Resting dairy cattle should have a dry bed. Stalls ordinarily should have bedding to allow for cow comfort and to minimize exposure to dampness or fecal contamination. When handled properly, many fibrous and granular bedding materials may be used, including long or chopped straw, poor-quality hay, sand, sawdust, shavings, and rice hulls. Inorganic bedding materials (sand or ground limestone) provide an environment that is less conducive to the growth of mastitis pathogens.. Bedding should be absorbent, free of toxic chemicals or residues that could injure animals or humans [13]. Compost barns have a concrete feed alley, a bedded pack area that is stirred two times a day, and a 1.2-m high wall surrounding the pack. The wall that separates the pack and feed alley has walkways to allow cow and equipment access to the stirred pack area. The stirred pack is sized to provide a minimum stirred bedded pack area of

7.4 m2/cow. Producers use dry fine wood shavings or sawdust for bedding. Fresh bedding is added when the bedded pack becomes moist enough to stick to the cows. The pack is stirred (aerated) at least two times each day to a producer recommended depth of 25 to 30 cm. Stirring aerates and mixes manure and urine on the surface into the pack to provide a fresh surface for cows to lie down on. The pack can provide manure storage for 6 to 7 months [11, 12]. Two of the most common methods of manure removal are the wheeled frontend loader and the box scraper. Both are effective at: 1) providing a smooth pen surface and 2) maintaining the integrity of the compacted protective hard pan under feedlot pens. A wheeled front-end loader requires a professional operator. A combination of a wheeled front-end loader for major manure removal and a scraper for final cleaning would be an effective compromise [9]. Evaluate different types of bedding materials included: pine sawdust (control) (SD), corn cobs (CC), and soybean straw (SS). material were collected twice a month and analyzed for dry matter. C:N ratios and pH . were measured weekly at various depths (15.2, 30.5, 45.7, and 61.0 cm) Moisture content of SD was 59.7; CC, 44.5 and SS, 60.6. ideal bedding material for compost barns should be dry, processed to parts length less than 2.5 cm, and have good water absorption and holding capacity [7] To produce high quality organic fertilizer in large scale using date palm trees mulch produced from local farms. Date palm trees mulch (DPM) was mixed with fresh farmyard manure (FYM) as nitrogen source. The mixtures were prepared by using 3 ratios (w/w) of 1:1, 2:1 and 3:1 (DPM : FCM).The results found for mixtures of 2 and 3 were better than mixture 1, therefore these mixtures are considered to be the best for composting date palm trees mulch. Regulated elements (N, P, K, Organic matter, pH, electrical conductivity, and TDS) were improved in all mixtures. The end product was free of salmonella, total coliform and faecal coliform bacteria [17]. The challenges of working with sand-laden manure are related to the physical properties of the sand. Sand has a density of 1,750–2,100 kg/m3 depending on its moisture content. Sand does not absorb moisture, This increase in weight, volume and density makes sand-laden manure extremely hard on manure-handling equipment, The simplest method of collecting sand-laden manure is to scrape with a tractor or skid steer Wear and tear on equipment is a major concern when handling sand-laden manure [8]. Manure is classified as one of the following: 1)Solid Manure is as-produced manure with a large amount of bedding, usually long stalk straw or hay. It is moved manually or by mechanical devices such as front-end loaders and skid-steers. To maintain a solid consistency during storage, 2) Semi-Solid Manure is as-produced manure with less bedding than solid. Conveying semi-solid manure with long stalk bedding from stanchion barns is the same as solid manure.3) Slurry Manure is as produced with limited bedding material, which allows the manure to flow and seek a level plane. It is handled as a liquid material, Conveying slurry manure from stanchion barns is usually by gutter cleaners or gravity flow gutters to a collection structure [15]. Organic bedding materials kind of straw, hay, saw dust, wood shavings, crop residue (corn stalks, cobs, etc.) shredded paper, paper pulp residue, composted or dried manure. They are used as bedding because their high moisture absorption. They are compatible with manure handling systems and are readily available [14]. The trailer box scraper is one of the equipment used for cleaning pens need professional operators. One of the problems is the hard of getting the corners of the pens along the feed bunk. Therefore, the front end loader, rear mounted scraper or a scraper on a tractor with a three point attachment are used to clean out the manure from the corners [14] .Solid manure is usually collected using scrapers, box scrapers, blades, front-end or skid-steer loaders or similar devices. Equipment sizes ranged from small blades suitable for tractors of 50 hp or less to large bucket loaders mounted on dedicated power units for operations generating large volumes of manure.[5]. Using semi trucks or tractors and manure tanks is the most common way today to transfer manure from one location to another. Tanks are readily available, may already part of a farms equipment package, and would not require any additional infrastructure [2,3] .

Our objectives for this study were:

1) Calculating the consumed power for all mechanical operation. (2) Estimating the machine costs for grinding date palm trees and mechanical operation to preparing the barn to distributing mulch bedding. (3) verifying the efficiency of using date palm trees mulch as bedding materials and its effect on livestock animals comfort ability and safety. At the time produce a suitable mix for further compost production. (4) Evaluate the physical, chemical and microbiological characteristics of the produced composite bedding material after 10 days.

Materials and methods

A. Animals and housing

The investigation on the ability of date palm trees mulch as a bedding material, to absorb dry heifers cow manure and urine was done at a large scale animal farm, Al Hasa City, K.S.A. Study consists of 120 dry heifers. Heifers sections were classified into 3 sections. Section A is an animal concrete feeding area (40 x3m). Section B is an animal comfort sandy area (40 x 5 m). Section C is an animal sandy floor movement area (40 x 19m). Total barn area of section A and SectionB (320 m^2) was bedded with date palm trees mulch. Integrated pest management program was selected as an optimum strategy of manure recycling to prevent the reproduction of house fly and change the medium of larva. First step, manure was removed from feeding area by tractor attached scraper 6 ft, to transfer the manure outside the area. Second step, feeding area was cleaned by using a high pressure water pumping machine in order to remove the remaining steak manure on the concret floor within trial areas, to ensure complete cleanliness of the barn. Third step, Water was sucked from concrete basin resulting assembly by using trailer Vacuum operating by tractor before date palm mulch was applied as a bedding material. Study area was sprayed using Larvacide. Finally, the mulch was distributed by using hand-fork. Fly density was measured within the bedding 10 days period. Movement of animals and bedding layer thickness were observed daily.

B. The bedding product is produced by grinding date palm trees using horizontal grinders machine from Al Hufof Stars Station, Al- AHSA K.S.A . The finished product is very dry compared to industry standard. It is lab tested at(IDAC LABE) to have a dry matter 95% and moisture content12-14%., Mulch denacity 160kg /m3 .The bedding has a fine texture consistency that is very absorbent with less dust, and it tends to hold up longer in high moisture without turning verss the traditional sawdust or shaving products

Calculate the Quantity of mulch required to covere the expremental areas Aand B according to [17] to produce high quality organic fertilizer in large scale using date palm trees mulch. Date palm trees mulch (DPM) and fresh farmyard manure (FYM), The mixing ratio (w/w) is 2:1 (DPM : FYM).The first stages of collecting information before starting the experiment found that wet manure samples were taken from the experiment site to determine the amount of daily production of fresh and dry manure from 120 heifers per day. Then calculate the quantity of mulch to caver expremental area.

C. Machine and equipments

1. Machine of preparing date palm trees mulch, as a bedding material from date palm trees waste.

The machine were 1- horizontal grinders machine Vermeer hg model 4000 which has an engine 384 kW, the capacity of grinding mulch is 10 ton per hour according to screen size 6 inch

2- grapple to loading date palm trees waste to grinding machine, Hitachi model ZAXIS 200 .has an engine 122 kw, and loader New Holaned handling mulch material model W170 has an engine 140 kw, bucket capacity 2.5 cubic meter.

2. Machine of preparing the parn area to putting the mulch. Integrated pest management program was selected as an optimum strategy of manure recycling using 3 steps

The first step is removing cow manure from feeding area (section A) by tractor 58.8 kw attach rear Box scraper 6ft to transferring manure outside the feeding area 2- Clean up the feeding area using a high pressure water pump machine to clean and remove sticky manure from concrete and corners .at the same time Use manual brushes to remove the remnants of the existing structures of manure to ensure complete cleanliness of the site. 3- tractor 58.8 kw A tractor is running Vacuum Trailer to suckeing liquide manure from concrete basin 4- ULV sprayer to spray the Larvacide.. The second step is removing and transfer soil with manure from rasting area B using Loader has an engine 140 kw,bucket capacity 2.5 cubic meter and trailer capacity 25 cubic meter. This step should apply before adding the mulch layer under the caw

The third step is Covering the area by mulch using dump truck has an engine power 255 kw, box capacityis 25m3, 30 ton. To distributing the mulch in areasA and B using hand-fork.

D. Controlling flies there are 3 steps to control flies 1- Spray area (A, B) using Larvicides Alsystin 050 UL to prevent larvae breeding 2- Controlling Adult flies using ULV machine 3- Install adhesive strips to measure the density of flies within the Contaminated site.

E. Estimating the horse power consumption for the different operation.

1. Machine of preparing the bedding material, to get out of machine as mulch by using Vermeer horizontal grinders machine model hg 4000 which has an engine 515 hp (384 kW), the capacity range from 6 to 10 ton per hour according to the size of screen, the fuel consumption was measured immediately after each treatment for screen 4 insh and 6insh by using fuel scale measurement. The following formula was used to estimate ending used power (EP) to grinding machine, grapple, and loader according to [6].

$$EP = \frac{f.c \times PE \times L.C.V \times 427 \times \eta thb \times \eta m}{3600 \times 75 \times 1.36} \qquad \text{(Eq. 1)}$$

Where :

EP = power consmipation (kW)

$f.c$ = The fuel consumption, (L/h)

PE = The density of fuel, (kg/L) (0.823 kg/L)

 L.C.V = The lower calorific value of fuel, (11000 k.cal/kg)

427 = Thermo-mechanical equivalent, (Kg.m/k.cal)

ηthb = Thermal efficiency of the engine, (35 % for Diesel)

ηm = Mechanical efficiency of the engine, (80 % for Diesel)

Fuel consumption: The rate of fuel consumption as quantity per time unit with load and without load, as shown in the following formula, according to [1]

$$FC = \frac{f}{t} \text{---------} 3.6 \qquad\qquad (\text{Eq. 2})$$

where: Fc = Fuel consumption, l/h;

f = volume of fuel consumption, cm3 and, t = time, s. =Calorific value of fuel (10000 kcal / kg),

2. The last formula in (Eq. 1) was used to estimate ending used power (EP for machine of preparing feeding area secation A to putting the mulch for tractor attach rear scraper, high pressure water pumping machine and trailer vacuum .and loder to remove cow manure from expremental area and estimate ending used power (EP for machine of remove bedding material after complet the expremental from secation Aand B

F. Cost analysis and economical evaluation:-

The cost analysis was calculated according to [16] It was performed to calculate the machine operating cost . Cost analysis inputs consist of both fixed and operating costs. Fixed machine costs included machine payments, insurance, taxes, and depreciation with interest rate. Operational costs for the machine included labor, fuel and lubrication, and maintenance and repair excluding the downtime costs.Also, a comparison between the traditional bedding manure handling cost and mechanical preparing date palm trees mulch, as a bedding material cost. These total costs (TC) include depreciation (D), annual capital interest taxes (I), housing and insurance cost (THI), repair and maintenance cost (R), fuel cost (F), lubrication cost, and labor cost (L) Machine age (n) .

$$T_C \text{---------------------------------} \frac{\{(D)+(L)+(THI\}+\{(R)+(F)(L_S)+(I)\}}{n} \qquad (\text{Eq. 3})$$

$$Tc = \frac{\left\{ \left[\left(\frac{Pc - Sv}{Y}\right) + \left(\frac{Pc + Sv}{2} * \frac{i}{100}\right) + 0.02\, Pc \right] + \left[\left(\frac{Pc * r_c}{Y}\right) + (0.25\, Pt * fc) + (Oc * c * n) + (N * L * n) \right] \right\}}{n}$$

where;

Tc = Total cost SR /h;

Pc = Machine price, LE;

Sv = Salvage value= Pc×Sp where, Sp=Salvage percentage = 56×0.885^n for mulching machine(%), and = $68 \times o.92^n$ for tractor.

n = Machine age = 5 years,10 years for tractor and 15years for grinding machine

i = Interest rate =13%;

rc =Coeficient of repair and maintenance = 1 for tractor, 0.6 for the machine;

Pt = Tractor and differant machine power (hp) ;

fc = Actual fuel consumption = measured (L l/h) ;

fp = Fuel price =3.o SR = for diesel fuel;

Lc = Lubrication cost =20% of fuel cost;

N1 = Number of labors

L = Labor cost = 15 SR /h

(2) Machine flied capacity (Fc): Actual field capacity was calculated as follows:

$$Fc = \frac{A}{t} \qquad (Eq.\ 4)$$

where; fc = machine capacity, m^2. / h; A = area m2. ; t = Machine operating time, h.

3. The specific energy was calculated by using the following equation:

$$Specific\ energy = \frac{Power\ requirement\ (kW)}{\frac{Effective\ field\ capacity\ (feddan)}{h}}\ kW.h/feddan \qquad (Eq.\ 5)$$

   G. Characterization of composite manure bedding material Raw materials, composite date palm trees mulch and manure bedding material were analyzed for pH, electrical conductivity, temperature, total nitrogen, total phosphorus, total potassium and ,moisture to determine the ability of bedding material to absorb moisture using standard procedure [4] Also initial raw materials were analyzed chemically and microbiologically (Tables 9, 10 and11). Microbial activity and temperature measurements of compost piles Composite date palm trees mulch and manure bedding material were biologically analyzed. Microbial analysis included total viable bacterial counts (cfu/g), total coliform (MPN/100g), bacterial pathogens detection using API kit identification. A nalysis was done according to [19]

Results and discussion
The choice of bedding material used on farms is dependent on many factors, including economics, animal health and manure management .Our study are required to elucidate the precise mechanisms by which the beneficial results of using date palm trees mulch as a bedding dry cows.

   1. Animals and housing Study consists of 120 dry heifers. Heifers barn was devided into 3 sections. 1-Section A is animal Feeding area (concrete floor) (40 x3m). 2-Section B is an animal comfortable area (Sandy floor) (Shadow area) (40 x 5 m). 3-Section C is an animal movement area (Sandy floor)(sunny area) (40 x 19m). Total barn area of section A and Section B is 320 m^2

   2. Calculate the Quantity of mulch required to covere area Aand B accouring to [17] to produce high quality organic fertilizer in large scale using date palm trees mulch, (DPM) and fresh farmyard manure (FYM), The mixing ratio (w/w) is 2:1 (DPM : FYM). The first stages of collecting information before starting the experiment found that, wet manure samples were taken from the experiment site to determine the amount of daily production of fresh manure from 120 heifers is 1067 kg/day wet manure and 490 kg/day dry manure . Calculate the quantity of mulch to caver expremental area [Section A 120 m^2, (concrete floor) and Section B 200 m2 (sandy floor)] total area is 320 m^2 and to fit the amount of manure for 10 days is 4840 kg .The equivalent quantity of mulch is 9680kg almost 10 ton of mulch enough to hold the experiment tim.

   3. Estimate power consumption for the different operation according to [6] .

3.1. power consumption kW, specific energy kW/ton and specific energy kW/m^3 for preparing the bedding material, to get out of machine as mulch . Data were tabulated in [table1] A total power consumption for horizontal grinder machine, grapple to loading the grinding machine and loader for handling maluch materal. Were 468 kW, and specific energy 46.8 kW/ton and 7.49 kW/m^3

Table 1: Power consumpation kW, specific energy kW/ton and specific energy kW/m3 for mechanical preparing the date palm trees mulch as bedding material .

	machine	Engine [kW]	Power consumpation [kW]	specific energy [kW/ton]	specific energy [kW/m^3]
1	Horizontal grinders machine	404 kW	283	28.3	4.5
2	Grapple	121 kw,	85	8.5	1.3
3	Loader	143 kw	100	10	1.6
	Total		468	46.8	7.4

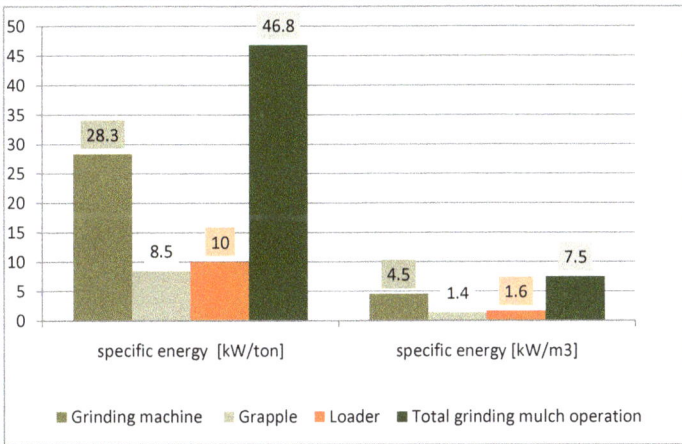

Fig. 1 Power consumpation kW, specific energy kW/ton and specific energy kW/m^3 for mechanical preparing the date palm trees mulch as bedding material .

3.2. Power consumption for Mechanical operation of preparing the barn to putting the mulch as shone in Table [2] this operation was comleted by 3 steps. 1-first step is removing manure from feeding area (section A). The machine which used in this operation were 1- removing manure from feeding area by tractor and attach rear scraper 6 ft . 2-complete cleaning feeding area by using a high pressure water pumping and remove sticky manure from concrete feeding area.3- sucking water from concrete basin of feeding area by tractor mounted vacuum trailer tank.

2-second step is removing soil with manure from rasting area (section B) to storage area

3-third step is covering the areas A and B by mulch.

4- transfer mulch from storage to the barn by dump truck capacity 30 m3. Total specific energy to preparing the barn to putting mulch for m^3 was 11.84 kW/m3 every 10 dayes

Table 2: Power consumpation kW and specific energy kW/m^3 for Mechanical operation of preparing the parn to putting the mulch every

.	Machine operation	Power onsumpation [kW]	specific energy [kW/m^3]
1	Tractor attach rear scraper 6 ft to cleane manure outside the feeding area	31	0.57
2	High pressure water pump to clean and remove sticky manure	3	0.37
3	Tractor mounted Vacuum Trailer tank 10m^3 to suckeing liquide manure	37	1.6
	loader removing soil and manure from area B	100	3.3
5	dump truck capacity 30m^3 transfer mulch to the barn	180	6
	Total in 10 dayes		11.84

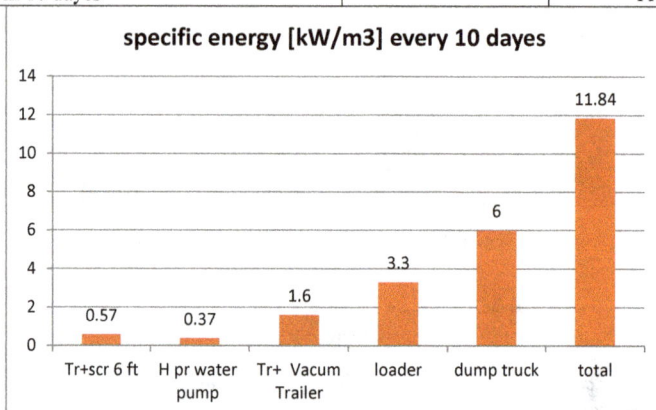

Fig. 2 Specific energy kW/m^3 for Mechanical operation of preparing the parn to putting the mulch every 10 days.

3.3. Machine removing the bedding with manur after expremintal completing using loader to removing and loading manure from areas A and B, and dump truck capacity 30 m^3 to transfer manure from expermintal area to composting yard.

Materials Research Forum LLC
doi: https://doi.org/10.21741/9781644900178-12

Table 3: Power consumpation kW and specific energy kW/m³ for removing manure from the barn every 10 dayes

.	Machine operation	Power consumpation kW	specific energy kW/m³
1	loader removing and loading manure from areas A and B	100	3.3
2	dump truck capacity 30 m3 transfer manure from expermintal area to composting yard	180	3.0
	Total	280	6.3

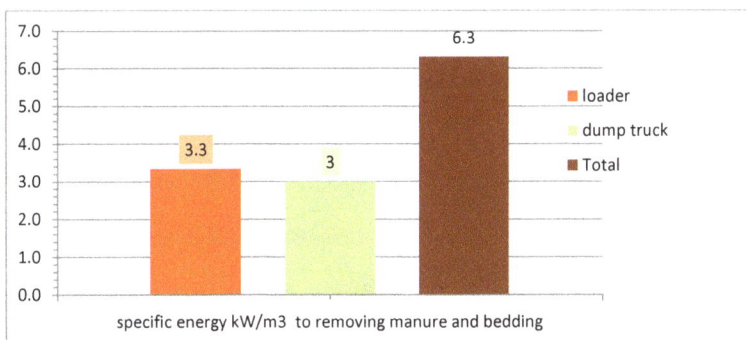

Fig. 3 Total specific energy kW/m3 for Mechanical operation of preparing the barn of new system to putting the mulch every 10 dayes 11.84+6.3 =18.14 kW/m3 - specific energy after 4 months (120 dayes) 18.14*12= 217.7 kW/m3

3.4. Power consumption for traditional manure management system in the farm, the following equipment was used to perform manure removal from barn.1- Cleaning feeding area by Tractor engine power 44kw attach rear scraper 6 ft to remove liquide manure to concrete basin and suckeing liquide manure from concrete basin by trailer vacuum 2- remove and transfer soil and manure from rasting area B to movment area C using tractor 58.8 kw attach rear scraper 10 ft and make cycle betwen area B and C Until remove all wet manure from area B and replac it by dray soil from area C . this operation made every day . as shone in table [4] . Total Power consumption of traditional manure handling system is 226 kW and specific energy 8.4 kW/m³ in one time and

Table [4] Power consumption kW and specific energy kW/m³ for mechanical operation of

Table 3: traditional manure handling system two times every day

	Machine operation	Power onsumption [kW]	specific energy [kW/m^3]
1	Tractor attach rear scraper 6 ft to cleane manure outside the feeding area.A	62	1.2
2	Tractor mounted Vacuum Trailer tank 10m^3 to suckeing liquide manure	74	3.2
3	Tractor attach rear scraper 10 ft to remove and transfer soil and manure from rasting area B to movment area C	90	4
	Total every day	226	8.4

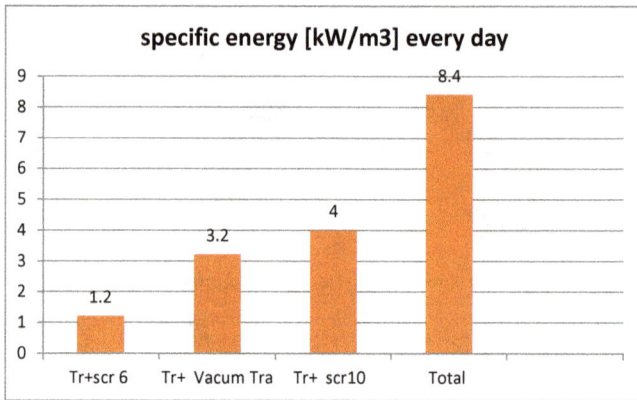

Fig. 4 After 4 months manure is completely removed using loader and dump truck capacity 30 m^3 to removing and loading manure from the barn to composting yard. as showen in table [5] total specific energy of traditional manure handling system after 120 days (4 months) is 1017. 3 kW/m^3

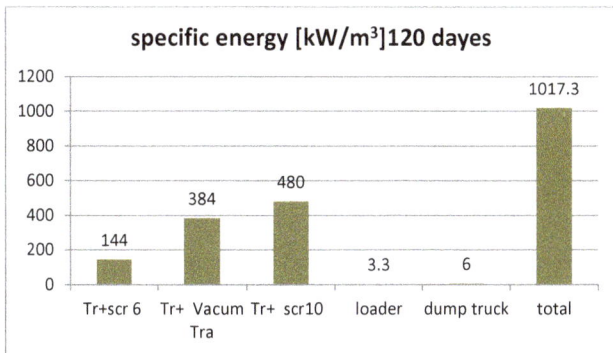

4. Cost analysis and economical evaluation included fixed and variable costs were calculated acording to [16]

4.1 Machine of preparing the bedding material, to get out of machine as mulch .As shown in Table 6, the total cost for mechanical preparing the date palm trees mulch as bedding material costs 1256 SR / h for the first stage.

Table 6: Cost analysis for mechanical preparing the date palm trees mulch as bedding material .

	Machine operation	Engen power kW	Machine capacity M³/h	Cost SR /m3	Cost SR /h
1	horizontal grinders	550	62	10.7	663.4
2	grapple	165	62	3.3	204.6
3	loader	195	62	6.3	388
	Total cost			20.3	1256

4.2. Cost analysis for mechanical preparing the barn area to putting mulch .the resaluts at table 7 for second step operation to remove manure at secation AandB costes 835.05 SR / hour and 26 SR / m3.

Table 7: Cost analysis for mechanical preparing the parn area to putting the mulch

.	Machine operation	Machine capacity M^3/h	Cost SR /m3	Cost SR /h
1	Tractor attach rear scraper 6 ft to cleane manure outside the feeding area	40	2.3	124
2	High pressure water pump to clean and remove sticky manure	13.5	1.1	14
3	Tractor mounted Vacuum Trailer tank to suckeing liquide manure	20	9.6	192
4	loader removing soil and manure from area B	60	6.3	388
5	dump truck capacity 30m3 transfer mulch from storage to the barn	60	5.4	324
			26	1042

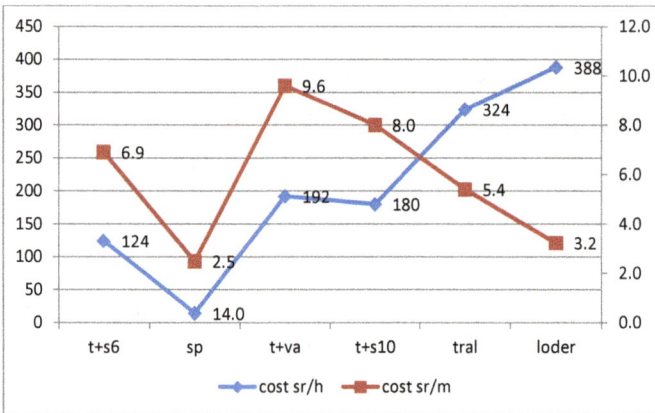

4.3. Removing the bedding with manur after the end of the experiment . As shown in Table 7b, the total cost of the manure removal process from barns is 8.6 SR /m3 and 712 SR /h for Removing the bedding

Materials Research Forum LLC
doi: https://doi.org/10.21741/9781644900178-12

Table 7b: Cost analysis for removing the bedding with manur after expremintal

.	Machine operation	Machine capacity M^3/h	Cost SR /m3	Cost SR /h
	loader remove manure and loading the truck from expermintal area	120	3.2	388
	dump truck capacity 30 m3 transfer manure from expermintal area to composting area	60	5.4	324
	Total cost		8.6	712

4.5. Total cost for using date palm trees mulch as shown in table 8, the total cost a bedding every 10 dayes were 3180.0 SR/m3 and total cost 28620 SR/m3 After 3 months .

Table 8: Total cost analysis and economical evaluation for using date palm trees mulch as a bedding

.	operation	Cost SR/m3	Total cost/barn (60m3)
1	Machine preparing mulch bedding material	20.2	1212.0
2	preparing the parn area to putting the mulch 60 m3	26	1560.0
3	removing the bedding with manur after expremintal	6.8	408.0
	Total cost by 10 dayes	53.0	3180.00
	Total cost After 3 months		28620

4.4 Cost analysis and economical evaluation. According to the traditional manure management system in the farm, the following equipment was used to perform manure removal from barn two times every day.

1- Cleaning feeding area by Tractor engine power 44kw attach rear scraper 6 ft to remove liquide manure to concrete basin and suckeing liquide manure from concrete basin by vacuum trailer 2- remove and transfer soil and manure from rasting area B to movment area C using tractor 58.8 kw attach rear scraper 10 ft and make cycle betwen sechan B and C antal remove all wet manure from area B and replac it by dray soil . this operation made every day . Every 3 months the whole manure is removed from each barn at a depth of about 50 cm . As shown in Table 9, feedinng area and the rest area are cleaned daily and the manure is completely removed every three months. Total cost for traditional manure management after 3 months is 57004 SR.

Table 9 Cost analysis for traditional manure handling system in the barn

.	Machine operation	Engine (kW)	Machine operation (h)	Cost SR /m^3	Cost SR /h	Cost/ One barn
1	tractor attach rear scraper 6ft to scrabing the manure outside the feeding area.A	44	0.30	6.9	124	37.2
	Tractor mounted Vacuum Trailer to suckeing liquide manure	58.8	0.50	9.6	192	96
	tractor attach rear scraper 10 ft to remove manure from area B to area C	58.8	0.75	8	180	135
	Total cost every day			24.5	506	268.2
	Total machine cost by 4monuth (120 day)			2938		32184
	sand Soil	400m3		50		20000
	loader remove manure and soil by truck from expermintal area		5		388	1940
	dump truck capacity 30 m3 transfer manure from expermintal area to composting area		5		324	1620
	dump truck capacity 30m3 transfer soil from storage to the barn		5		324	1260
	Total cost After 4 months					57004

5. The choice of bedding material used on farms is dependent on many factors, including economics, animal health, manure management, and animal well- being. Data presented in table 3 showed that in section A date palm trees mulch absorb and filter manure and urine highly till day10 (Moisture of bedding material was 30.0%) which could indicate that date palm trees mulch is a good bedding material as it behave as a filter. On the other hand in section B bedded material of date palm trees and sand has higher absorption, moisture content at day 10 (52.19%). In the present study, although elevated relative to Heifers bedded with date palm trees mulch was within range considered normal. Concentration of regulated elements in composite bedded material of date palm trees mulch mixed with fresh heifers manure in section A and section B are present in table 4. Bedding in section A and section B were closer in elements content. Indeed, N, P and K concentrations were higher in section A than section B and within the normal standard concentration ranges observed in healthy cattle [20] Date palm trees mulch mixed with fresh heifers manure denacity is 620 kg /m3.

Table 10: Chemical Composition of raw materials used for composting

Characteristics	Date palm trees mulch	Fresh cow manure
pH	6.66	7.22
Moisture (%	**4.5**	**41.8**
Organic Matter (%)	**88.70**	**38.00**
Organic Carbon (%)	**44.45**	**19**
C/N Ratio	76.63: 1	15: 1
NDF (%)	**61.38**	**67.38**
ADF (%	**41.85**	**62.05**
Ash (%)	18.01	70.75
Nitrogen (%)	**0.58**	**1.26**
Phosphorus (%)	**0.25**	**0.54**
Potassium (%)	**1.70**	**1.55**
Calcium (%)	**2.25**	**2.80**
Sodium (%)	**0.55**	**0.54**
Magnesium (%)	**0.26**	**0.27**

Table 11: Microbiological Composition of raw materials used in composting

Characteristics	Date palm trees mulch	cow manure
Total viable bacterial counts (cfu/g)	TNTC	TNTC
Total coliform (MPN/g)	Nil	TNTC
Bacterial detection	Bacillus sps	Salmonella, E.coli

Table 12: Chemical properties of the composite bedding materials

Time	Section A (Date palm trees mulch in concrete area)					Section B (Date palm trees mulch and sand)				
Days	pH	EC (us cm-1)	Moisture (%	NDF %	ADF %	pH	EC (us cm-1)	Moisture (%	NDF %	ADF %
Day1	7.21	15.57	57.23	65.14	57.53	7.34	15. 7	56.37	61.6	44.50
Day2	7.33	20.24	55.38	63.51	46.32	7.53	18.6	54.17	68.51	42.03
Day3	7.37	25.31	62.60	63.61	48.06	7.25	22.1	54.61	63.93	47.97
Day5	6.72	19.99	37.73	63.42	51.62	6.60	18.8	54.21	66.10	49.56
Day7	7.16	22.57	39.70	69.62	60.22	7.03	18.7	53.30	66.78	56.29
Day10	7.53	16.88	30.00	60.00	49.40	7.34	27.3	52.19	68.27	56.38

Table 13 Major elements properties of the composite samples.

Time	Section A (Date palm trees mulch in concrete area)						Section B (Date palm trees mulch and sand)					
Days	N%	P%	K%	Na%	Ca%	Mg %	N%	P%	K%	Na%	Ca%	Mg%
Day1	1.50	0.81	1.67	0.65	2.46	0.26	2.03	o.49	1.11	o.35	2.51	0.03
Day2	1.51	0.65	1.59	0.53	2.18	0.27	1.38	o.49	1.32	0.48	2.33	0.29
Day3	2.00	0.53	1.71	0.54	2.36	0.30	1.66	o.61	1.61	0.55	2.40	0.03
Day5	1.14	0.16	1.73	0.60	2.11	0.26	1.35	0.18	1.37	o.53	2.14	0.26
Day7	1.12	0.12	1.11	0.37	1.93	0.20	o.96	0.34	0.98	o.55	1.87	0.69
Day10	1.85	0.50	1.42	0.59	2.56	0.26	0.38	0.38	1.09	0.43	2.35	0.24

Summary

Date palm trees mulch can be safely and effectively used as a bedding material for cow feedlots. Feedlot managers will need to adjust bedding rates according to facilities, environment, and cow comfort. Feedlot managers interested in using date palm trees mulch as bedding will recognize that absorbency of date palm trees mulch is lower than that of sand. However, utilizing of date palm trees mulch eliminates costs of harvesting sand. A total power consumption for horizontal grinder machine, grapple to loading the grinding machine and loader for handling maluch materal. Were 468 kW/h, specific energy were 46.8 kW/ton and 7.49 kW/m3. Total Power consumpation to preparing the barn to use mulch bedding material for m3 was 7.84 kW/m3 every day and 78.4 kW/m3 after finish experiment time 10 days. Total cost for using date palm trees mulch as a bedding every 10 dayes were 3180.0 SR. Bedding management every 120 days were 28620 SR

Total cost for traditional manure management every 120 days were 57004 SR for all operation

Acknowledgment

The researchers are thankful to Department of Plant Protection, Ministry of Agriculture in Saudi Arabia for their research participation. Also thanks for Al Hofuf Stars Establishment for research participation and financial support. Thanks are extending to the animal farm for supplying animals and area

References

[1] A.E. Suliman, G.E.M. Nasr, W.M.I. Adawy, Energy requirements for land preparation of peas crop under Egyption conditions, Misr. J. Agric. 10-2 (1993) 190-206

[2] A.C. Lenkaitis, Dairy Freestall Manure Collection and Transfer Systems: Energy and Operational Comparisons, Oral Presentation, Louisville, Kentucky, American Society of Agricultural and Biological Engineers 1111819 (2011).

[3] A.C. Lenkaitis, MANURE COLLECTION AND TRANSFER SYSTEMS IN LIVESTOCK OPERATIONS WITH DIGESTERS, GEA Farm Technologies Inc. (Houle USA), Naperville, 2012.

[4] APHA American Public Health Association, Compendium of methods for the microbiological examination of foods Pub. 3rd edition, 1992.

[5] Charles Fulhage, Joe Harner , Solid Manure Collection and Handling Systems America's land-grant universities enabled by eXtension.org, 2015.

[6] Donnell Hunt, Farm Power and Machinery Management, Iowa State University Press, 1983, pp. 28 – 29.

[7] E.M. Shane, M. Endres, D.G. Johnson, J.K. Reneau, Bedding options for an alternative housing system for dairy cows: A descriptive study: Applied Engineering in Agriculture 26-4 (2010) 659-666. https://doi.org/10.13031/2013.32062

[8] H.K. House, Sand-Laden Manure Handling and Storage, OMAFRA, www.omafra.gov.on.ca, 2010.

[9] J.G. Davis, T.L. Stanton, T. Haren, Feedlot Manure Management no. 1.220,Colorado State University Cooperative Extension. 5-97(1997).

[10] J.I. Sprague, Fly Control: Manure Clean-up, Feedlot Magazine, 2013.

[11] K.A. Janni, M.I. Endres, J.K. Reneau, W.W. Schoper, Compost barns: An alternative dairy housing system in Minnesota, ASABE Annual International Meeting 9 -12 July 2006, American Society of Agricultural and Biological Engineers, 2006. https://doi.org/10.13031/2013.20909

[12] K.A. Janni, M.I. Endres, J.K. Reneau, W.W. Schoper, Compost Dairy Barn Layout and Management Recommendations- Applied Engineering in Agriculture, American Society of Agricultural and Biological Engineers 23-1 (2007) 97-102. https://doi.org/10.13031/2013.22333

[13] MWPS. Dairy Freestall Housing and Equipment, 7th ed., Iowa State Univ., Ames., 2000.

[14] New York State Cattle Health Assurance Bedding Materials and Udder Health, Fact Sheet, Program Mastitis Module, 2008.

[15] NRCS manure transfer, code 634, natural resources conservation service, conservation practice standard, 2006.

[16] A. Oida, Using personal computer for agricultural machinery management, Kyoto University, Japan, JICA publishing, 1997.

[17] M.W. Sadik, A.O. Al Ashhab, M.K. Zahran, F.M. Alsaqan, Composting mulch of date palm trees through microbial activator in Saudi Arabia International Journal of Biochemistry and Biotechnology, Available online at http://internationalscholarsjournals, 1-5 (2012) 156-161.

[18] Stephen Herbert, Masoud Hashemi, Carrie Chickering-Sears, Sarah Weis, Factsheets, UMass Extension Crops, Dairy, Livestock, www.umass.edu/cdl, 2008.

[19] S.Y. Wong, S.S. Lin, Composts as soil supplement enhances plant growth and fruit quality of straw berry, J. Plant Nutr., 25 (2002) 2243-2259. https://doi.org/10.1081/pln-120014073

[20] W.G. Bickert, Ventilation: The Merck Veterinary Manual, 9th ed., Merck & Co., Whitehouse Station, 2005, pp. 1609-1702.

[21] S.Y. Wong, S.S. Lin, Composts as soil supplement enhances plant growth and fruit quality of straw berry, J. Plant Nutr., 25 (2006) 2243-2259.

By-Products of Palm Trees and Their Applications   Materials Research Forum LLC
Materials Research Proceedings 11 (2019) 186-192   doi: https://doi.org/10.21741/9781644900178-13

Effect of Natural Additives as Coconut Milk on the Shooting and Rooting Media of *in vitro* Barhi Date Palm (*Phoenix dactylifera L.*)

H.S. Ghazzawy[1,2,a*], S.F. El-Sharabasy[1]

[1]Date palm Research Center of Excellence, King Faisal University, Saudi Arabia

[2]Central Laboratory for Date palm Research and Development, Agriculture Research Center, Giza, Egypt

[a]hishamdates@hotmail.com

Keywords: natural additives, Barhi, Casein Hydrolysate, coconut milk, yeast extract

Abstract. The objective of the research study was to determine the effect of addition of different concentrations of three types of natural additives on Date Palm cv. Barhi: (1.25g/l, 2.5g/l, 5.0g/l for Casein Hydrolysate and 10%, 20%, 30% for (Coconut Milk and Yeast Extract), in addition to the control (0.05 BA mg/l) for shooting stage and (0.1 NAA mg/l, 3 g/l AC) for rooting stage. The results show that the use of 30% Coconut Milk achieved a high number of shoots and the highest shoot length was recorded with 10% Coconut Milk. In the date palm rooting stage, the results show that the use of 30% Coconut Milk increased the number of roots, shoot thickness and rooting percentage. However, root length was increased with 10% Coconut Milk. The lowest values were recorded with using Yeast Extract in this stage.

Introduction

Date palm (*Phoenix dactylifera*. L.) has a great economical importananc and agricultural uses throughout human's history. Also, it is one of the oldest cultivated fruit trees in the world. Date palm is a very important crop in the Middle East, since it can grow well in both semi-dry desert areas and the newly cultivated land. The production of Arab world of dates is about 80% of the total production of the world. Egypt is the world largest date producing country i.e. more fruitful female palms (1.5M tonnes/annum) produce 1.694.813 tons of dates [9], [7]. In Egypt, date palm trees distribution covers a large area extends from Aswan to north Delta, beside the Oasis of Siwa, Bahriya, Farafra, Kharga, Dakhla. Egypt is one of the most productive countries of dates in the world, the number of fruitful female palms in Egypt is about 15 million produce 1.694.813 tons of dates [9]. Date palm is commonly propagated by ground offshoots; however, a female date palm produces only 10-20 offshoots in its entire life [20], which is a limiting factor for the propagation of commercial cultivars. A non-conventional technique of in vitro culture is widely used in many species including date palm [14]. The production of plants through in vitro culture is successfully introduced in many species [23]. The technique of tissue culture for propagation date palm, also called in vitro propagation, has many advantages as larg scale multiplication troughout the year, production healthy female cultivars, (disease and pest-free), or males having superior pollen; production of genetically uniform plants [19]. Recently, the natural products is using Yeast and plant extracts *in vitro* which have been discovered. Some undefined components such as Yeast Extracts, Frnit Juices and Protein Hydrolysate were frequently used in nutrient media as opposed to defined amino acids or vitamins as a further supplementation [4]. In addition, some other natural additives as Coconut Milk is frequently used as a popular addition to the media of orchid cultures in the floral industry of tissue culturing [5]. Natural extract could be

Materials Research Forum LLC
doi: https://doi.org/10.21741/9781644900178-13

used at a 6% concentration as a replacement for sucrose [7] the utilization of natural additives compounds instead of hormones in culture media may decrease the possibility of genetic instability in plants. Organic additives such as Coconut Water and Casein Hydrolysate have been used to rise embryogenic callus growth and somatic embryogenesis in several plant species as well as date palm [6].

The aim of the research study was to determine the effect of different concentrations of combination natural additives such as Coconut Milk, Casein Hydrolysate and Yeast Extract, with the goal of enhancing the in vitro date palm cv. Barhi shoot and root proliferation.

Materials and methods
Explant and sterilization: The experiments were carried out in the Tissue Culture Laboratory for Date Palm Research and Development, Agriculture Research Center, Giza, Egypt. Four-year-old female offshoots of date palms *cv.* Barhi were collected and used as explants. Preparation of explants was done by removing the roots and outer green mature leaves from the offshoots, then reducing the size to less than 25 cm. remaining mature leaves were removed gradually from the bottom offshoot to the top in the laboratory [14]. The gradual removal of white young leaves and surrounding white fibrous leaf sheath resulted in 5 cm shoot tips, which were further trimmed to 2 cm for explant use. All excised shoot apexs were stored temporarily in an anti-oxidant solution (150 mg/l ascorbic acid and 100 mg/l citric acid) prior to surface-sterilization. Under aseptic conditions, shoot apexs were soaked in 70% ethanol alcohol solution for 30 seconds, followed by immersion in (1.0 g/l) of mercuric chloride for 5 min and thoroughly washed with sterilized distilled water for one-time. After that additional leaf primordial were removed from sterilized explants and then these explants were sterilized in 50%(v/v) commercial bleach (Clorox) 5.25% w/v, sodium hypochlorite NaOCl plus 1 drop Tween 20 for 15 min with rotary agitation, rinsed three times with sterilized distilled water.

Effect of different natural additives on shooting and rooting stages:
Shoots clusters which havd been received from indirect somatic embryogenesis as recomnded by (El-Dawayati et al,2018) were used as explants in this experiment.
Different concentrations of three natural additives as follows: (1.25g/l, 2.5g/l, 5.0g/l for Casein Hydrolysate and 10%, 20%, 30% for Coconut Milk and Yeast Extract), were supplemented to a standard nutrient growth medium (control treatment without natural additives) for shooting and rooting, Control (treatments) were prepared by culturing the same explants on the same media under the same conditions without any supplements to study their effects on shoots development during shooting and rooting stages. All refined techniques were completed under aseptic conditions. Standerd growth media preparation for shooting stage was composed of ¾ MS basal nutrient medium according to Murashige and Skoog with vitamins [16, 22], with addition of 100 mg/l Myo-Inositol; 80 Adenine Sulfate; 170 mg/l NaH2PO4.2H2O; 0.3 mgl/l Ca panthothianic acid; 0.4 mg/l thiamine- HCl; 2 glycine; 0.5 mg/l nicotinic acid; 0.5 mg/l pyridoxin-HCl; 100 myo-inositol; 30g/l Sucrose; 0.05 mg/l (BA) and 0.1 NAA mg/l growth regulators and 6 g/l Agar; 7000 [Agar-agar/Gum agar] (Sigma Chem. Co., St. Louis, MO) (in mgl⁻¹) [1]. Standerd growth media preparation for rooting stage Also the same different three natural additives at different concentrations were added to rooting media which consist of the same components of previous standerd growth media of shooting but supplemented only with 0.1 NAA mg/l growth regulator, with the addition of 1.5 g/l activated charcoal . and. 0.3 mgl/l Ca panthothianic acid; 0.4 mg/l thiamine- HCl; 2 glycine; 0.5 mg/l nicotinic acid; 0.5 mg/l pyridoxin-HCl; 100 myo-inositol; 200 glutamine; 1g; 30000 sucrose; and 6000 agar.

By-Products of Palm Trees and Their Applications   Materials Research Forum LLC
Materials Research Proceedings 11 (2019) 186-192   doi: https://doi.org/10.21741/9781644900178-13

The pH value was adjusted at 5.7 before adding agar gerlite and autoclaving the medium at 1.2 Kg.cm^{-2} equivalent to 121°C for 20 min. The nutrient media was dispensed into small jars twenty-five ml of media for shooting stage. The plantlets were cultured in tube size (25 x 250 mm) each tube contained 25 ml for rooting stage. Explants of each treatment and control treatments were transferred and repeatedly recultured for 2 recultures every 8 weeks into fresh medium of the same compostion. [10]. all samples were incubated for 16 hours under 1500 lux light conditions shooting stage and 3000 lux light conditions for rooting stage. They were then subjected to 8-hr dark conditions at 27 ± 2°C for the shoot multiplication stage. Subculturing was performed twice on the control samples and three times for the natural additives with their three different concentrations [4]. All procedures were carried out in a decontaminated horizontal laminar flow hood. The experimental design was completely randomized with three replicates in each treatment. Data recorded of 10 treatments were first analyzed as a whole using the aforementioned statistical design and then it was divided into groups as follows [14]:

Table 1, Different concentrations of three types of natural additives (1.25, 2.5, 5.0 mg.L^{-1}) for Casein Hydrolysate and (10%, 20%, 30% for Coconut Milk and Yeast Extract)

T$_1$ Control 0. 5 BA mg.L^{-1} (Shooting stage).

T$_2$ Control 0. 1 NAA mg.L^{-1} +3AC g.L^{-1} (Rooting stage).

T$_3$ Casein Hydrolysate 1.25 mg.L^{-1}.

T$_4$ Casein Hydrolysate 2.5 mg.L^{-1}.

T$_5$ Casein Hydrolysate 5 mg.L^{-1}

T$_6$ Coconut Milk 10%.

T$_7$ Coconut Milk 20%.

T$_8$ Coconut Milk 30%.

T$_9$ Yeast Extract 1.25 mg.L^{-1}.

T$_{10}$ Yeast Extract 2.5 mg.L^{-1}.

T$_{11}$ Yeast Extract 5 mg.L^{-1}

Collected data for shooting were calculated by estimated the number of shooting, shoot length per cluster in cm and shoot thickens per cluster, number of roots formed rooting % and the length of roots per cluster in cm.

Statistical Analysis: A factorial design in completely randomized arrangement was used and data were subjected to analysis of variance. Difference of means among treatments was determined using L.S.D. test at the 5% significance level according to Smith et al. [11].

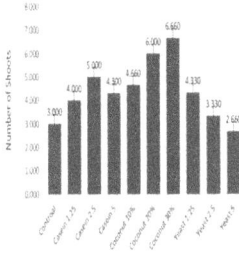

Fig (1): Effect of Coconut milk, Casein Hydrolysate and Yeast extract on No. of shoots of in vitro Barhi CV. (Phoenix dactylifera L.).

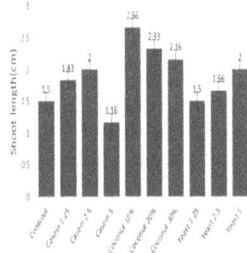

Fig (2): Effect of Coconut milk, Casein Hydrolysate and Yeast extract on shoot length (cm) of in vitro Barhi CV. (Phoenix dactylifera L.).

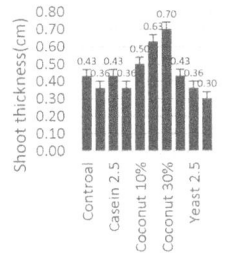

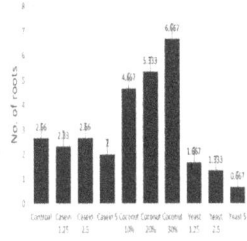

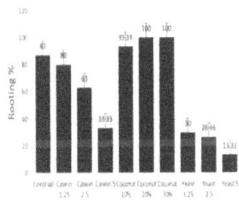

Fig (3): Effect of Coconut milk, Casein Hydrolysate and Yeast extract on (shoot thickness cm) of in vitro Barhi CV. (Phoenix dactylifera L.).

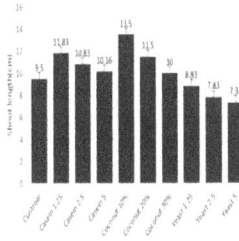

Fig (4): Effect of Coconut milk, Casein Hydrolysate and Yeast extract on shoot length (cm) of in vitro Barhi CV. (Phoenix dactylifera L.).

Fig (5): Effect of Coconut milk, Casein Hydrolysate and Yeast extract No. of roots on of in vitro Barhi CV. (Phoenix dactylifera L.).

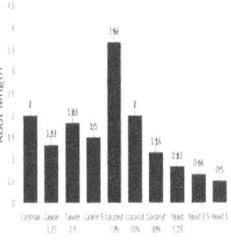

Fig (6): Effect of Coconut milk, Casein Hydrolysate and Yeast extract on root lenght (cm) of in vitro Barhi CV. (Phoenix dactylifera L.).

Fig (7): Effect of Coconut milk, Casein Hydrolysate and Yeast extract on rooting *(%) of in vitro Barhi CV. (Phoenix dactylifera L.).*

By-Products of Palm Trees and Their Applications   Materials Research Forum LLC
Materials Research Proceedings **11** (2019) 186-192   doi: https://doi.org/10.21741/9781644900178-13

Results and discussion

Data in Fig (1) show the effect of addition of different concentrations of three types of natural additives on Date Palm cv. Barhi: (1.25g/l, 2.5g/l, 5.0g/l for Casein Hydrolysate and 10%, 20%, 30% for Coconut Milk and Yeast Extract), in addition to the control (0.05 BA mg/l mg/l) among different treatments regarding the number of shoots, Maximum increase in number of shoots with (6.66) was observed when Coconut Milk 30% (T7) added as addition to the control (0.05 BA mg/l) for shooting stage, followed by the same materials (T6) of Coconut Milk 20% and (T3) Casein Hydrolysate 2.5 g/l (5.00) *i.e.* T5, T4, T8, T2, T9, T1 and T10 (2.66). Data regarding *in vitro* shoot length (cm) showed that T5 (Coconut Milk 10%) had highest value (2.66) followed by T6, T7 were quite close (Coconut Milk 20% & 30%) among various means of different concentrations of three types of natural additives on Date Palm cv. Barhi the three sources (Casein Hydrolysate, Coconut Milk and Yeast Extract) when compared to the control samples. The lowest values were attributed to T4 (Casein Hydrolysate 5g/l) (1.16) then followed by T1, T8 (1.50) (Fig 2). Concerning the shoot thickness (cm) the highest results were with (T7) of Coconut Milk 30% (0.70) and (T6) of Coconut Milk 20%, then (T5) of Coconut Milk 10% (0.63 &0.50), respectively as shown in Fig (3).

Shoot length (cm) (Fig 4) clarified gave demonstrate efficiency with T5 (Coconut Milk 10%) which recorded (13.50) followed by T2 (Casein Hydrolysate 1.25 g/l) and T6 (Coconut Milk 20%) (11.83 & 11.50) and the lowest results were viewed the Yeast Extract 5g/l (7.33). Higher value recorded for number of roots (6.67) was founded in T7 (Coconut Milk 30%) followed by T6 and T5 (5.33&4.66), respectively as shown in Fig (5). Results presented in Figure (6) when assessing root lengths in (cm), the highest results were acquired and identical in Coconut Milk 10% (T 5) and Coconut Milk 20%, control were quite close (2 cm), on the hand, lowest results were in Yeast Extract 5g/l (0.50). Regarding to Fig (7), the highest same values were cleared observed with three treatments (T1, T6 and T7) which recorded (100%), the lowest results were in Yeast Extract 5g/l (13.33). using of natural additives instead of plant growth regulators when added to culture medium may be gave minimum or reduce the possibility of genetic instability in plants [4]. Our results showed the potential use of natural additives to stimulate proliferation. Medium composition, genotype and plant hormones some factors, which affected on multiplication. [12] date palm cv. Maktoom showed higher shoot-bud multiplication in MS medium with a hormone combination of 1 mg L-1 NOA, 1 mg L-1 NAA, 4 mg L-1 2iP and 2 mg L- BAP. Half-strength MS medium improved with 0.5 mg L^{-1} NOA and 0.5 mg L^{-1} Kin produced (23.5) shoot buds per explant after 3 months of multiplication in cv. Najda [13]. Average of production an of 18.2 buds per culture in cv. Hillawi, in the MS medium containing 1 mg L-1 BAP and 0.5 mg L-1 TDZ [2]. Many researchers [20] studied the effects of using plant extracts and Yeast *in vitro* culture. In media undefined components such as fruit juices, Yeast Extract and Casein Hydrolysate were frequently used in place of defined vitamins or Amino Acids, or even as further supplements. As it is essential that a medium should be the same each time it is prepared, materials, which can differ in their composition, are best avoided if at all possible, although improved results are sometimes obtained by their addition [4, 5, 15]. High protein content was founded in Coconut Milk, while high Amino Acid and vitamin were in Casein Hydrolysate, so this confirms that these natural additives increase cell division. Additionally, both Casein hydrolysate and Coconut milk act as cytokine, so they both affect the growth of shoots. These results are in accordance with [8]. Duhamet and Gautheret [26] declared that Coconut Milk are frequently used as a stimulator of cell division; this is due to the high Amino Acid content in Coconut Milk as mentioned with [17]. [18] Suggested 1 mg L^{-1} NAA induces optimum and better rooting at the same concentration IBA or IAA. Mejhoul cv. [3] reported that the shoots grew an average of (13.4 cm) with an average 4.6 roots number per shoot with wide and green leaves from (3 months) old hormone-free 1/2 MS medium strength. Yeast Extract showed an inversely proportional relationship with indoles, which could be an indicator to least efficacy being attributed to them, where they acquired the lowest results in number and length of roots; [19] corroborates these findings. [17] declared that the most efficient secondary somatic embryo formation in association with coconut milk was the most effective component and 5.00g/l casein hydrolysate that for the growth vigor,. The Yeast Extract [2] produced the lowest readings in all assessed concentrations. In addition, chemical analysis was performed that tested Chlorophyll' A' & 'B', Amino Acid, total Carbohydrate, Protein, Indole and Phenol. In addition, the results showed that Coconut Milk 30% and Casein Hydrolysate 2.5g/l gaved the best results, both in test responsiveness and regenerative abilities.

Summary
Our findings validate the results in the supplementation of nutrient media with natural additives as growth regulators. It revealed Coconut Milk 30%, 20%, 10% then Casein Hydrolysate 2.5g/1 as the most successful inducers and are recommended for the *in vitro* culturing of Barhi Date Palm (*Phoenix Dactylifera* L.).

References

[1] A.A. Abul-Soad, Z.E. Zayed, and R.A. Sedqi, Improved method for the micropropagation of date palm (Phoenix dactylifera L.) through elongation and rooting stages, Bulletin of Faculty of Agriculture, Cairo University 57-4 (2006) 789-802.

[2] A.M. AL-MAYAHI, W. Thidiazuron-induced in vitro bud organogenesis of the date palm (Phoenix dactylifera L.) cv. Hillawi, African Journal of Biotechnology 13 (2014) 3581- 3590. https://doi.org/10.5897/ajb2014.13762

[3] S.A. BEKHEET, Direct organogenesis of date palm (Phoenix dactylifera L.) for propagation of true-to-type plants. Scientia Agriculturae 4 (2013) 85-92.

[4] 1. Beshir, S. El Sharbassy, G. Safwat, A. Diab, The effects of some natural materials in the development of shoot and root of Banana (Musa spp.) using tissue culture technology, New York Science Journal 5-1 (2012) 132-138.

[5] L. Duhamet, R.G. Gautheret, Strncture Anatomique de Fragments de Tubercules de Topinambour Cultives en Presence de Lait de Coco. Compo Rendus de la Soc, de Biol. 144 (1950) 177-184. https://doi.org/10.1111/j.1438-8677.1955.tb00343.x

[6] A.A. El-Khateeb, Comparison effects of sucrose and Date Palm syrup on somatic embryogenesis of Date Palms Phoenix Dactylifera L., Am. J. Biotech Biochem 4-1 (2008) 9-23. https://doi.org/10.3844/ajbbsp.2008.19.23

[7] S.F. El-Sharabasy, H.S. Ghazzawy, and M. Munir, In-vitro application of silver nanoparticles as explant disinfectant for date palm cultivar Barhee. J. Appl. Hort. 19-2 (2017) 106-112.

[8] FAO, FAO Production Year Book. Food and agriculture Organization of the United Nation, Rome Italy, 2016.

[9] F.E. George, Plant tissue Culture Technique, The Components of Culture Media, Printed in the UK by Butler and Tanner Ltd, Frome Somerset, (1993) 273.

[10] H.S. KHIERALLAH, M.S. BADER, Micropropagation of date palm (Phoenix dactylifera L.) var. Maktoom throughorganogenesis. Acta Horticulturae 736 (2007) 213-223. https://doi.org/10.17660/actahortic.2007.736.19

[11] H.S.M. Khierallah, N.H. Hussein, The Role of coconut water and casein hydrolysate in somatic embryogenesis of Date Palm and genetic stability detection using RAPD markers, Date Palm Research Unit, College of Agriculture, University of Baghdad, Iraq (2013).

[12] M. A. MAZRI,; R. MEZIANI, An improved method for micropropagation and regeneration of date palm (Phœnix dactylifera L). Journal of Plant Biochemistry and Biotechnology, 22:176-184, 2013. https://doi.org/10.1007/s13562-012-0147-9

[13] R. MEZIANI, J.M. Mouaad, A. Mohamed, A.C. Mustapha, E.f. Jamal, A. Chakib, Effects of plant growth regulators and light intensity on the micropropagation of date palm (Phoenix dactylifera L) cv. Mejhoul. Journal of Crop Science and Biotechnology, 18 (2015) 325-331. https://doi.org/10.1007/s12892-015-0062-4

[14] I.A. Mohammed, S.M. Alturki; F.S. Wael, H.S. Ghazzawy, Effect of potassium Nitrate on Antioxidants Production of Date Palm (Phoenix dactylifera L.) in vitro. Pakistan journal of Biological sciences 17-12 (2014) 1209-1218. https://doi.org/10.3923/pjbs.2014.1209.1218

[15] T. Murashige, F.A Skoog, A Revised medium for rapid growth and bioassays with Tobacco tissue cultures. Physiology Planternm 15 (1962) 473-497. https://doi.org/10.1111/j.1399-3054.1962.tb08052.x

[16] SAS, SAS/STAT Usera Guide. SAS, Cary, NC. USA (2005).

[17] M.H. Shereen, H. Gehan, S.F. El Sharbasy, Z. Zayed, Effect of Coconut Milk, Casein Hydrolysate and Yeast Extract on the Proliferation of in vitro Barhi Date Palm (Phoenix dactylifera L.), Journal of Horticultural Science & Ornamental Plants 8-1 (2016) 46-54.

[18] A. Zaid, P.F. De Wet, Chapter I Botanical and Systematic Description of Date Palm. FAO Plant Production and Protection Papers, 1999, pp. 1-28.

[19] J. Mc Cubbin, A. Zaid, 1. Van Stade, A southern african survey conducted for off types on date palm produced using somatic embryogenesis, EmirJ. Agric. Sci., 16-1 (2004) 8-14. https://doi.org/10.9755/ejfa.v12i1.5213

[20] K.M. Sudipta, M. Kumara Swamy, M. Anuradha, Influence of Various Carbon Sources and organic Additives on In vitro Growth and Morphogenesis of Leptadenia reticulata (Wight & Am), A Valuable Medicinal Plant of India. Int. 1. Pharm. Sci. Rev. Res. 21-2 (2013) 174-179.

[21] F.Abraham, A. Bhatt, C.L. Keng, G. Indrayanto, S. Sulaiman, Effect of yeast extract and chitosan on shoot proliferation, morphology and antioxidant activity of curcuma mangga in vitro plantlets, African Journal of Biotechnology 10-40 (2011) 7787-7795. https://doi.org/10.5897/ajb10.1261

[22] A. Baque, Y.K. Shin, T. Elshmari, E.J. Lee, K.Y. Pack, Effect of light quality, sucrose and coconut water concentration on the micropropagation of canthe hybrids. Australian Journal of Crop Science, 5-10 (2011) 1247-1254.

[23] H.S. Ghazzawy, M.R. Alhajhoj, M. Munir, In-vitro somatic embryogenesis response of date palm cv. Sukkary to sucrose and activated charcoal concentrations. Journal of Applied Horticulture 19-2 (2017) 91-95.

By-Products of Palm Trees and Their Applications
Materials Research Proceedings 11 (2019)

Materials Research Forum LLC
doi: https://doi.org/10.21741/9781644900178

Bio-Composites

By-Products of Palm Trees and Their Applications
Materials Research Proceedings 11 (2019) 195-200

Materials Research Forum LLC
doi: https://doi.org/10.21741/9781644900178-14

Evaluation of Coconut (*Cocos nucifera*) Husk Fibre as a Potential Reinforcing Material for Bioplastic Production

Omoniyi A. Babalola[1,a] and Abel O. Olorunnisola[2,b*]

[1]Department of Agricultural & Environmental Engineering, University of Ibadan, Nigeria

[2]Department of Wood Products Engineering, University of Ibadan, Nigeria

[a]niyibarbz@gmail.com, [b]abelolorunnisola@yahoo.com

Keywords: coconut husk, bioplastic, cassava starch, biodegradable materials

Abstract. In this study the potential use of coconut husk (*Cocos nucifera*) husk fibre for the reinforcement of bio-plastic produced with cassava (*Manihot utilissima*) starch was investigated. Five compositions of the bioplastics were formulated containing 0% (control), 5%, 10%, 15% and 20% of coconut husk fibre. The tensile strength, modulus of elasticity, impact energy, water absorption, and biodegradability of the fibre-reinforced bioplastic samples were then determined in accordance with standard methods. Results obtained showed that the tensile strength values ranged from 0.36 to 0.68MPa; while the modulus of elasticity ranged from 2.7 x10^6 to 4.9 x10^6 N/m^2. The impact energy range was 1.73 - 3.7 J. Analysis of variance showed that coconut husk fibre content had a significant effect on the tensile strength. The impact energy increased with an increase in fibre content up to 15%. Also, water absorption (27.3 - 42.9%) increased with an increase in fibre content. The bioplastics were biodegraded within one month of grave yard test. The optimum fibre reinforcement level was found to be 10%. This may, however, be increased to 15% for impact resistance improvement.

Introduction

Plastic is a material consisting of a wide range of synthetic or semi-synthetic organics that are malleable and can be moulded into solid objects of diverse shapes. Plastics are used in an enormous and expanding range of products due to their long life and attractive properties including relatively low cost, ease of manufacture, versatility, and imperviousness to water. The world's annual consumption of plastic materials has increased from around 5 million tons in the 1950s to nearly 100 million tons; thus, 20 times more plastic is produced today than 50 years ago [1]. This implies that on one hand, more resources are being used to meet the increased demand for plastics, and on the other hand, more plastic waste is being generated. Most of the plastic waste is neither collected properly nor disposed of in appropriate manner to avoid its negative impacts on the environment and public health in many African countries. Due to extremely long periods required for their natural decomposition, waste plastic is often the most visible component in waste dumps and open landfills.

The increased use of synthetic plastics in developing countries is a particular concern as their waste management infrastructure are seldom able to deal effectively with the increasing levels of plastic waste [2,3]. In spite of this daunting challenge, however, the use of plastics has significantly replaced leaves, glasses and metals as a cheaper and more efficient means of packaging in many African societies, except in Kenya and few other countries that have recently banned the use plastic bags in shops and supermarkets.

A group of more environmental friendly alternative materials worthy of consideration in Africa is collectively known as bio-plastics, i.e., plastics derived from renewable biomass sources, such as vegetable fats and oils, corn starch, straw, woodchips, food waste, etc. Bioplastics have been used in a variety of consumer products, such as food containers, grocery bags, biodegradable utensils, and food packaging. These are called commodity plastics. Bioplastics can, however, also be used for engineering grade applications, such as electrical and electronic housings and enclosures. The greatest advantages of bioplastics are a smaller carbon footprint and a less polluted ecosystem. The problem of overflowing landfills and floating islands of trash in Africa may be addressed through increased use of bioplastics.

Starch and cellulose are two of the most common renewable feedstocks used to create bioplastics and these typically come from corn and sugarcane. In Nigeria, cassava starch is of interest as a candidate for producing bioplastics, given the fact that *Nigeria* is the largest producer of cassava (tapioca) in the world, accounting for up to 20 % of the global, about 34% of Africa's and about 46 % of West Africa's cassava production. Annual *production* in *2017 was* conservatively put at 50,000 metric tonnes [4]. It has been shown that reinforcement of bioplastics with a natural fibre may enhance the tensile strength [5]. The typical biofibre sources for bioplastics include cotton, flax or hemp, recycled wood, waste paper, crop processing byproducts, etc. However, a major potential reinforcement material in Nigeria is coconut (*Cocus nucifera*) husk fibre which has relatively high tensile strength, is available in abundant quantities, but largely treated as a waste material [6,7]. *Nigeria* produces about 267,500 metric tonnes of *coconuts* annually and the country occupied the 18th position on the world *coconut production* index as at 2017 [8].

There are two types of coconut fibres, i.e., brown fibre extracted from matured coconuts and white fibres extracted from immature coconuts. Brown fibres are thick, strong and have high abrasion resistance, while white fibres are smoother and finer but weaker. There are many general advantages of brown coconut husk fibres, the object of this study: they are moth-proof, resistant to fungi and rot, provide excellent insulation against heat and sound, not easily combustible, flame-retardant, unaffected by moisture and dampness, tough and durable, resilient, springs back to shape even after constant use, totally static free, and easy to clean [8].

The aim of this study was to investigate the effects of coconut fibre reinforcement on selected properties of a cassava starch-based bioplastic.

Methodology

Bioplastic specimens (a sample is shown in Figure 1) were produced with cassava starch and varied coconut husk fibre contents of 0, 5, 10, 15 and 20% (by weight). The fibres were cut into 2 mm length followed by chemical treatment to reduce lignin and hemi-celluloses contents involving soaking in 1 molar solution of NaOH at 50°C for 4 hours. The fibres were then washed, dried at 80°C in an oven and mixed with cassava starch, water, glycerin and acetic acid in predetermined proportions in a container. The blended mixture was then transferred to a mould, heated up to 65°C for 4 h and air cooled at room temperature. Tensile test was performed using 3 mm thick dog-bone shaped specimens. Three replicate samples were tested for each composition at a fixed crosshead speed of 5 mm/ min in accordance with ASTM D638 standard test methods for plastic properties in tension.

Fig. 1: A sample of the coconut husk fibre-reinforced bioplastic

Impact resistance tests were performed on 60 x 60 x 2 mm specimens using an adaptation of ISO 6603-1. A known weight of 0.616kg was raised to a known height and allowed to fall on each specimen. The height travelled by the ball when the first visible crack had developed on the face was recorded. The total energy of fracture was determined using equation 1:

$$T= mgh. \tag{1}$$

Where T is the total energy, m is the mass, g is the gravitational acceleration, and h is the height.

Water absorption properties were determined by weighing samples in air before and after immersion in water for 24 hours. The percentage amount of water absorbed by samples was calculated using equation 2:

$$W (\%) = \frac{Wt-Wo}{Wo} \times \frac{100}{1}. \tag{2}$$

Where W is percent water absorption, Wo and Wt are the initial and final masses of the specimen before and after immersion in water respectively.

Biodegradability of the samples was determined using the standard grave yard test that involved burying the samples in a soil at a depth of about 900 mm for one month. The degree of biodegradation was monitored for a period of one month.

Results and Discussion

Tensile strength
Table 1 shows the tensile strength values of the bioplastics reinforced with coconut husk fibres at different levels. The values obtained were relatively low compared to those reported for bioplastics produced with cassava starch and reinforced with recycled newspaper pulp [9]. Also, the tensile strength increased as the coconut husk fibre content increased from 0 to 10%, but decreased thereafter. The significant increase (p<0.05) in tensile strength (0.36 to 0.68MPa) in the 5 to 10% fibre content ranges was most likely a result of good interaction between the coir fibre and matrix and a strong interface created which led to a strong bonding as noted in previous studies [9,10].

By-Products of Palm Trees and Their Applications   Materials Research Forum LLC
Materials Research Proceedings 11 (2019) 195-200   doi: https://doi.org/10.21741/9781644900178-14

Table 1: Tensile Strength of the Bioplastic Specimens

Specimen	Mean Tensile Strength [N/m^2]	Elastic Modulus [x 10^5 N/m^2]
0% fibre content	0.36	26.7
5% fibre content	0.62	42.7
10% fibre content	0.68	49.1
15% fibre content	0.45	33.4
20% fibre content	0.37	33.2

Elastic Modulus

As shown in Table 1, modulus of elasticity of the bioplastics ranged from 26.7 x 10^5 to 49.1 x 10^5 N/m^2. These values are quite high, suggesting that the material a lot of stress with minimal strain. Again, as the coconut husk fibre content increased up to 10%, the modulus of elasticity also increased significantly perhaps due to the good interaction between the coir fibre and matrix. Beyond this point, there was a considerable decrease in modulus of elasticity attributable to relatively poor interface between the fibre and the matrix.

Impact Resistance

The impact test results are presented in Table 2. The values ranged from 1.73 to 3.7 J. The relatively low impact resistance suggests that the material may not be used in applications where impact resistance is critical such as electrical and electronic housings and enclosures.

Table 2: Impact Resistance of the Bioplastics

SAMPLES	MEAN IMPACT ENERGY [J]
0% fibre content	1.7308
5% fibre content	2.3948
10% fibre content	2.9381
15% fibre content	3.7024
20% fibre content	3.6420

However, the impact resistance increased significantly with increasing coconut husk fibre content up to 15, an indication of the fact that 15% is about the upper limit of the quantity of fibre that could be incorporated in the bioplastics to improve their desirable properties.

Water absorption

The water absorption (WA) after 24hrs for the different bioplastic specimens is shown in Figure 2. The values ranged between 27.2 and 42.9%. These values are much lower than the WA values reported for bioplastics produced with cassava starch and reinforced with recycled newspaper pulp [9], but higher than the WA values reported for polypropylene reinforced with alkaline-treated coconut husk fibres [10]. However, in line with a previous finding on polypropylene reinforced with alkaline-treated coconut husk fibres, but contrary to another finding cassava starch-based bioplastics reinforced with recycled newspaper, the control samples (without fibre reinforcement) had the lowest WA and the WA increased with increasing fibre content. The behaviour of the bioplastic specimens tested in this study is, perhaps, due to micro void creation in the matrix with the introduction of the coconut husk fibres and/or the hydrophilic nature of fibre [5,11].

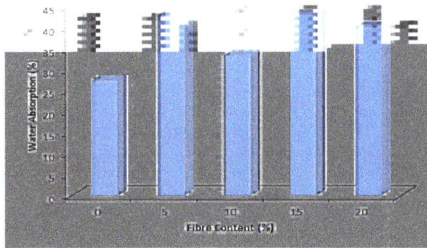

Figure 6: Mean water absorption of the bioplastic specimens

Biodegradability

A complete degradation of all the bioplastic samples was observed after one month burial in the soil, indicating that the material is biodegradable and can be used in a variety of consumer products, such as food containers, grocery bags, biodegradable utensils, and food packaging. Technically, all materials are biodegradable, but not all bio-based plastics are biodegradable. For practical purposes, only bioplastics that degrade within a relatively short period of time (weeks to months) are usually considered biodegradable; those that do not degrade within a few months or years are said to be durable [12].

Conclusion

It was concluded that coconut husk fibre could be used as reinforcement in a cassava starch-based bioplastic. An an inclusion of up to 10 % of coconut husk fibre could result in significant improvement in the tensile strength and modulus of elasticity. The fibre quntity may, however, be increased to 15% for impact strength improvement.

References

[1] United Nations Environmental Programme, Developing integrated solid waste management plan. Training manual volume 3: Targets and issues of concern for ISWM. UNEP, Nairobi, Kenya, 2009.

[2] W.R. Yu, B.H. Lee, H.J. Kim, Fabrication of long and discontinuous natural fibre reinforced polypropylene biocomposites and their mechanical properties, Fibres and Polymers 10-1 (2009) 83-90. https://doi.org/10.1007/s12221-009-0083-z

[3] Information on http://www.onlinesciences.com/health/danger-and-bad-effects-of-burning-plastics-and-rubber-on-humans-and-global-warming.

[4] Information on http://www.fao.org/nigeria.

[5] F.U. Felix, A. Adetifa, Characteristics of kenaf fiber-reinforced mortar composites. IJRRAS 12-1 (2012) 18-26.

[6] R.D. Filho, N.P. Barbosa, K. Ghavami, Application of sisal and coconut fibres in adobe blocks. In: HS Sobral (ed.), Vegetable plants and their fibres as building materials, Chapman and Hall, London (1990) 139-149.

Materials Research Proceedings **11** (2019) 195-200   doi: https://doi.org/10.21741/9781644900178-14

[7] A.O. Olorunnisola, Strength and water absorption characteristics of cement-bonded particleboard produced from coconut husk, Journal of Civil Engineering Research and Practice 3-1 (2006) 41 – 49. https://doi.org/10.4314/jcerp.v3i1.29150

[8] Information on: https://agriculturenigeria.com/farming.

[9] R.A.M. Sujuthi, K.C. Liew, Properties of bioplastic sheets made from different types of starch incorporated with recycled newspaper pulp, Transactions on Science and Technology, 3-1&2 (2016) 257 – 264.

[10] O.G. Agbabiaka, I.O. Oladele, O.O Daramola, Mechanical and Water Absorption Properties of Alkaline Treated Coconut (cocosnucifera) and Sponge (acanthus montanus) Fibers Reinforced Polypropylene Composites, American Journal of Materials Science & Technology, 4-2 (2015) 84-92. https://doi.org/10.7726/ajmst.2015.1007

[11] F. J.Aranda-Garcia, R. Gonzalez-Nunez, C.F Jasso-Gastinel, E. Mendizabal, Water absorption and thermomechanical characterization of extruded starch/poly (lactic acid)/agave bagasse fiber bioplastic composites. International Journal of Polymer Science, article ID 343294. https://doi.org/10.1155/2015/343294

[12] Information on: https://www.creativemechanisms.com/blog/everything-you-need-to-know-about-bioplastics.

By-Products of Palm Trees and Their Applications   Materials Research Forum LLC
Materials Research Proceedings 11 (2019) 201-210   doi: https://doi.org/10.21741/9781644900178-15

Enhancement of the Mechanical Behavior of Starch-Palm Fiber Composites

H. Megahed[1, 4, a *], M. Emara[2, b], Mahmoud Farag[3, c], Abdalla Wifi[4, d], and Mostafa El. Shazly[5, e]

[1]Department of Materials Engineering, Faculty of Engineering and Materials Science, German University in Cairo, New Cairo, 11835, Egypt

[2]Department of Manufacturing Engineering, School of Engineering, Canadian International College in Cairo, New Cairo 11835, Egypt

[3]Department of Mechanical Engineering, School of Sciences and Engineering, American University in Cairo, New Cairo, 11835, Egypt

[4]Department of Mechanical Design and Production, Faculty of Engineering, Cairo University, Giza, 12316, Egypt

[5]Department of Mechanical Engineering, Faculty of Engineering, The British University in Egypt, Cairo, Egypt, Al-Sherouk, 11837, Egypt

[a]hebatullah.shwa@gmail.com, [a]hebatullah.alieldin@guc.edu.eg, [b] mohamed_emara@cic-cairo.com, [c]mmfarag@aucegypt.edu, [d]aswifi@yahoo.com, [e]mostafa.shazly@bue.edu.eg

Keywords: hybrid composites, date palm fiber, chopped fibers, polymer-matrix composites

Abstract. This study discusses the fabrication of starch- based hybrid composite reinforced with chopped randomly oriented flax, sisal, and date palm fibers. The tensile properties, before and after chemical treatment, as well as the morphology of the fibers were evaluated. The hybrid composites were fabricated using hot compaction technique at 5MPa and 160°C for 30min. Fracture surface investigations using field emission scanning microscopy showed a good adhesion between fibers and matrix. The fracture surface revealed the presence of matrix micro cracks as well as fibers fracture and pullout. Hybrid composites containing 20 vf % sisal, and 5 vf % flax at 25 vf % date palm as well as 35vf% sisal, and 5 vf % flax at 10 vf % date palm had the optimum mechanical properties and consequently can serve as competitive eco-friendly candidates for various applications. A finite element (FE) approach was developed to simplify the treatment of random orientation of chopped fibers and predict elastic modulus using Embedded Element technique. Analyses based on rule of hybrid composite (ROHM), COX rule, and Leowenstein rule are presented to validate both experimental and FE numerical results. The FE results compared favorably with the experimental results.

Introduction

The construction of natural fiber bio-composite may have very good applications in the automotive and transportation industry such as car door panels which may save up to 45% from door panel carrier weight, bio-based cushions, the driver's seat back rest, etc. Moreover, reducing cost of bio-composites will be more desirable to industrial economic development [1].

Biodegradable composite materials based on natural fibers and starch had attracted attention over the past several years. Starch is one of polysaccharide matrices. Owing to its low cost, availability as a renewable resource, biodegradable and nontoxic degradation products, it is one of the important raw materials used for packaging, biomedical applications, and in some automotive parts. Starch, however, has some drawbacks such as poor melting process ability, high water solubility, difficulty of processing, and brittleness. Gelatinization process converts starch to thermoplastic starch (TPS) and improves those draw backs [2-3].

Date-Palm fiber (DPF) is a low cost material with mechanical properties that depend on the place of extraction. DPF can be considered one of the best types of fibers regarding several evaluation criteria such as specific strength to cost ratio if compared to other fiber types [4]. Sisal fiber (SF) is known by its high strength but it has some limitations such as high cost and is not cultivated in Egypt [5]. Flax fiber (FF) has mechanical properties near to SF; however, the cultivation of Flax has been diminished in Egypt as it can be replaced by other imported materials [1, 6].

Several fiber types are incorporated into hybrid composites and such composites can be tailored to meet various design requirements in a more economical way than conventional composites. Their behavior depends on the characteristics and the mechanical properties of the incorporated fibers [7]. Several factors will affect the composite mechanical properties such as fiber type, length, orientation, characterization, resin type, and volume fraction of the reinforcements [8].

The objective of the present work is to study the behavior of starch- based hybrid composites containing three types of fibers, namely, DPF, FF, and SF, and to compare the mechanical properties obtained to flax/date palm hybrid composite at 1:1 matrix/ fiber volumetric ratio. The present work involves both experimental and numerical investigations. Composite preparation stage was performed by using different mixtures of fibers with different volume fraction as shown in Table 1. The composite analysis first stage was based on measuring mechanical properties, examining the fracture surfaces, and applying the morphological characterization of the materials. Finally the finite element analysis (FEA) stage; where different models were implemented using ABAQUS software.

Mixed FE-analytical approaches are suggested for the prediction of the Young's Modulus of reinforced composite having randomly oriented chopped fibers. An attempt is suggested to overcome the difficulty of representing random orientation of chopped fiber across composite in finite element representation. The attempt is based on unidirectional fibers having 33.3% volume fraction to represent the actual fiber volume fraction of randomly oriented chopped fiber in the composite.

Experimental Work

Materials Preparation, Characterization, and Mechanical Testing: Corn Starch was purchased from Aro Sheri Company in Egypt with an average particle size of 16μm. Glycerin with 99.7% purity was used as a plasticizer. Gelatinization process of starch following Ref. [3] methodology is used to form TPS by mixing native starch with 30Wt. % glycerin and 20Wt. % distilled water in temperature range from 60–80°C. Adding glycerin improves process ability and reduces embrittlement by inhibiting the retro gradation process. The TPS was kept in polyethylene bags over night to enhance its flow properties before being used.

Flax and sisal strands were donated by the Egyptian Industrial Center E.I.C. DPFs were extracted from the stem of date palm trees at the American University in Cairo. Sodium hydroxide (NaOH) with molecular weight 40g/mol. was used for alkaline treatment of fibers. The three fibers (DPF, FF, and SF) were chemically treated using the following procedure: 1) Dipping in 5% NaOH for 3 hours at room temperature. 2) Rinsing the treated fibers in cold water. 3) Dipping the fibers in 5% acetic acid to remove any excess NaOH from fibers surface. 4) Rinsing in cold water and oven drying at 120°c for 3hrs. 5) The treated fibers were cut manually into short fibers with average length varies from 15 to 30mm according to the aspect ratio.

Characterization and testing were performed using the following procedures: 1) Measuring fibers diameter before and after chemical treatment by Leica stereoscopic microscope using 10 samples with a μm divisions scale lens. 2) Measuring density of TPS and fibers using the Mittler Toledo densitometer for 10 samples (Xylene was used as the immersing liquid with relative density is 0.86). 3) Tensile testing using Instron 3382 universal testing machine at 50% RH, 18°C and strain rate of 0.01per min. with a gauge length of 50mm at strain rate of 0.01/min. 4) Fracture surface study of fibers Using ZEISS scanning electron microscope (SEM) operated at a vacuum pressure 1e-4 mbar and 8KV.

Hybrid Random Composite Preparation: The different composites were prepared using 1:1 fiber to matrix volume fraction according to Eqs. 1- 4. The DPF was used with 50vf% to 20vf% of fiber at different SF and FF volume fraction percentages. The fiber cutting length was based on fiber aspect ratios. Stearic acid with concentration 98% was used as a mold coating releasing agent. The fiber mixture was uniformly distributed in a die cavity (120X80X10mm) to form ten different fiber volume fractions of hybrid composites as shown in Table 1.The emulsified TPS was poured on the random mixed fibers. The mixture was then pre-heated at 140±3°c for 30min to remove excess water from the mixture. This was followed by hot pressing at 5MPa and 160°c for another 30min then cooling at a rate of about 2°C/min.

$$V_T = V_f + V_m = \frac{Wt_f}{\rho_f} + \frac{Wt_m}{\rho_m} = (\frac{\pi}{4} * d_f{}^2 * l_f * n_f) + (W_m * l_m * h_m) \qquad (1)$$

$$\sum_i^n v_i = 1 \qquad (2)$$

$$v_f = V_f/V_T \qquad (3)$$

$$v_m = V_m/V_T \qquad (4)$$

Where; v_i is the volume fractional for constituent i, v_f and v_m are fiber and matrix volume fracture, V_f and V_m are fiber and matrix volume, V_T is Total Volume, Wt_f and Wt_m are fiber and matrix weights, ρ_f and ρ_m are fiber and matrix density, d_f is the fiber diameter, l_f is the fiber length, n_f is the number of the different fibers used in a composite, w_m, h_m, l_m are matrix width, thickness, and length respectively.

Hybrid Composite Characterization and Mechanical Testing: Density measurement of composite using the densitometer for five samples per each hybrid composite is very important as indicator for measuring void fraction percentage. Voids can be as a result of an imperfection from the processing of the material and is generally deemed undesirable. Its presence can affect the mechanical properties and lifespan of the composite. Voids can also act as a crack nucleation site as well as allow moisture to penetrate the composite [9]. Void percentage was calculated using Eq. 5, and Eq. 6.

$$\rho_{theoretical} = \rho_c = \rho_f V_{f+} \rho_m (1 - V_f) \qquad (5)$$

$$void\ fraction = \frac{\rho_{theoretical} - \rho_{exp}}{\rho_{ttheoretical}} \qquad (6)$$

By-Products of Palm Trees and Their Applications   Materials Research Forum LLC
Materials Research Proceedings 11 (2019) 201-210   doi: https://doi.org/10.21741/9781644900178-15

Tension tests were carried out at a strain rate of 2mm/min for 5 samples for the different volume fraction composites. Test specimens were cut manually in the form of rectangular bars 80x8x2mm according to ASTM D5083-10 with working gauge length equals 50mm.

In case of analytical calculation for UTS; Kelly and Tyson [10] proposed an approach to deal with discontinuous unidirectional fiber and to overcome the problem of unequal strain in fibers and matrix using Eq. 7 and Eq. 8. It is accomplished by assuming perfect bonding between fibers and matrix, besides both fibers and matrix behave as linear elastic material so the stress on the fibers start from zero at ends to reach the maximum stress within fiber length. If the fibers are shorter than critical length, they cannot be loaded to their failure stress.

$$L_c = \frac{\sigma_f d_f}{2\tau} \tag{7}$$

$$\sigma_c = \sum \left(\frac{3}{8}\left(1 - \frac{L_c}{2L}\right)\sigma_f V_f\right) + \sigma_m V_m \tag{8}$$

Where; L_c is the critical length, τ is the shear strength of fiber/matrix bond which is set to be about 0.5 of matrix strength, and L is the fiber length.

In case of Elastic modulus prediction; ROHM [11] was used to present the elastic modulus for random fibers oriented composite (E_c) by substituting Eq. 10 and Eq. 11 in Eq. 9. Cox [12] implemented an analytical equation, where its concept is based on averaging the elastic constants over all possible orientations by integration as shown in Eq. 12. Leowenstein [13] governing rule is as shown in Eq. 13. Also Leowenstein rule [13] is based on ROHM [11] criteria, whereas; the most effective volume fraction is fiber. These predictions are only good at very low fiber volume fractions. At high fiber volume fractions, the predicted modulus is much higher than measured.

$$E_c = \frac{3}{8}E_1 + \frac{5}{8}E_2 \tag{9}$$

$$E_1 = E_f V_f + E_m V_m = E_f v_{ff} + E_s v_{fs} + E_D v_{fd} + E_m v_m \tag{10}$$

$$E_2 = E_m \left(\frac{1 + \xi(\sum \eta_f V_f)}{1 - (\sum \eta_f V_f)}\right) \tag{11}$$

$$E_c = \frac{1}{3}E_1 \tag{12}$$

$$E_c = \frac{3}{8}E_1 \tag{13}$$

Where; η_f is the fiber stress portioning parameter in transverse direction, and ξ is the curve fitting parameter.

Finite Element Proposed Mixed Technique

In the present work finite element analyses (FEA) using ABAQUS 6.12 commercial software were conducted to predict the Elastic Modulus of fiber reinforced composites using embedded element technique. In order to make an adequate prediction of Elastic Modulus for natural fiber reinforced composites using FEM, it is essential to carry out an appropriate analysis which requires both the correct mechanical characterization of the materials and composite structure.

Analytical or numerical micro mechanical analysis of fiber reinforced composites had involved the study of the Representative Element Volume (RVE).

In an attempt to avoid the difficulties associated with modeling randomly oriented chopped fibers, two simplified FE approaches were used. In the first approach, a mixed FE / ROHM model was developed for the prediction of Elastic Modulus based on Eq. 9 to represent final hybrid composite Elastic Modulus. Fibers were placed in the matrix using array as longitudinal beam element in unidirectional and with approximate equal spaces. The load was applied in longitudinal and transverse directions. The output data E1 and E2 are substituted in Eq. 9. Based on this analysis, mixed FE/Cox and mixed FE/Leowenstein was made by substitution of E1 results in Eq. 12 and Eq. 13. The results were compared to those obtained from the mixed FE/ROHM model.

In the second approach; two modified mixed FEMs were examined. The first modified mixed model followed Leowenstein approach in which fibers RVE in the model constitute 3/8 from the fibers actual volume fractional. The second modified mixed FE model followed Cox approach in which fibers RVE in model constitute 1/3 of the fibers actual volume fractional. The results of two modified mixed FEMs were used directly to predict the Elastic Modulus of randomly oriented chopped fiber hybrid composite.

In all analyses, the composite structure was assumed to be a rectangular prism consisting of fibers surrounded by matrices in which both were created as 3-D elements. The matrix structure was 3D deformable solid model and was shaped to the required thickness based on the RVE. The fibers were modeled as 3D deformable wires lay in a plane, as the wire represented the beam elements immersed inside matrix. The fiber length must be equal to the length of the hosting matrix to avoid the end point factor. The area of matrix is controlled by matrix volume fraction in which thickness and width are assumed to be equal. In the present model; the structure of meshing used in case of fiber was beam element with 2 nodes but the matrix was eight-node hexahedral element shape. .

Results and Discussion

Physical and Mechanical Properties of Fibers and Matrix: The experimental measurements for FF, DPF, and SF diameter ranges were 9-20μm, 19.6-30 μm, and 27-187.5 μm after chemical treatment stage, and the average densities were 1.44, 1.352, and 1.43 gm/cm^3, respectively. The ultimate tensile strength (UTS) for FF, DPF, and SF were 351 MPa, 195 MPa, and 565.7 Mpa. And Young's moduli were 21.5 GPA, 8.81 GPA, and 41.2 GPA, respectively.

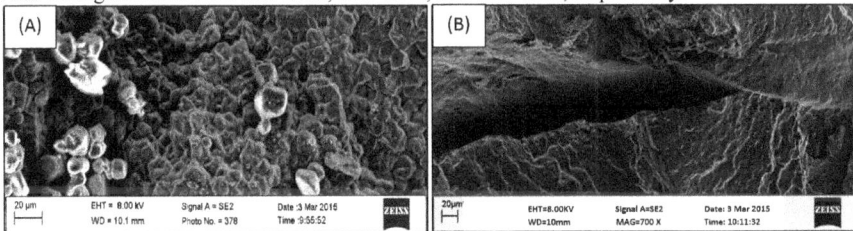

Fig. 1: (A) SEM investigation for plasticized corn starch; agglomerated and connected corn starch particles, (B) SEM fracture surface investigation of hot-pressed matrix (TPS emulsion technique); smooth.

The matrix granules size investigated by SEM varied from 4.9 to 33 μm with average size being about 16 μm. The presence of dimpled structure in TPS blend increased the ductility of material as shown in Fig.1A. This effect was more noticeable in composite fracture SEM in Fig.1B. The hot pressed TPS matrix had an average density of 1.445 gm/cm^3, UTS of 3.6 MPa, and Young's modulus of 378 MPa.

Physical Properties of Composites: The densities of the implemented hybrid composites with different volume fractions of DPF are summarized in Table 1. The void percentage ranged from 1.2% to 4.6%. The variation in diameter along the length of the fibers had an effect on the void contents as well as acting as stress concentration sites that may weaken the final composite mechanical properties and this difference in diameters was more obvious in Fig. 2.

Table 1: Different implemented hybrid composites at 10% and 25% volume fractions DPF and void percentage results.

| Composites | Fibers volume fractional | | | Theoretical Density results (gm /cm³) | Experimental Density results (gm /cm³) | Void (%) |
	DPF (%)	FF (%)	SF (%)			
SFD0520R	25	5	20	1.422	1.357	4.6
SFD1015R		10	15	1.423	1.405	1.3
SFD1313R		12.5	12.5	1.423	1.404	1.3
SFD1510R		15	10	1.423	1.405	1.2
SFD2005R		20	5	1.424	1.404	1.4
SFD0535R	10	5	35	1.435	1.413	1.5
SFD1525R		15	25	1.436	1.373	4.3
SFD2020R		20	20	1.436	1.374	4.3
SFD2515R		25	15	1.437	1.377	4.1
SFD3505R		35	5	1.437	1.383	3.7

Mechanical Behavior of Composites: The surface fracture morphology of the implemented hybrid composites is shown in Fig. 2. Fig. 2A shows the presence of good adhesion in matrix/fibers and also good wettability. The brittle matrix surface cracking behavior and voids may be as result of the relatively fast cooling rate of composite after compression molding.

Fig. 2B shows fibers failure and unexpected bad adhesion at the matrix/fibers interface, due to the presence of untreated fibers lignin. The fracture of a SF in tension as a result of the inhomogeneity in fibers crosses section would lead to unequal stress distributions. Shear took place across the fiber due to the lateral strains produced accompanied with the axial stresses.

Good fibers wettability is also obvious in Fig.2C, unfortunately; the presence of brittle matrix surface cracking is also detected. This was all due to difficulty of distribution of mixture across composite.

By-Products of Palm Trees and Their Applications   Materials Research Forum LLC
Materials Research Proceedings 11 (2019) 201-210   doi: https://doi.org/10.21741/9781644900178-15

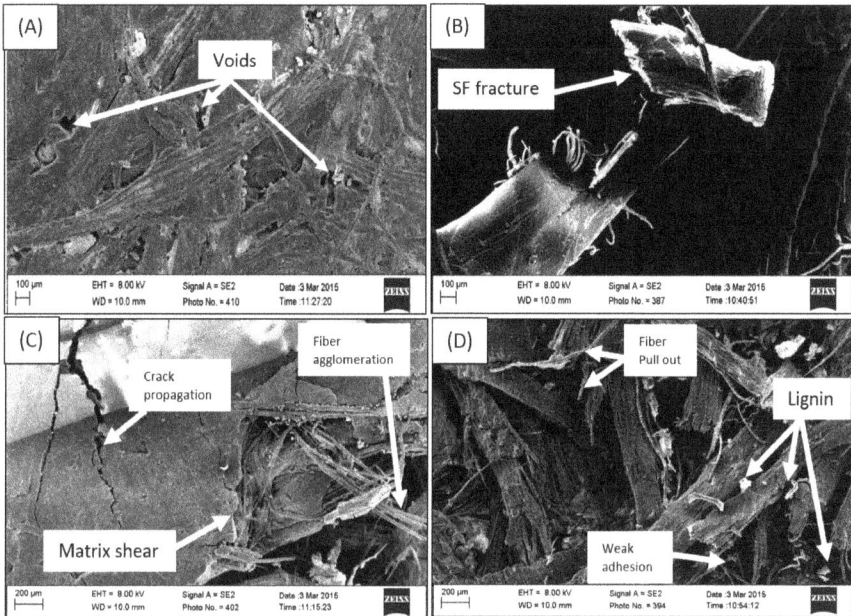

Fig. 2: A) SFD3505R hybrid composite - Matrix surface voids, B) SFD0520R hybrid composite - Fiber cracking, C) SFD2515R hybrid composite – Matrix crack propagation as a result of stress concentration as well as fiber agglomeration, D) SFD2020R hybrid composite-Multiple failure sources.

Fiber aspect ratio, orientation, and agglomeration are other sources of premature failures. With increasing SF vf% at low DPFs vf%, the composite suffered from a combined matrix and fiber shear. The agglomeration of date palm fibers acted as internal stress raisers where cracks were initiated and consequently propagated across the composite as shown in Fig. 2C. In Fig. 2D, fibers pull out and fracture is also detected. The fiber pull out were more obvious with the increase of the FF vf%.

Table 2 shows the UTS results achieved with 25vf% and 10 vf% DPF content in comparison with Kelly and Tyson [10] analytical results. It was found that increasing FF coupled with decreasing SF volume fractional led to the decrease of composite strength. The increase of SF will increase the strength and consequently the cost since SF is more expensive than FF. Knowing that the SF strength is the highest among the three fibers, the experimental results were unexpected. This can be as result of miss-distribution of SF in the composite and/or fabrication problems leading to a large void percentage in the composite. Hybrid composite of 5vf% FF and 35vf% SF shows the highest strength for 25vf% and 10vf% Date-palm. Based on strength and Young's modulus results, this composite is considered to be the best amongst other compositions in the study. Ibrahim et al. [3] observed an average strength of 26.77 MPa using FF and DPF chopped randomly hybrid composite with equal weight fractions. The strength trends show good

By-Products of Palm Trees and Their Applications   Materials Research Forum LLC
Materials Research Proceedings **11** (2019) 201-210   doi: https://doi.org/10.21741/9781644900178-15

agreement to that of Kelly and Tyson results. At equal volume fractions of SF and FF with 10vf% Date-palm; the composite behaves differently than the rest of composites.

Table 2: Experimental UTS results versus Kelly and Tyson analytical results.

Composites	UTS (Mpa)	
	Experimental results	Kelly and Tyson rule [10]
SFD0520R	16.37	62
SFD1015R	8.83	59
SFD1313R	6.68	57
SFD1510R	6.51	56
SFD2005R	6.1	53
SFD0535R	18.56	79
SFD1525R	14.86	73
SFD2020R	18.13	70
SFD2515R	10.12	67

Table 3 shows a comparison of Young's modulus results using ROHM, Leowenstein and COX. The empirical analytical results show higher values than experimental ones. This can be due to the effect of void percentage, compression molding technique, and the difference in fibers diameters on the composite elastic modulus. The Cox results are the closest results to experimental ones.

The experimental results showed favorable agreement with those of Cox and Leowenstein approaches and show lower agreement than the ROHM approach based on both longitudinal and transverse solutions. It must be noticed also that the voids effect is not taken into consideration in analytical analyses.

The results of the suggested FE/ROHM mixed model show good agreement with the analytical results, but are higher as compared to experimental results. This is expected in view of the presence of large void percent and the lower strength of the fabricated composites compared to that of Ibrahim et al [3] as discussed in the previous section.

Both mixed FE/ Leowenstein and FE/Cox models show a higher results compared to experimental results, with the mixed FE/Cox approach showing closer results to the experimental ones. Modified Cox FEM shows closer agreement to experimental results than Leowenstein based model. In fact, such results are consistent with the fact that less fraction of fibers are used in the Cox approach leading to lower predictions of the Elastic Modulus. Generally speaking, the FEM results compare well with most of the analytical solutions.

In view of the results of the various attempted models, the modified mixed FEM model based on Cox approach was adopted to predict Elastic Modulus of various hybrid composites.

Table 3: Comparison between experiment, analytical and FE Young's Modulus results.

Composites	Experimental results	Analytical Results			1st approach			2nd approach		Final FE adopted approach
		ROHM rule [11]	Leowenstein rule [13]	Cox rule [12]	FEM mixed ROHM	FEM mixed Leowenstein rule	FEM mixed COX	FEM based on Leowenstein rule approach	FEM based on COX approach	Adopted Modified FEM/COX
SFD0520R	2.1-2.69	5.23	4.39	3.90	4.9	4.5	4.05	4.02	2.59	2.59
SFD1015R	1.3-1.91	4.84	4	3.6	4.35	3.99	3.55	2.29	1.6	1.6
SFD1313R	1.16-1.22	4.68	3.84	3.4				3.24	2.29	2.29
SFD1510R	1.74-2.33	4.48	3.64	3.2						3.62
SFD2005R	1.06-1.32	4.10	3.27	2.91						3.75
SFD0535R	4.2-4.7	7.12	6.24	5.54				4.69	4.62	4.62
SFD1525R	2.9-3.26	6.4	5.5	4.9	4.91	4.54	4.038	4.28	4.13	4.13
SFD2020R	2.27-2.68	6.00	5.12	4.6						4.46
SFD2515R	2.03-2.13	5.62	4.75	4.22						4.83
SFD3505R	1.2-2.6	4.87	4.00	3.56						4.33

Concluding Remarks

It should be emphasized that the experimental results generally show lower Elastic Modulus values as compared to both the analytical and the FEM results. This could be attributed to the manual compression molding process used in the present work leading to high void percentage, and the lower efficiency in preparing the composite with no guarantee that the actual volume fraction of fibers is necessarily equal to that in the RVE.

Furthermore, as detailed in Ref. [14] and indicated previously in microscopic and SEM investigations, fiber cross sectional area changes along the fiber length, and fiber lengths are not necessarily cut to equal lengths which could affect the strength of the implemented composite mixture. This is not the case in analytical and FEM solutions where the fibers are assumed to be of uniform cylindrical cross section of equal lengths to overcome the end point effect.

However, it should be noticed that a perfect mixing and bond between matrix and fibers are assumed in analytical and the mixed FEMs leading to results of strength and Elastic Modulus that are generally higher and could be out of range of realistic experimental work. The suggested

Mixed FE results indicate favorable agreement with those of Cox and Leowenstein approaches and show lower predictions than the ROHM approach based on both longitudinal and transverse solutions.

Ref [14] discussed in details the relative cost of the implemented hybrid composites under considerations, and that DPFs are observed to be the cheapest among the three fibers in both markets. The cost of the hybrid composites based on the local market increases with an increase in SF volume fraction.

Reference

[1] Akampumuza, Obed, et al., Review of the applications of biocomposites in the automotive industry, Polymer Composites 38-11 (2017) 2553-2569.

[2] Guleria, Ashish, Amar Singh Singha, and Raj K. Rana, Mechanical, Thermal, Morphological, and Biodegradable Studies of Okra Cellulosic Fiber Reinforced Starch-Based Biocomposites, Advances in Polymer Technology 37-1 (2018) 104-112. https://doi.org/10.1002/adv.21646

[3] H. Ibrahim, M. Farag, H. Megahed, S. Mehanny, Characteristics of starch-based biodegradable composites reinforced with date palm and flax fibers, Carbohydrate polymers 101 (2014) 11–19. https://doi.org/10.1016/j.carbpol.2013.08.051

[4] M. Asadzadeh et al., Bending Properties of Date Palm Fiber and Jute Fiber Reinforced Polymeric Composite, International Journal of Advanced Design and Manufacturing Technology 5-4 (2013) 59–63.

[5] Xie, Qi, et al., A new biodegradable sisal fiber–starch packing composite with nest structure, Carbohydrate polymers 189 (2018) 56-64. https://doi.org/10.1016/j.carbpol.2018.01.063

[6] L. Yan, N. Chouw, and K. Jayaraman, Flax fibre and its composites–A review, Composites Part B: Engineering 56 (2014) 296–317. https://doi.org/10.1016/j.compositesb.2013.08.014

[7] Ashori, Alireza., Hybrid thermoplastic composites using nonwood plant fibers, Hybrid Polymer Composite Materials 3 (2017) 39-56. https://doi.org/10.1016/b978-0-08-100787-7.00002-0

[8] Chauhan, P. Chauhan, Natural Fibers Reinforced Advanced Materials" Journal of Chemical Engineering & Process Technology 6 (2013) 417-421. https://doi.org/10.4172/2157-7048.s6-003

[9] D647, standard test methods for void content of reinforced plastics, ASTM D647, 2013.

[10] Wongpajan, Rutchaneekorn, et al., Interfacial shear strength of glass fiber reinforced polymer composites by the modified rule of mixture and Kelly-Tyson model, Energy Procedia 89 (2016) 328-334. https://doi.org/10.1016/j.egypro.2016.05.043

[11] R. F. Gibson, Principles of composite material mechanics, CRC press, 2011.

[12] H. Cox, The elasticity and strength of paper and other fibrous materials, British journal of applied physics, 3-3 (1952) 72.

[13] D. Agarwal, L. J. Broutman, and K. Chandrashekhara, Analysis and performance of fiber composites. John Wiley & Sons, 2006.

[14] H. Megahed, Experimental and numerical study of starch matrix hybrid biodegradable composites reinforced with chopped randomly oriented fibers, MSc. Thesis, Faculty of Engineering, Cairo University, 2016.

Materials Research Forum LLC
doi: https://doi.org/10.21741/9781644900178

Biomedicine and Biotechnology

By-Products of Palm Trees and Their Applications
Materials Research Proceedings 11 (2019) 213-218

Materials Research Forum LLC
doi: https://doi.org/10.21741/9781644900178-16

Effect of some Micro-Elements on Steroids Production from Embryogenic Callus of *in vitro* Date Palm Sakkoty and Bartamuda Cultivars

Sherif F. El sharabasy[1,a*], Abdel-Aal W. B.[1], Hussein A. Bosila[2],
Bayome M. Mansour[2] and Abdel-Monem A. Bana[1]

[1]The Central Lab of Date Palm Researches and Development, ARC, Egypt.

[2]Floriculture, Medicinal & Aromatic Plants. Dept., Fac. of Agric., Al-Azhar University, Egypt.

[a]sharabasydates@yahoo.com

Keywords: micro-elements, steroids, embryogenic callus, manganese sulfate, zinc sulfate, cupric sulfate, date palm cultivars

Abstract. The ability of plant cell, tissue, and organ cultures to produce and accumulate many of the same valuable chemical compounds as the parent plant in nature has been known almost since the inception of *in vitro* technology. Date palm has been recognized as an important crop containing high valuable secondary metabolism. Some microelements such as, manganese sulfate ($MnSO_4 2H_2O$), zinc sulfate ($ZnSO_4 7H_2O$) and copper sulfate ($CuSO_4 5H_2O$) were used as precursor to produce steroids from embryonic callus two date palm dry cvs. In this study, embryogenic callus explants were cultured on MS nutrient medium supplemented with different concentrations of ($MnSO_4 2H_2O$), (22.3, 44.6 and 66.9 mg/l), $ZnSO_4 7H_2O$ (8.6, 17.2 and 25.8 mg/l) and $CuSO_4 5H_2O$ (0.025, 0.050, 0.075 mg/l). The highest significant value of total steroids (0.94 mg/g dry weight) was recorded when embryogenic callus of Sakkoty cv. was cultured on medium contained (22.3mg/l) $MnSO_4 2H_2O$. Where embryogenic callus of Bartamuda cv. cultured on nutrient medium supplemented at (17.2 mg/l) $ZnSO_4 7H_2O$ gave the highest significant value of total steroid (0.92 mg/g dry weight).

Introduction

Date palm (*Phoenix dactylifera* L.) is a member of (Arecaceae) family it is a heterozygous and dioecious tree it was known in ancient Egypt since 4000 years ago and this fact can be simply indicated from date palm inscriptions appearing on the walls of ancient Egyptian temples. In general, the importance of this tree all over in its cultivation region in North Africa and the Middle East was referred to the numerous advantages from its fruits and from the tree as a whole [1]. Biotechnology approach has a great deal for the production of chemicals and pharmaceuticals from *in vitro* plant cell culture [2].

Steroids are a set of cholesterol derivative lipophilic that are low molecular weight and may found in synthetic sources. They include sterols, hormones gonadal and adrenal ones, hydrocarbons and bile acids. Steroids family plays an important role in the biochemistry and composition of organisms [3]. Steroids are used as anti-cancer agents, antibiotics, and anti-inflammatory, and anti-hormones drugs [4]. First study for the steroids production in date palm tissues was documented by El Sharabasy *et al.*, [5] who found that total steroids in tow Egyptian cultivars tissues demonstrated higher values in pollen grain and shoot tip of in vivo tissues, and also in leaf and roots of the *in vitro* tissues. Also, the separation and identification of cholesterol and ß-sitosterol from callus cultures by Thin Layer Chromatography (TLC), was detected. El-Sharabasy [6] indicated also, that the precursors have great effect in the biosynthesis of steroids in date palm callus and embryogenic callus cells. Enhancement in secondary metabolite production can be obtained by selection of high-producing and medium optimizations [7,8]. Microelements are required in trace amounts for plant growth and development and have many diverse roles [9]. Manganese, iodine, copper, cobalt, boron, molybdenum, iron, and zinc usually comprise the microelements, although other elements, such as nickel and aluminum, are frequently found in some formulations. Iron is usually added as iron sulphate, although iron citrate can also be used. The aim of this work is to study the effect of Microelements on steroids production content (mg/ g dry weight) in embryonic callus stage of *in vitro* date palm Bartamuda, Sakkoty cultivars in order to optimize strategy for enhancing steroids production from *in vitro* date palm tissues by targeting manipulation of culture media composition.

Materials And Methods

Preparation of plant material
Callus explants of two cultivars Bartamuda and Sakkoty were produced from indirect protocol of date palm micropropagation described by [10,11].

In this study received embryonic callus explants for both cultivars were cultured on basic nutrient medium for callus formation which composed of MS basal medium [12], supplemented 30 g/l sucrose and 3.0 g/l activated charcoal with 40 mg/l adenine – sulfate, 200 mg/l glutamine, 100 mg/l myo-inositol, 0.1 mg/l biotin, 170 mg/l Na $_2HPO_4$,0.1 mg/l thiamine HCl 0.5 mg/l pyridoxine,0.5 mg/l nicotinic acid, 3.0 mg/l 2- isopentenyl adenine (2iP) + 10.0 mg/l 2,4 –D dichlorophenoxy acetic acid (2,4 – D). Pyruvic acid was added at 0.01 mg/l to induce steroids compounds production [13]

Micro elements compounds, manganese sulfate ($MnSO_42H_2O$), zinc sulfate heptahydrate ($ZnSO_47H_2O$) and copper sulfate ($CuSO_4.5H_2O$) were added to previous basic nutrient medium for both Bartamuda and Sakkoty cv. callus cultures, in three different separated treatments for each as follows:-

1-Manganese sulfate ($MnSO_42H_2O$) were added at (22.3, 44.6 and 66.9 mg/l)

2- Zinc sulfate Heptahydrate ($ZnSO_47H_2O$) were added at (8.6, 17.2 and 25.8 mg/l)

3- Copper sulfate ($CuSO_4.5H_2O$) were added at (0.025, 0.050 and 0.075 mg/l)

6.0 g/L agar were used to solidified Culture medium which were distributed in culture jars (250 ml); each jar contained 25 ml of culture nutrient medium. Culture jars were immediately capped with polypropylene closure autoclaved at 121°C at 1.05 kg/cm^2 for 20 min. The cultured jars were incubated under total darkness at 27±1°C and data were recorded every (6 weeks) for three subcultures on total steroids content (mg/g dry weight).

Callus samples were collected from all studied treatments of the Micro elements compounds, Manganese sulfate ($MnSO_42H_2O$), Zinc sulfate Heptahydrate ($ZnSO_47H_2O$) and copper sulfate ($CuSO_4.5H_2O$) for both Bartamuda and Sakkoty cv. for total steroids assay

Determination of total steroids (mg/g dry weight)
Total steroids were calculated as β-sitosterol and determined by spectrophotometer according to the methods described by [11] as follows:

Test solution preparation: - 0.5 g weight of embryogenic callus sample is dried in an oven at 75 °C for 48 h. dried embryogenic callus sample is placed in a clean flask, with addition of 100 ml of 5% potassium hydroxide solution in alcohol (90% v/v) and are heated on a water bath at 50°C to smooth reflux for 2 hours, then are cooled for 5 min, then the flask contents are transferred to a separator funnel. The residual contents of flask were washed for two times, firstly with 100 ml water followed by 100 ml diethyl ether then the washings were transferred

By-Products of Palm Trees and Their Applications   Materials Research Forum LLC
Materials Research Proceedings **11** (2019) 213-218   doi: https://doi.org/10.21741/9781644900178-16

into the same separator funnel and they are shacked altogether slowly by hand for 3 min. To separate the formed layer the aqueous phase was removed from separator funnel. This layer was washed in a separator funnel four times with 100 ml diethyl ether then, is placed in a clean flask. The received ethereal extracts are washed with three successive portions of 40 ml water (shaking was gently to avoid emulsions), 40 ml 5% w/v hydrochloric acid, and 40 ml 3% w/v potassium hydroxide aqueous solution. Successive portions of 40 ml water (each wash) are edited until the washings become neutral to phenolphthalein solution (2 drops 1% phenolphthalein in 70% ethanol and 2 N NaOH until rose color is stable). One drop of 0.1 N HCl is added to sample and rapidly mix until the rose color disappears. Hundred mg anhydrous sodium sulfate powder is added to the sample with well shacking, then the mixture is filtered through folded Whatman filter paper. The resulted solution is evaporated in water bath at 50°C until fully dry. 100 ml glacial acetic acid is added to the residue with stirring for 30 min in small glass bowl.

Test solution: - 2ml of previous resulted solution is transferred to a 20 ml volumetric flask and dilute to 20 ml with glacial acetic acid.

The reference solution preparation:- 40 mg β-sitosterol is dissolved in 100 ml glacial acetic acid then 5 ml of this solution is taken then diluted to 50 mL with glacial acetic acid.

The deniges reagent preparation: - This reagent is consisting of mixing of two solutions (solution A) is prepared by adding 100 ml sulfuric acid to 50 ml glacial acetic acid. (solution B) is prepared by dissolving 5g mercury oxide (HgO_2) and 20 ml sulfuric acid into 100 ml water. 100 ml of solution (A) is added to 1 ml of solution (B), then are mixed and filtered through a sintered glass filter (grade G4) before use.

Finally, 5 ml of Deniges reagent mixture solutions is added to test tube filled with 1 ml (Test solution) and 1 ml (Reference solution) for evaluation of β-sitosterol amount.

The blank is carried out by 1 ml glacial acetic acid instead of the sample in a test tube. Both tubes are lifted on the stand under the dark for 15 min. The absorbance is read using a spectrophotometer at 510 nm against the blank reading. The amount of steroids is calculated as β-sitosterol from a standard curve prepared by dissolving 40 mg of β-sitosterol in 10 ml glacial acetic acid. Series of standards are prepared as 5, 10, 20, and 40 mg/100 ml, respectively; 1 ml of each is mixed with 5 ml deniges reagent and read at 510 nm against the blank. The absorbance of each concentration is plotted against the absorbance obtained from the standard curve.

Statistical analysis
The obtained data were subjected to analysis of variance. The mean values were compared using LSD test at the 5% level of probability. The data were tabulated and statistically factorial analyses according to the randomized complete block design with three replicates [14].

Results and Discussion
1. Effect of manganese sulfate on total steroids content (mg/g dry weight).
Data in Table 1 clearly showed that no significant differences were found between the two cultivars under investigation (0.53, 0.61 mg/g dry weight) was for Bartamuda and Sakkoty respectively). The manganese sulfate concentration 22.3 mg/l was the most effective forming the highest significant value (0.83 mg/g dry weight), concerning the interaction between cultivars and manganese sulfate concentrations, the results illustrated that the highest significant value(0.94mg/g dry weight)of total steroids was for Sakkoty cultivar embryogenic callus grown on medium contained 22.3mg/l manganese sulfate. The lowest value (0.25mg/g dry weight) was for Bartamuda cultivar embryogenic callus grown on medium contained 66.9 mg/l.

Table 1: Effect of manganese sulfate on total steroids content (mg/g dry weight).

Cultivar (A)	Manganese sulfate mg/ l (B)			
	22.3	44.6	66.9	Mean (A)
Bartamuda	0.72	0.61	0.25	0.53
Sakkoty	0.94	0.59	0.31	0.61
Mean (B)	0.83	0.60	0.28	
L.S.D 0.05: A=0.057, B=0.070, AB=0.099				

2. Effect of zinc sulfate on the total steroids content (mg/g dry weight).

Table 2: Effect of zinc sulfate on total steroids content (mg/g dry weight).

Cultivar (A)	zinc sulfate mg/ l (B)			
	8.6	17.2	25.8	Mean (A)
Bartamuda	0.70	0.92	0.52	0.71
Sakkoty	0.59	0.83	0.44	0.62
Mean (B)	0.64	0.88	0.48	
L.S.D 0.05: A=0.033, B=0.040, AB=0.057				

Data in Table 2 showed that significant differences were found between the two cultivars under investigation (0.71, 0.62 mg/g dry weight respectively), the zinc sulfate concentration(17.2mg/l) was the most effective, forming the highest significant value (0.88 mg/g dry weight), concerning the interaction between cultivars and zinc sulfate concentrations, the highest significant value(0.92 mg/g dry weight)was for Bartamuda cultivar embryogenic callus grown on medium contained 17.2 mg/l zinc sulfate. The lowest value (0.44 mg/g dry weight) was for Sakkoty cultivar embryogenic callus grown on medium contained 25.8 mg/l zinc sulfate.

3. Effect of copper sulfate on total steroids content (mg/g dry weight).

Table 3: Effect of copper sulfate on total steroids content (mg/g dry weight).

Cultivar (A)	cupric sulfate mg/l (B)			
	0.025	0.050	0.075	Mean (A)
Bartamuda	0.92	0.61	0.32	0.62
Sakkoty	0.85	0.63	0.43	0.63
Mean (B)	0.88	0.62	0.38	
L.S.D 0.05: A=N.S, B=0.040, AB=0.057				

Materials Research Forum LLC
doi: https://doi.org/10.21741/9781644900178-16

Data in Table 3 showed that no significant differences were found between the two cultivars under investigation (0.62, 0.63 mg/g dry weight respectively), cupric sulfate concentration 0.025mg/l was the most effective as it produced the highest significant value (0.88 mg/g dry weight), There are some reports showing effects of some trace minerals on the production of secondary metabolites, i.e. Cu^{2+} had a positive effect on ginseng saopnin and polysaccharide production in cell cultures of *Panax notoginseng*. [15]. concerning the interaction between cultivars and cupric sulfate concentrations, the highest significant value (0.92 mg/g dry weight) was for Bartamuda cultivar embryogenic callus grown on medium contained 0.025 mg/l copper sulfate. The lowest value (0.32 mg/g dry weight) was for Bartamuda cultivar when the embryogenic callus grown on medium contained 0.075 mg/l cupric sulfate. There are many studies made on the method that could be used for the enhancement of the production of valuable secondary metabolites. Microelements are required in trace amounts (Manganese, iodine, copper, cobalt, boron, molybdenum, iron, and zinc) usually comprise the microelements for plant growth and development and have many diverse roles The effects of the medium employed in various processes have been reported. It has been reported that proper concentration microelements have been considered as nutrient factors or as abiotic elicitors, which trigger the formation of secondary metabolites [8,9,16]. Andrijany *et al.* [17] found in callus cultures of *Agave amaniensis* the relatively high concentration of magnesium, cobalt and copper ions simultaneously inhibited the sapogenin steroid formation while the absence of calcium ions in media induced the increasing in the sapogenin steroid content. It seems to be that the productions of specific useful secondary metabolites by plant cell cultures have intensive researches for economic implications, but until now only a few studies addressing the possibility of date palm steroids production.

Summary
Concentration of microelements have significant role in enhancing steroids accumulation in date palm callus. It could be suggested that medium components at certain concentrations have been considered as nutrient factors or precursors, which trigger the in vitro formation of steroids in date palm metabolites. More intensive studies are needed in this approach.

References

[1] S. Gantait, M.M. El-Dawayati, J. Panigrahi, C. Labrooy, S.K. Verma, The retrospect and prospect of the applications of biotechnology in (Phoenix dactylifera L.), App. Microbial. Biotech. 102 (2018) 8229–8259. https://doi.org/10.1007/s00253-018-9232-x

[2] K. Jamwal, S. Bhattacharya, S. Puri, Plant growth regulator mediated consequences of secondary metabolites in medicinal plants, Journal of applied research on medicinal and aromatic plants (2018). https://doi.org/10.1016/j.jarmap.2017.12.003

[3] A. Sultan, A.R. Raza, Steroids: a diverse class of secondary metabolites Med Chem, 5-7 (2015), 310–317.

[4] S.S. Jovanovi Santa, E.T. Petri, O.R. Klisuric, M. Szecsi, R. Kovacevic, Antihormonal potential of selected D-homo and Dsecoestratriene derivatives Steroids, 97 (2015) 45–53. https://doi.org/10.1016/j.steroids.2014.08.026

[5] S.F. El-Sharabasy, H. Bosila, S.Mohamed, I. Ibrahim, K. Refay, Production of some secondary products from date palm (Phoenix dactylifera) tissue cultures, Sewi cultivar using

some precursors during Embryogenesis stage, Second Emirates conference of date palm. Abu Dhabi (2001) 540-545.

[6] S.F. El-Sharabasy, Effects of some precursors on development of secondary products in tissues and media of embryogenic callus of date palm cv. Sewi. Arab J. Biotech. 7(2004) 83-90.

[7] M.I. Dias, M.J. Sousa, R.C. Alves, I.C. Ferreira, Exploring plant tissue culture to improve the production of phenolic compounds: A review, Industrial Crops and Products 1-82(2016) 9-22. https://doi.org/10.1016/j.indcrop.2015.12.016

[8] I. Smetanska, Production of secondary metabolites using plant cell cultures, In Food biotech Springer, Berlin, Heidelberg (2008) 187-228.

[9] W. Yue, Q.L. Ming, B. Lin, K. Rahman, C.J. Zheng, T. Han, L.P. Qin, Medicinal plant cell suspension cultures: pharmaceutical applications and high-yielding strategies for the desired secondary metabolites, Critical reviews in biotechnology 36-2 (2016) 215-32. https://doi.org/10.1016/j.indcrop.2015.12.016

[10] Z.E. Zayed, Enhanced Indirect Somatic Embryogenesis from Shoot-Tip Explants of Date Palm by Gradual Reductions of 2, 4-D Concentration. In Date Palm Biotechnology Protocols, Humana Press, New York 1 (2017) 77-88. https://doi.org/10.1007/978-1-4939-7156-5_7

[11] M.M. El-Dawayati, H.S. Ghazzawy, M. Munir, Somatic embryogenesis enhancement of date palm cultivar Sewi using different types of polyamines and glutamine amino acid concentration under in-vitro solid and liquid media conditions, Int J Biosci, 12 (2018) 149-159. https://doi.org/10.12692/ijb/12.1.149-159

[12] T. Murashige, F.A. Skoog, revised medium for rapid growth and bioassays with tobacco tissue cultures, Physiol Plant 15 (1962) 473-497. https://doi.org/10.1111/j.1399-3054.1962.tb08052.x

[13] El-Sharabasy, M.M. El-Dawayati, Bioreactor steroid production and analysis of date palm embryogenic callus. in: J.M. Al-Khayri, S.M. Jain, D. Johnson (Eds), Date palm biotechnology protocols, Tissue culture and applications, Humana Press, New York 1 (2017) 309–318. https://doi.org/10.1007/978-1-4939-7156-5_25

[14] G.W. Snedecor, W.G. Cochran, "Statisical Methods". Oxford and J.B.H. Publishing Co., 6th edition, 1980, pp. 507.

[15] X.W. Pan, Y.Y. Shi, X. Liu, X. Gao, Y.T Lu, Influence of inorganic microelements on the production of camptothecin with suspension cultures of Camptotheca acuminate, Plant growth regulation 44-1 (2004) 59-63. https://doi.org/10.1007/s10725-004-1654-z

[16] D. Pavokovic, M. Krsnik-Rasol, Complex biochemistry and biotechnological production of -betalains, Food Tech. Biotech 49-2 (2011) 145.

[17] V.S. Andrijany, G. Indrayanto, L.A. Soehono, Simultaneous effect of calcium, magnesium, copper and cobalt ions on sapogenin steroids content in callus cultures of Agave amaniensis, Plant cell, tissue and organ culture 55-2 (1998) 103-108. https://doi.org/10.1023/a:1006119600153

By-Products of Palm Trees and Their Applications
Materials Research Proceedings **11** (2019) 219-228

Materials Research Forum LLC
doi: https://doi.org/10.21741/9781644900178-17

Steroids Production of Embryogenic Callus Cultures of Date Palm under the Effect of Vitamins (Pyridoxine Hydrochloride, Nicotinic acid) Thiamine Hydrochloride and Myo- Insitol

[1,a*]Sherif F. El Sharabasy [1]Abdel-Aal W. B., [2]Hussein A .Bosila
[2]Bayome M. Mansour and [1]Abdel-Monem A. Bana

[1]The Central Lab of Date Palm Researches and Development, ARC, Egypt

[2]Floriculture, Medicinal & Aromatic Plants. Dept., Fac. of Agric., Al-Azhar University, Egypt

[a]sharabasydates@yahoo.com

Keywords: steroids, vitamins, Pyridoxine hydrochloride, Nicotinic acid, Thiamine hydrochloride, Myo- insitol, embryogenic callus, tissue culture, date palm

Abstract. A number of chemical and physical factors influence biomass accumulation and synthesis of secondary metabolites in plant cell and organ cultures. The medium composition is a basic and critical factor affecting the cell physiology and metabolism. Date palm tissues were abundance with valuable secondary products. Steroids production in embryogenic callus stage of in vitro date palm (Sakkoty and Bartamuda cultivar) was studied with the effect of addition of vitamins Pyridoxine hydrochloride , Nicotinic acid Thiamine hydrochloride at (0.5, 1.0, 2.0 mg/l) and Myo- insitol (25, 50, 100 mg/l) to MS nutrient medium. The pyridoxine concentration (0.5mg/l) was the most effective as it resulted in the highest significant value of mean (0.78 mg/g dry weight of embryonic callus of Sakkoty cv. and the highest significant value (0.80 mg/g dry weight) embryonic callus of Bartamuda cv. According to vitamin of nicotinic acid data showed that concentration at (0.5mg/l) was the most effective as it induced the highest significant value of steroid content (0.82 mg/g dry weight) of embryonic callus of Bartamoda and the highest significant value of steroid content (0.91 mg/g dry weight) of embryonic callus of Sakkoty cv..On other hand data showed that thiamine concentration at (2.0mg/l) was the most effective in inducing steroid content in embryonic callus of sakkoty cv. (0.83 mg/g dry weight) and Bartamuda cv. (0.87 mg/g dry weight).Results indicated also that The myo-inositol concentration at (25mg/l) gave the highest significant value of steroid content (0.71 mg/g dry weight) of embryonic callus of Bartamoda and the highest significant value of steroid content (0.80 mg/g dry weight) of embryonic callus of Sakkoty cv.

Introduction

Plant cells and callus cultures have been extensively used to explore the possibility of producing useful secondary metabolites through biotechnology methods and many studies have been undertaken to search for the most effective protocols for valuable secondary bioactive compounds production. In plant cells, secondary metabolism can be induced by using elections process or through targeting cell metabolism by specific precursors or controlling of chemicals and physiological of growth conditions [1-3] .Steroids are a set of cholesterol derivative lipophilic that are low molecular weight and may found in synthetic sources. They include sterols, hormones gonadal and adrenal ones, hydrocarbons and bile acids. Steroids family plays an important role in the biochemistry and composition of organisms. Steroids are used as anti-cancer agents, antibiotics, and anti-inflammatory, and anti-hormones drugs [4-5] . Date palm tree *Phoenix dactylifera* L. possess its importance not only for high nutrition's value of fruits but also for the multipurpose of the whole tree, the cultivation of this crop was distributed in North Africa and Middle East specially in Arabian Peninsula [6] .

Date palm tissues are abundance with valuable secondary products[7.] Steroids considered to be an important secondary product were localized in date palm tissues[8-10]   Recently there are some studies had paid attention for date palm ability for producing steroids *in vitro*

El Sharbasy et al.,[9] demonstrated the existence of total steroids in two Egyption cultivars tissues at higher values in pollen grain and shoot tip of in vivo tissues also they found in leaf and roots of the *in vitro* tissues. Cholesterol and ß-sitosterol have been also,separated and identified from date palm callus cultures by Thin Layer Chromatography (TLC). El-Sharabasy [10] found that precursors as pyruvic acid and cholesterols have a great effect on the biosynthesis of steroids in date palm callus and embryogenic callus cells. Enhancement in secondary metabolite production can be obtained by changing chemicals or physicals factors of growth conditions of plant cells such as culture medium components, pH, light, temperature,*etc.*, since the optimization of nutrient medim composition was the most important factor could strongly targting the plant cells metabolism[11-12]. Precursor feeding has been a normal and a popular approachto increase the yield of secondary compounds in plants cells.This is upon on the theory that any compound, which is an intermediate involved in the biosynthetic pathway [13].The aim of this work is to study Effect of some vitamins (Pyridoxine hydrochloride , Nicotinic acid , Thiamine hydrochloride , Myo- inositol) on steroids production in embryogeinic callus stage of in vitro date palm (Sakkoty and Bartamuda cultivar).

Materials and Methods

Callus explants of tow cultivars Bartamuda and Sakkoty were produced from indirect protocol of date palm micropropagation discribed by Zayed and El-Dawayati, et al. [14,15 .

In this study recived embryonic callus explants for both cultivars were cultured on basic nutrient medium for callus formation which composed of MS basal medium Murashige and Skoog [16],with the addition of 30 g/l sucrose ,3.0 g/l activated charcoal ,40 mg/L adenine – sulfate , 200 mg/l glutamine, 100 mg/l myo-inositol, 0.1 mg/l biotin , 170 mg/L NaH_2PO_4,0.1 mg/l thiamine HCl 0.5 mg/l pyridoxine ,0.5 mg/l nicotinic acid, 3.0 mg/l 2- isopentenyl adenine (2iP), and 10.0 mg/l 2,4 –D dichlorophenoxy acetic acid (2,4 – D). Pyruvic acid was added at 0.01 mg/l to induce steroids compounds production [10]. Four treatment were conducted separatly for both callus cultures of the tow cultivars (Bartamuda and Sakkoty) to induce the total steroid production by:-

a-Vitamins

1-Pyridoxine hydrochloride at(0.5 1.0 and 2.0 mg/l) concentrations

2-Nicotinic acid at(0.5 1.0 and 2.0 mg/l) concentrations

b-Thiamine hydrochloride at(0.5 1.0 and 2.0 mg/l) concentrations

c-Myo- insitol concentration at (25,50 and 100 mg/l) concentrations

Data in every treatment were recorded about the determination of total steroids (mg/g dry weight) in embryonic callus explants for both cultivars.

Total steroids were calculated as β-sitosterol and determined by spectrophotometer according to the protocol described by El-Sharabasy, and El-Dawayati [17] as follows:-

Test solution preparation: - 0.5 g weight of embryogenic callus sample is dried in an oven at 75 _C for 48 h. dried embryogenic callus sample is placed in a clean flask, with addition of 100 ml of 5% potassium hydroxide solution in alcohol (90% v/v) and are heated on a water bath at 50_°C to smooth reflux for 2 hours, then are cooled for 5 min, then the flask contents are transferred to a separator funnel. The residual contents of flask were washed for two times, firstly with 100 mL water followed by 100 mL diethyl ether then the washings were transferred into the same separator funnel and they are shacked altogether slowly by hand for 3 min. To separate the formed layer the aqueous phase was removed from separator funnel. This layer was washed in a separator funnel four times with 100 mL diethyl ether then, is placed in a clean flask. The received ethereal extracts are washed with three successive portions of 40 mL water (shaking was gently to avoid emulsions), 40 mL 5% w/v hydrochloric acid, and 40 mL 3% w/v potassium hydroxide aqueous solution. Successive portions of 40 mL water (each wash) are edited until the washings become neutral to phenolphthalein solution (2 drops 1% phenolphthalein in 70% ethanol and 2 N NaOH until rose color is stable). One drop of 0.1 N HCl is added to sample and rapidly mix until the rose color disappears. 100 mg anhydrous sodium sulfate powder is added to the sample with well shacking, then the mixture is filtered through folded Whatman filter paper. The resulted solution is evaporated in water bath at 50 °C until fully dry. 100 mL glacial acetic acid is added to the residue with stirring for 30 min in small glass bowl.

Test solution: - 2ml of previous resulted solution is transferred to a 20 mL volumetric flask, and dilute to 20 mL with glacial acetic acid.

The reference solution preparation: - 40 mg β-sitosterol is dissolved in 100 mL glacial acetic acid then 5 mL of this solution is taken then diluted to 50 mL with glacial acetic acid .

The Deniges reagent preparation: - This reagent is consisting of mixing of two solutions (solution A) is prepared by adding 100 mL sulfuric acid to 50 mL glacial acetic acid. .(solution B) is prepared by dissolving 5g mercury oxide (HgO_2) and 20 mL sulfuric acid into 100 mL water. 100 mL of solution (A) is added to 1 mL of solution (B), then are mixed and filtered through a sintered glass filter (grade G4) before use.

Finally, 5 mL of Deniges reagent mixture solutions is added to test tube filled with 1 mL (Test solution) and 1 mL (Reference solution) for evaluation of β-sitosterol amount.

The blank is carried out by 1 mL glacial acetic acid instead of the sample in a test tube. Both tubes are lifted on the stand under the dark for 15 min. The absorbance is read using a spectrophotometer at 510 nm against the blank reading. The amount of steroids is calculated as β-sitosterol from a standard curve prepared by dissolving 40 mg of β-sitosterol in 10 mL glacial acetic acid. Series of standards are prepared as 5, 10, 20, and 40 mg/100 mL, respectively; 1 mL of each is mixed with 5 mL Deniges reagent, and read at 510 nm against the blank. The absorbance of each concentration is plotted against the absorbance obtained from the standard curve.

Statistical analysis:
The obtained data were subjected to analysis of variance. The mean values were compared using LSD test at the 5% level of probability. The data were tabulated and statistically factorial analyses according to the randomized complete block design with three replicates [18].

Results and Discussion

1- Effect of Pyridoxine HCl on total steroids content (mg/g dry weight).
Data in Table (1) and Fig. (1) clearly showed that no significant differences were observed between the two cultivars under investigation (0.61, 0.64 mg/g dry weight respectively) the pyridoxine concentration (0.5mg/l) was the most effective as it resulted in the highest significant value (0.78 mg/g dry weight), concerning the interaction between cultivars and pyridoxine concentrations, the highest significant value (0.80 mg/g dry weight) was for Bartamuda cultivar

embryogenic callus grown on medium contained(0.5 mg/l) Pyridoxine. The lowest value (0.32 mg/g dry weight) was for Bartamuda cultivar embryogenic callus grown on medium contained (2.0 mg/l) pyridoxine.

Table1 Effect of Pyridoxine HCl on total steroids content (mg/g dry weight).

Cultivar (A)	Pyridoxine HCL mg/l(B)			
	0.5	1.0	2.0	Mean (A)
Bartamuda	0.80	0.71	0.32	0.61
Sakkoty	0.76	0.63	0.52	0.64
Mean (B)	0.78	0.67	0.42	
L.S.D 0.05: A=N.S, B=0.040, AB=0.575				

*Fig. 1 Effect of Pyridoxine **HCl** on total steroids content (mg/g dry weight).*

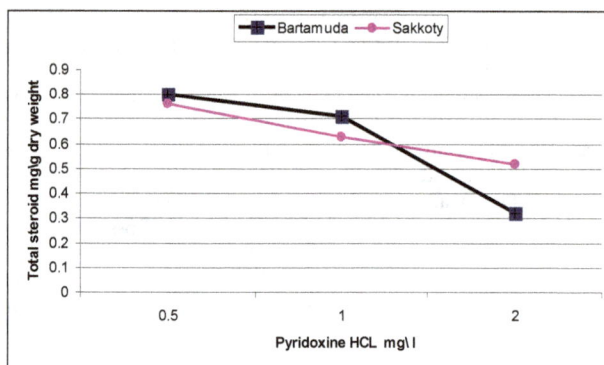

2- Effect of Nicotinic acid on total steroids content (mg/g dry weight).

Table2 Effect of Nicotinic acid on total steroids content (mg/g dry weight).

Cultivar (A)	Nicotinic acid mg/l (B)			
	0.5	**1.0**	**2.0**	**Mean (A)**
Bartamuda	0.74	0.30	0.19	0.41
Sakkoty	0.91	0.72	0.40	0.67
Mean (B)	0.82	0.51	0.29	
L.S.D 0.05: A=0.046, B=0.057, AB=0.081				

Fig. 2 Effect of Nicotinic acid on total steroids content (mg/g dry weight).

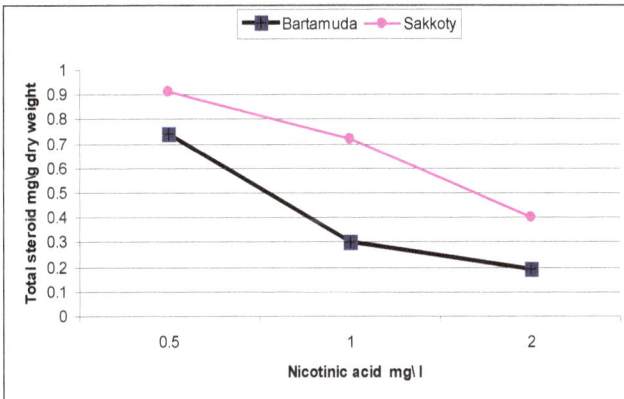

Data in Table (2) and Fig. (2) clearly showed that, significant differences were found between the two cultivars under investigation (0.41, 0.67 mg/g dry weight respectively), the nicotinic acid concentration(0.5mg/l) was the most effective as it induced the highest significant value (0.82 mg/g dry weight), concerning the interaction between cultivars and nicotinic acid concentrations, the highest significant value(0.91 mg/g dry weight) was for Sakkoty cultivar embryogenic callus grown on medium contained(0.5 mg/l) nicotinic acid. The lowest value (0.19 mg/g dry weight) was for Bartamuda cultivar embryogenic callus grown on medium contained (2.0 mg/l) nicotinic acid.

3- Effect of Thiamine on total steroids content (mg/g dry weight).

Table3 Effect of Thiamine HCl on total steroids content (mg/g dry weight).

Cultivar (A)	Thiamine HCL mg/l (B)			
	0.5	**1.0**	**2.0**	**Mean (A)**
Bartamuda	0.52	0.73	0.87	0.70
Sakkoty	0.42	0.48	0.80	0.56
Mean (B)	0.47	0.60	0.83	
L.S.D 0.05; A=0.033, B=0.040, AB=0.057				

Fig. 3 Effect of Thiamine HCl on total steroids content (mg/g dry weight).

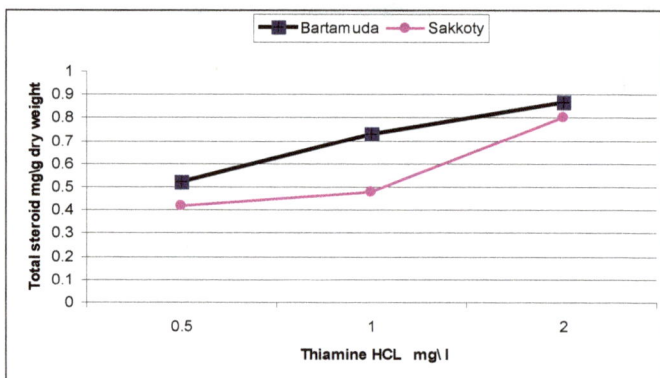

Data in Table (3) and Fig. (3) clearly showed that significant differences were found between the two cultivars under investigation (0.70, 0.56 mg/g dry weight respectively), the thiamine concentration(2.0mg/l) was the most effective as it resulted in the highest significant value (0.83 mg/g dry weight)then came (0.60 and 0.47 mg/g dry weight respectively), concerning the interaction between cultivars and thiamine concentrations, the highest significant value (0.87 mg/g dry weight) was for Bartamuda cultivar embryogenic callus grown on medium contained(2.0 mg/l) thiamine. The lowest value (0.42 mg/g dry weight) was for Sakkoty cultivar embryogenic callus grown on medium contained (0.5 mg/l) thiamine.

4. Effect of myo-insitol on total steroids content (mg/g dry weight).

Table.4 Effect of myo-inositol on total steroids content (mg/g dry weight).

Cultivar (A)	Myo-inositol mg/l (B)			
	25	50	100	Mean (A)
Bartamuda	0.62	0.40	0.31	0.44
Sakkoty	0.80	0.56	0.31	0.56
Mean (B)	0.71	0.48	0.31	
L.S.D: 0.05, A=0.033, B=0.040, AB = 0.057				

Fig. 4 Effect of Myo-insitol on total steroids content (mg/g dry weight).

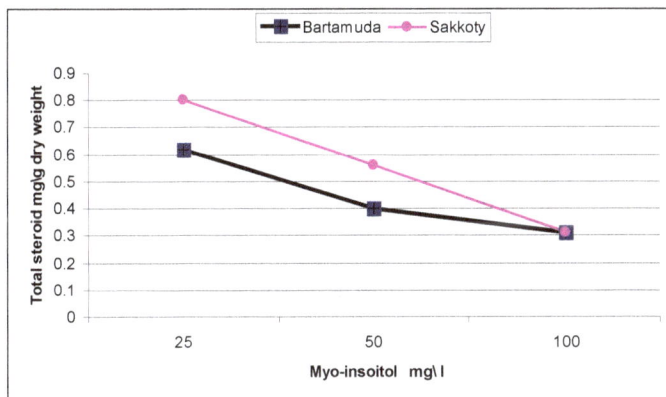

Data in Table (4) and Fig. (4) clearly showed that, there are significant differences were found between the two cultivars under investigation (0.44, 0.56 mg/g dry weight respectively), the myo-inositol concentration (25mg/l) was the most effective as it resulted in the highest significant value (0.71 mg/g dry weight), concerning the interaction between cultivars and myo-inositol concentrations, the highest significant value(0.80 mg/g dry weight) was for Sakkoty cultivar embryogenic callus grown on medium contained(25 mg/l) myo-inositol. The lowest value (0.31 mg/g dry weight) was for Bartamuda and Sakkoty cultivars embryogenic callus grown on medium contained (100 mg/l) myo-inositol. The basal MS medium includes vitamins such as pyridoxine-Hcl, nicotinic acid also, myo –inositol, , and thiamine-Hcl. Vitamins stimulated the growth of cells. The requirements of cells for vitamins addition is vary according to the nature of the plant and the type culture. Thiamine is essential for many plant cells, it is also involved in cell biosynthesis and metabolism. Myo-inositol has been described as a natural constituent of plant which involved in cell membrane permeability. It stimulated the cell division when added at low concentrations to the culture medium[19,20].On the light of our results these compounds additives have induced the production of steroids in date palm callus dependent on the concentration. There are few studies about the effect of these medium components on secondary metabolites production in plants cells. Some studies are reported in this investigation to support our finding, Jacob and Malpathak [21] found that the increased in vitamin concentration reduced fresh growth index in *Solanum khisanium* but solasodine (steroidal compound) production was increased by 2 times of concentration. In pine apple culture the proteolytic activity of the culture significantly decreased when the myo-inositol added at higher concentration than 1.10 Mm [22], while high concentrations of myo-inositol were favorable for secondary metabolite production as shown in callus culture of *Rheum ribes*, myo-inositol which increased the anthraquinone content [23],. However , addition of myo-inositol at high concentrations may induce the formation of calcium- inositol and ferrous- inositol complex in the culture medium. Plant cells can't easily up take these complexes, because of increasing in the osmotic potential of culture medium which may limit the plant growth and biosynthesis processes [24,25] .

Summary
Constituents of culture medium are important determinants of production of secondary metabolites as steroids compounds from date palm callus cultures. Much More studies are needed in these concerns.

References
[1] H.N. Murthy, V.S. Dandin, J.J. Zhong, K.Y. Paek, Strategies for enhanced production of plant secondary metabolites from cell and organ cultures, In Production of biomass and bioactive compounds using bioreactor technology, Springer, Dordrecht (2014) 471-508. https://doi.org/10.1007/978-94-017-9223-3_20

[2] A. Venugopalan, S. Srivastava, Endophytes as in vitro production platforms of high value plant secondary metabolites, Biotech. Adv. 33-6 (2015) 873-87. https://doi.org/10.1016/j.biotechadv.2015.07.004

[3] W. Yue, Q.L. Ming, B. Lin, K. Rahman, C.J. Zheng, T. Han, L.P. Qin, Medicinal plant cell suspension cultures: pharmaceutical applications and high-yielding strategies for the desired secondary metabolites, Critical reviews in biotechnology 36-2 (2016) 215-32. https://doi.org/10.3109/07388551.2014.923986

[4] A. Sultan, A.R. Raza Steroids: a diverse class of secondary metabolites Med Chem 5-7 (2015) 310–317.

[5] S. Jovanovic, S.Santa, E.T. Petri, O.R. Klisuric, M.Szecsi, R.Kovacevic Antihormonal potential of selected D-homo and Dsecoestratriene derivatives, Steroids 97 (2015) 45–53. https://doi.org/10.1016/j.steroids.2014.08.026

[6] S. Gantait, M.M. El-Dawayati, J. Panigrahi, C. Labrooy, S.K. Verma, The retrospect and prospect of the applications of biotechnology in (Phoenix dactylifera L.), App. Microbial. Biotech. 102 (2018) 8229–8259. https://doi.org/10.1007/s00253-018-9232-x

[7] R. Al-Alawi, J. Al-Mashiqri, J. Al-Nadabi, B. Al-Shihi, Y. Baq, Date palm tree (Phoenix dactylifera L.) natural products and therapeutic options, Front Plant Sci 8 (2017) 1–12. https://doi.org/10.3389/fpls.2017.00845

[8] S.F. El-Sharabasy, Studies on the production of secondary metabolites from date palm by using tissue culture technique: Ph. D. Thesis, Fac. Agric., Al-Azhar University, 2000.

[9] S.F. El-Sharabasy, H. Bosila, S.Mohamed, I. Ibrahim, K. Refay, Production of some secondary products from date palm (Phoenix dactylifera) tissue cultures, Sewi cultivar using some precursors during Embryogenesis stage Second Emirates conference of date palm. Abu Dhabi, (2001) 540-545.

[10] S.F. El-Sharabasy, Effects of some precursors on development of secondary products in tissues and media of embryogenic callus of date palm cv. Sewi, Arab J. Biotech. 7 (2004) 83-90.

[11] I.Smetanska, Production of secondary metabolites using plant cell cultures. In Food biotech Springer, Berlin, Heidelberg (2008) 187-228.

[12] M.I. Dias, M.J. Sousa, R.C. Alves, I.C. Ferreira, Exploring plant tissue culture to improve the production of phenolic compounds: A review, Industrial Crops and Products 1-82 (2016) 9-22. https://doi.org/10.1016/j.indcrop.2015.12.016

[13] N.A. Fadzliana, S. Rogayah, N.A. Shaharuddin, O.A. Janna, Addition of L-Tyrosine to Improve Betalain Production in Red Pitaya Callus. Pertanika J. Tropical Agr. Sci. 40-4 (2017) 521-532.

[14] Z.E. Zayed, Enhanced Indirect Somatic Embryogenesis from Shoot-Tip Explants of Date Palm by Gradual Reductions of 2, 4-D Concentration. In Date Palm Biotechnology Protocols, Humana Press, New York 1 (2017) 77-88. https://doi.org/10.1007/978-1-4939-7156-5_7

15- M.M. El-Dawayati, H.S. Ghazzawy, M. Munir, Somatic embryogenesis enhancement of date palm cultivar Sewi using different types of polyamines and glutamine amino acid concentration under in-vitro solid and liquid media conditions, Int J Biosci 12 (2018) 149-159. https://doi.org/10.12692/ijb/12.1.149-159

[16] T. Murashige, F. Skoog, A revised medium for rapid growth and bioassays with tobacco tissue cultures. Physiol. Plant 15 (1962) 473-497. https://doi.org/10.1111/j.1399-3054.1962.tb08052.x

[17] S.F. El-Sharabasy, M.M. El-Dawayati, Bioreactor steroid production and analysis of date palm embryogenic callus. in: J.M. Al-Khayri, S.M. Jain, D. Johnson (Eds), Date palm biotechnology protocols, Tissue culture and applications, Humana Press, New York 1(2017) 309–318. https://doi.org/10.1007/978-1-4939-7156-5_25

[18] G.W. Snedecor, W.G. Cochran, Statisical Methods, Oxford and J.B.H. Publishing Co., 6th edition, 1980, pp. 507.

[19] T. Thorpe, S.E.A. Yeung, G.J. de Klerek, A. Robert, E.F. George, The component of plant tissue culture media II, Organic additives ,osmotic and pH effects and support system, In: George EF, Hall MA, de Klerek GJ (Eds) Plant propagation by tissue culture the background (3rd Edn), Vol. (1), Springer The Netherland, 2010, pp.115-174. https://doi.org/10.1007/978-1-4020-5005-3_4

[20] A.P. Ling, S.L. Ong, H. Sobri, Strategies in enhancing secondary metabolites production in plant cell cultures, Med Aromat Plant Sci Biotechnol. 5 (2011) 94-101.

[21] A. Jacob, N. Malpathak, Manipulation of MS and B5 components for enhancement of growth and solasodine production in hairy root cultures of Solanum khasianum Clarke, Plant cell, tiss org cult. 80-3 (2005) 247-57. https://doi.org/10.1007/s11240-004-0740-2

[22] A. Pérez, L. Nápoles, C. Carvajal, M. Hernandez, JC. Lorenzo, Effect of sucrose, inorganic salts, inositol, and thiamine on protease excretion during pineapple culture in temporary immersion bioreactors, In Vitro Cellular & Developmental Biology-Plant. 40-3 (2004) 311-316. https://doi.org/10.1079/ivp2004529

[23] B.K. Drøbak, P.A.C. Watkins, Inositol(1,4,5) trisphosphate production in plant cells: an early response to salinity and hyperosmotic stress, FEBS Lett. 481 (2000) 240–244. https://doi.org/10.1016/s0014-5793(00)01941-4

[24] F.M. Sepehr, M. Gohrbanli, Effect of nutritional factors on the formulation of anthraquinone content in callus cultures of Rheum ribes, Plant Cell tissue culture and organ culture 68 () 171-175.

[25] J.K. Zhu, Plant salt tolerance, Trends in plant science, 6-2 (2001) 66-71.

By-Products of Palm Trees and Their Applications
Materials Research Proceedings 11 (2019) 229-234

Materials Research Forum LLC
doi: https://doi.org/10.21741/9781644900178-18

Effect of Murashige and Skoog Salts Strength Medium (MS) on Steroids Production and Total Amino Acids Content of Date Palm Embryonic Callus (Sakkoty and Bartamuda cultivar)

[2]Hussein A.Bosila, [1,a*]Sherif F. El sharabasy, [1]Abdel-Aal W. B.,
[2]Bayome M. Mansour and [1]Abdel-Monem A. Bana

[1]The Central Lab of Date Palm Researches and Development, ARC, Egypt

[2]Floriculture, Medicinal&Aromatic Plants. Dept., Fac. of Agric., Al-Azhar University, Egypt

[a]sharabasydates@yahoo.com

Keywords: MS, steroids, amino acid, embryogenic callus, date palm

Abstract. The potential to use tissue culture technique for the production of some bioactive compounds is immense, since it allows the manipulation of the biosynthetic routes to increase the production and accumulation of specific compounds. This study was conducted to investigate the effect of MS salt strength on steroids production and total amino acids content in embryonic callus cultures of two cultivars of date palm (Sakkoty and Bartamuda). embryonic callus explants were cultured on MS (Full), ¾ MS, ½ MS and ¼ MS), date was recorded every 6 weeks for three subculture. It obviously displays the superiority full MS over the three other investigated levels (¾ MS, ½ MS and ¼ MS) of steroids production (0.55, 0.38, 0.32 and 0.44 45mg/g dry weight respectively). Also the full MS level was the most effective of amino acids content (0.95 mg/g fresh weight). Bartamuda cv. was the superior of steroids production (0.45 mg/g dry weight) and amino acid content (1.13 mg/g fresh weight) compared with Sakkoty cvs.

Introduction

Date palm, *Phoenix dactylifera* L. is a heterozygous and dioecious tree belongs to (Arecaceae) family. It is considered to be as the most significant fruit crop in the Arabian Peninsula and North Africa countries, where it is closely related to the life and culture of the people since ancient times [1]. In addition to the high nutritional value of fruits also there are economic benefits of the parts of the whole tree. It was found that date palm tissues are rich of phytosterols compounds [2,3]. It has been discovered by Arabs and Egyptians that date palm grains are considered as a cure for sterility or antsterility agent [4]. Steroids are a set of cholesterol derivative lipophilic that are low molecular weight and may found in synthetic sources. They are essential for standard growth, development and differentiation of multicellular of organisms. The animal sterols are coprostanol and cholesterol, and plant ones such as campestrol, ergosterol, and B-sitosterol [5,6]. Cholesterol is the chief animal sterol, that made to be in certain amounts in plants and found in oil, date palm [7]. Extraction of secondary metabolites for industrial application has become an attractive solution by biotechnological approaches [8]. El-Sharabasy [9] separated and identified cholesterol and ß-sitosterol from callus cultures by Thin Layer Chromatography (TLC). El-Sharabasy [10] found that embryogenic callus cells of date palm were stimulated greatly for steroids biosynthesis by the addition of the precursors. A number of chemical and physical factors influence biomass accumulation and synthesis of secondary metabolites in plant cell and organ cultures. Nutrient media components, growth regulators, pH, temperature, light, etc… are considered to be the most important factors to enhance secondary metabolite accumulation in plant cell [11,12]. For the first time in this work the effect of MS basal medium (Murashige and Skoog) [13] salts strength on total steroids production and total amino acids content of embryonic callus stage of *in vitro* date palm Bartamuda, Sakkoty cultivars were studied in order to determine the optimal nutrient medium for enhancing accumulation of important secondary metabolites in date palm callus.

Materials and Methods

Preparation of plant material

Callus explants were produced from indirect protocol of date palm micropropagation discribed by Zayed [14] and El-Dawayati et al. [15].

Received embryonic callus explants for both Bartamuda. and Sakkoty cvs. were cultured on different levels of MS [13] salts strength, full strength of MS salts (Control), three fourth strength of MS salts, half strength of MS salts, and one fourth strength of MS salts.The basic nutrient medium components for all treatments are 30 g/l sucrose and 3.0 g/ l activated charcoal with 40 mg/L adenine – sulfate, 200 mg/l glutamine, 100 mg/l myo-inositol, 0.1 mg/l biotin, 170 mg/L NaH2PO4,0.1 mg/l thiamine HCL 0.5 mg/l pyridoxine,0.5 mg/l nicotinic acid, 3.0 mg/L 2- isopentenyl adenine (2iP) + 10.0 mg/l 2,4 –D dichlorophenoxy acetic acid (2,4 – D). Pyruvic acid was added at 0.01 mg/L to induce steroids compounds production [4]. 6.0 g/L agar were used to solidified Culture medium which were distributed in culture jars (250 ml); each jar contained 25 ml of culture nutrient medium. Culture jars were immediately capped with polypropylene closure autoclaved at 121_C at 1.05 kg/cm2 for 20 min. The cultured jars were incubated under total darkness at 27±1_C and data were recorded every (6 weeks) for three subcultures on total steroids content (mg/g dry weight).

1-Determination of total steroids (mg/g dry weight)

Total steroids were calculated as β-sitosterol and determined by spectrophotometer according to protocol described by El-Sharabasy and El-Dawayati [4] as follows:-

Test solution preparation: - 0.5 g weight of embryogenic callus sample is dried in an oven at 75 _C for 48 h. dried embryogenic callus sample is placed in a clean flask, with addition of 100 mL of 5% potassium hydroxide solution in alcohol (90% v/v) and are heated on a water bath at 50_C to smooth reflux for 2 hours, then are cooled for 5 min, then the flask contents are transferred to a separator funnel. The residual contents of flask were washed for two times, firstly with 100 mL water followed by 100 mL diethyl ether then the washings were transferred into the same separator funnel and they are shacked altogether slowly by hand for 3 min. To separate the formed layer the aqueous phase was removed from separator funnel. This layer was washed in a separator funnel four times with 100 mL diethyl ether then, is placed in a clean flask. The received ethereal extracts are washed with three successive portions of 40 mL water (shaking was gently to avoid emulsions), 40 mL 5% w/v hydrochloric acid, and 40 mL 3% w/v potassium hydroxide aqueous solution. Successive portions of 40 mL water (each wash) are edited until the washings become neutral to phenolphthalein solution (2 drops 1% phenolphthalein in 70% ethanol and 2 N NaOH until rose color is stable). One drop of 0.1 N HCl is added to sample and rapidly mix until the rose color disappears. 100 mg anhydrous sodium sulfate powder is added to the sample with well shacking, then the mixture is filtered through folded Whatman filter paper. The resulted solution is evaporated in water bath at 50 _C until fully dry. 100 mL glacial acetic acid is added to the residue with stirring for 30 min in small glass bowl.

Test solution: - 2ml of previous resulted solution is transferred to a 20 mL volumetric flask, and dilute to 20 mL with glacial acetic acid.

The reference solution preparation:- 40 mg β-sitosterol is dissolved in 100 mL glacial acetic acid then 5 mL of this solution is taken then diluted to 50 mL with glacial acetic acid.

The deniges reagent preparation: - This reagent is consist of mixing of two solutions (solution A) is prepared by adding 100 mL sulfuric acid to 50 mL glacial acetic acid..(solution B) is prepared by dissolving 5g mercury oxide (HgO_2) and 20 mL sulfuric acid into 100 mL water. 100 mL of solution (A) is added to 1 mL of solution (B), then are mixed and filtered through a sintered glass filter (grade G4) before use.

Finally 5 mL of Deniges reagent mixture solutions is added to test tube filled with 1 mL (Test solution) and 1 mL (Reference solution) for evaluation of β-sitosterol amount.

The blank is carried out by 1 mL glacial acetic acid instead of the sample in a test tube. Both tubes are lifted on the stand under the dark for 15 min. The absorbance is read using a spectrophotometer at 510 nm against the blank reading. The amount of steroids is calculated as β-sitosterol from a standard curve prepared by dissolving 40 mg of β-sitosterol in 10 mL glacial acetic acid. Series of standards are prepared as 5, 10, 20, and 40 mg/100 mL, respectively; 1 mL of each is mixed with 5 mL Deniges reagent, and read at 510 nm against the blank. The absorbance of each concentration is plotted against the absorbance obtained from the standard curve.

Determination of total amino acids content (mg/g fresh weight)

Total amino nitrogen or free amino acids were determined according to Rosein [16].

Statistical analysis

The obtained data were subjected to analysis of variance. The mean values were compared using LSD test at the 5% level of probability. The data were tabulated and statistically factorial analyses according to the randomized complete block design with three replicates [17].

Results and Discussion

Effect of different levels of MS salt strength on total steroids content (mg/g dry weight) embryonic callus date palm Sakkoty and Bartamuda cultivar in vitro.

Data in Fig (1) showed that there are significant differences between different cultivars under investigation. The highest amount of total steroids was recorded was for Bartamuda cv. (0.45mg/g dry weight), while Sakkoty cv. recorded (0.40 mg/g dry weight) of total steroids..Referring to the specific effect of MS salts level it obviously displays the superiority of the highest level (full MS) over the three other investigated levels (¾ MS, ½ MS and ¼ MS). Concerning, the interaction between MS salt levels and cultivars, it was noticed that, Bartamuda cultivar grown on full MS salt strength formed the highest significant value (0. 57 mg/g dry weight) of total steroids. On the light of our results it has been reported that medium manipulation is the most fundamental approch in plant tissue culture technology [11]. Murashig and Skoog is one of the most commonly medium were used for plant *in vitro* cultures, the high concentrations of nitrate,potasium and ammonium is the significant feature of Murashig and Skoog medium, nitrogen concentration was found to affect the amount of secondary metabolits in plant cell cultures as it has an regulation role for the expresstion of specific protiens through mechanisms affecting m RNA stability and transcription. Also, phosphour have important effect on the production of secondary metabbolites in plant cell cultures [18]. Which is confirmed with our results that the full MS salt strength of high concentration of total nitrogen and phosphour gave the highest signficant results of total steroids production in date palm callus.

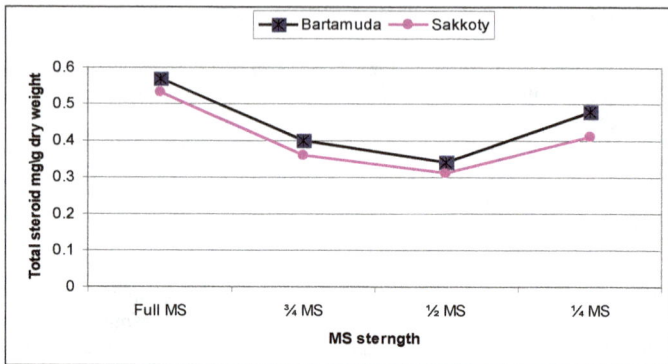

Fig. 1: Effect of different levels of MS salt strength on total steroids content (mg/g dry weight) of embryonic callus date palm Sakkoty and Bartamuda cultivar.

Effect of different levels of MS salt strength on total amino acids content (mg/g fresh weight) of embryonic callus of date palm Sakkoty and Bartamuda cultivars.

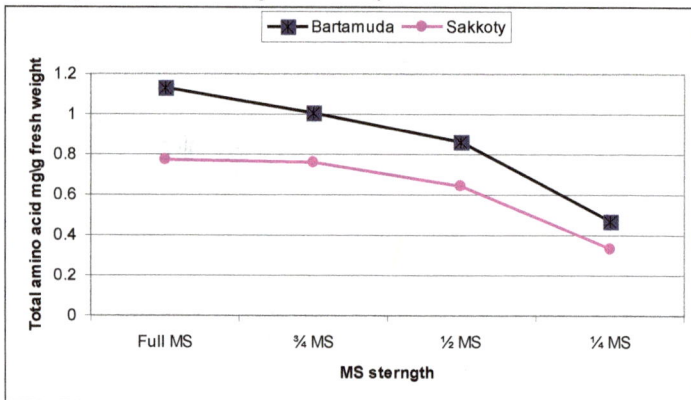

Fig. 2: Effect of MS salt strength on total amino acids content (mg/g fresh weight) in embryonic callus stage of date palm (Sakkoty and Bartamuda cultivar).

Data in Fig. (2) indicated that, there were significant differences between the two cultivars under investigation in total amino acid from embryogenic callus (0.87 and 0.62 mg/g fresh weight respectively). It was also observed that the full MS concentration was the most effective as it induced 0.95 mg/g fresh weight of amino acids while other concentrations (¾ MS, ½MS and¼ MS respectively) formed 0.88, 0.75 and 0.40 mg/g fresh weight respectively.

Concerning the interaction between cultivars and MS salt concentrations, the results illustrated that the highest significant value was of total amino acids (1.13 mg/g fresh weight) was for Bartamuda cultivar was formed by embryogenic callus grown on full MS medium. The lowest value (0.33 mg/g fresh weight) of total amino acids was formed by Sakkoty cultivar embryogenic callus grown on medium contained ¼ MS. The present results regarding the influence of MS medium on total free amino acids content was in general agreement with the finding of Mohamed [19] who decided that total amino acids significantly differed between cultivars and the highest amino acid content was detected with Bartamuda followed by Sakkoty.

Summary

Date palm steroids production from callus cultures seems to be an important approch for various studies depend on the manipulation of nutrient medium.

References

[1] D.V. Johnson, J.M. Al-Khayri, S.M. Jain, Introduction: Date production status and prospects in Asia and Europe. In Date palm genetic resources and utilization, Springer, Dordrecht, 2015, pp. 1-16. https://doi.org/10.1007/978-94-017-9707-8_1

[2] S.F. El-Sharabasy, Studies on the production of secondary metabolites from date palm by using tissue culture technique: Ph.D. Thesis, Fac Agric, Al-Azhar University, Cairo, 2000, pp 200.

[3] S. Besbes, C. Blecker, C. Deroanne, N. Bahloul, G. Lognay, N.E. Drira, H. Attia, Date seed oil: phenolic, tocopherol and sterol profiles. Journal of food lipids 11-4 (2004) 251-65. https://doi.org/10.1111/j.1745-4522.2004.01141.x

[4] S.F. El-Sharabasy, M.M. El-Dawayati, Bioreactor steroid production and analysis of date palm embryogenic callus, In: J.M. Al-Khayri, S.M. Jain, D. Johnson (Eds), Date palm biotechnology protocols, Tissue culture and applications, Humana Press, New York, 1 (2017) 309–318. https://doi.org/10.1007/978-1-4939-7156-5_25

[5] D. Lednicer, Steroid chemistry at a glance. John Wiley & Sons, 2011.

[6] A. Sultan, A.R. Raza, Steroids: a diverse class of secondary metabolites, Med Chem 5-7 (2015) 310–317.

[7] Z.G. Brill, M.L. Condakes, C.P. Ting, T.J. Maimone, Navigating the chiral pool in the total synthesis of complex terpene natural products, Chem. Rev. 117-18 (2017) 53-95. https://doi.org/10.1021/acs.chemrev.6b00834

[8] F. Biglari, A.F. AlKarkhi, A.M. Easa, Antioxidant activity and phenolic content of various date palm (Phoenix dactylifera) fruits from Iran. Food chem 107-4 (2008) 1636-41. https://doi.org/10.1016/j.foodchem.2007.10.033

[9] S.F. El-Sharabasy, H. Bosila, S.Mohamed, I. Ibrahim, K. Refay, Production of some secondary products from date palm (Phoenix dactylifera) tissue cultures, Sewi cultivar using some precursors during Embryogenesis stage, Second Emirates conference of date palm, Abu Dhabi (2001) 540-545.

[10] S. F. El-Sharabasy, Effects of some precursors on development of secondary products in tissues and media of embryogenic callus of date palm cv. Sewi. Arab J. Biotech 7 (2004) 83-90.

[11] A.P. Ling, S.L. Ong, H. Sobri, Strategies in enhancing secondary metabolites production in plant cell cultures, Med Aromat Plant Sci Biotechnol. 5 (2011) 94-101.

[12] E.M. Abd El-Kadder, I.I. Lashin, M.S. Aref, E.A. Hussian, E.A. Ewais, Physical elicitation of (Dillenia indica) callus for production of secondary metabolites, New York Sci J, 7-10 (2014) 48-57.

[13] T. Murashige, F. Skoog, A revised medium for rapid growth and bioassays with tobacco tissue cultures, Physiol. Plant 15 (1962) 473-497. https://doi.org/10.1111/j.1399-3054.1962.tb08052.x

[14] Z. E., Zayed, Enhanced Indirect Somatic Embryogenesis from Shoot-Tip Explants of Date Palm by Gradual Reductions of 2, 4-D Concentration, In Date Palm Biotechnology Protocols, Humana Press, New York 1 (2017) 77-88. https://doi.org/10.1007/978-1-4939-7156-5_7

[15] M.M. El-Dawayati, H.S. Ghazzawy, M. Munir, Somatic embryogenesis enhancement of date palm cultivar Sewi using different types of polyamines and glutamine amino acid concentration under in-vitro solid and liquid media conditions, Int J Biosci, 12 (2018) 149-159. https://doi.org/10.12692/ijb/12.1.149-159

[16] H. Rosein, A modified ninhydrincoloremetric analysis for amino acids, Archives of Biochemistry and Biophysics 67 (1957) 10-15.

[17] G.W. Snedecor, W.G. Cochran, Statistical Methods, Oxford and J.B.H. Publishing Co. 6 (1980) 507.

[18] W. Yue, Q.L. Ming, B. Lin, K. Rahman, C.J. Zheng, T.Han, L.P. Qin. Medicinal plant cell suspension cultures: pharmaceutical applications and high-yielding strategies for the desired secondary metabolites, Critical reviews in biotechnology, 36-2 (2016) 215-32. https://doi.org/10.3109/07388551.2014.923986

[19] A.M. Mohamed, Resistant cellulars selection of some date palm varieties through new genetically techniques: Ph.D thesis, Department of Agricultural Botany Fac. Agric., Al-Azhar University, 2008, pp. 53.

By-Products of Palm Trees and Their Applications
Materials Research Proceedings **11** (2019) 235-243

Materials Research Forum LLC
doi: https://doi.org/10.21741/9781644900178-19

The Effect of Some Micro-Elements on Free Amino Acids, Indols and total Phenols Production from Embryogenic Callus of Tow Date Palm Cultivars (Sakkoty and Bartamuda)

[1,a]*Sherif F. El sharabasy, [1]Abdel-Aal W. B., [2]Hussein A.Bosila,
[1]Abdel-Monem A. Bana and [2] Bayome M. Mansour

[1]The Central Lab of Date Palm Researches and Development, ARC, Giza, Egypt

[2]Floriculture, Medicina l& Aromatic Plants. Dept., Fac. of Agric., Al-Azhar University, Cairo Egypt

[a]sharabasydates@yahoo.com

Keywords: microelements, manganese sulfate ($MnSO_4$ 2 H_2O), zinc sulfate ($ZnSO_4 7H_2O$), copper sulfate ($CuSO_4 5H_2O$), amino acids, phenols, indols, embryogenic callus, date palm

Abstract. Effect of microelements on some chemicals analysis of secondry metabolits such as free amino acids, total indols content and total phenols content of date palm cultivars (Sakkoty and Bartamuda) were study in this work. Different concentrations of manganese sulfate ($MnSO_4$ 2 H_2O) (22.3, 44.6 and 66.9 mg/l), zinc sulfate ($ZnSO_4 7H_2O$) (8.6, 17.2 and 25.8 mg/l) and copper sulfate ($CuSO_4 5H_2O$) (0.025, 0.050, 0.075 mg/l) were added into nutrient medium of embryogenic callus stage. The results illustrated that, addtion of manganese sulfate at (22.3 mg/l) to culture medium of embryonic callus of Bartamoda cv. gave the highest significant values of total free amino acids (1.75 mg/g fresh weight) and (0.33 mg/g fresh weight) of total indols. Where the addition of manganese sulfate at (66.9 mg/l) to nuutrient medium of growing embryogenic callus of Bartamuda cv. gave the highest significant value of total phenols (1.17 mg/g fresh weight). The addititon of zinc sulfate at (17.2 mg/l) to culture medium of embryogenic callus of Sakkoty cv., recorded the highest significant values of total amino acids (1.64 mg/g fresh weight) and Indoles(0.40 mg/g fresh weight). While the highest significant values of total phenol content was (1.24 mg/g fresh weight) when embryonic callus of Sakkoty cv. grown on medium contained zinc sulfate at (25.8 mg/l). Data showed also the highest significant values of total free amino acids and total indols content (1.36 and 0.40 mg/g fresh weight respectively) were achived when embryogenic callus of Bartamuda cv. was grown on medium containing of copper sulfate at (0.025 mg/l),wherase the highest significant value of total phenols content (1.83 mg/g fresh weight) were recorded when embryonic callus of Sakkoty cv. was grown on nutrient medium supplemented with copper sulfate at (0.075 mg/l).

Introduction

Date palm has indispensable utilization in the economy and domestic life of growing countries. It is considered one of the most important commercial crops in the Middle East and Arab World [1].

Secondary metabolites are considered as chemicals that are produced by plants and these chemicals are diverse, Identification of them made into many classes. Each species or plant family has its own mixture of secondary metabolites and that's considered a main advantage in classification of plants. These chemicals could be used for medicinal purposes for humans [2]. Date palms can accumulate many chemicals in their tissues, as a primary metabolites containing carbohydrates and proteins, and secondary metabolites which are produced from primary ones

By-Products of Palm Trees and Their Applications   Materials Research Forum LLC
Materials Research Proceedings 11 (2019) 235-243   doi: https://doi.org/10.21741/9781644900178-19

such as phenolics [3].Secondary metabolite production can be induced by medium optimizations [4,5]. Microelements have many diverse roles and they are required in trace amounts for plant growth and development [6]. Culture conditions play an important role in the quality and quantity of the material obtained through secondary metabolites [7]. Optimization of the culture condition is effective in improving the accumulation of the desired product. External factors such as carbon source, nitrogen source, growth regulators, medium pH, temperature, light and oxygen are considered easy to regulate the expressions of plant secondary metabolite pathways [8]. Constituents in plant cell culture medium are determinants of growth and production of secondary metabolites. The specific roles for essential micronutrients in the production of active principles is due to their function as components or activators of enzymes of the secondary metabolism. Moreover, metals can quelate certain phytochemicals in plant tissues [9]. The aim of this work is to study the effect of microelements on production of (free amino acids, total indols content and total phenols content) in embryogenic callus stage of in vitro date palm (Bartamuda and Sakkoty cultivar).

Materials and Methods
Callus explants of tow cultivars Bartamuda and Sakkoty were produced from indirect protocol of date palm micropropagation discribed by [10,11].

In this study recived embryonic callus explants for both cultivars were cultured on basic nutrient medium for callus formation which composed of MS basal medium [12], supplemented 30 g/l sucrose and 3.0 g/l activated charcoal with 40 mg/L adenine – sulfate , 200 mg/l glutamine, 100 mg/l myo-inositol, 0.1 mg/l biotin , 170 mg/L NaH2PO4,0.1 mg/l thiamine HCL 0.5 mg/l pyridoxine ,0.5 mg/l nicotinic acid, 3.0 mg/L 2- isopentenyl adenine (2iP) + 10.0 mg/l 2,4 –D dichlorophenoxy acetic acid (2,4 – D).

Micro elements compounds, Manganese sulfate (MnSO4.4H2O), Zinc sulfate Heptahydrate (ZnSO4.7H2O) and Cupric Sulfate (CuSO4.5H2O) were added to previous basic nutrient medium for both Bartamuda and Sakkoty cv. callus cultures, in three different separated treatments for each as follows:-

1-Manganese sulfate (MnSO4.4H2O) were added at (22.3, 44.6 and 66.9 mg/l)

2- Zinc sulfate Heptahydrate (ZnSO4.7H2O) were added at (8.6, 17.2 and 25.8 mg/l)

3- Cupric Sulfate (CuSO4.5H2O) were added at (0.025, 0.050 and 0.075 mg/l)

6.0 g/L agar were used to solidified Culture medium which were distributed in culture jars (250 ml); each jar contained 25 ml of culture nutrient medium. Culture jars were immediately capped with polypropylene closure autoclaved at 121_C at 1.05 kg/cm2 for 20 min. The cultured jars were incubated under total darkness at 27±1_C and data were recorded every (6 weeks) for three subcultures on total steroids content (mg/g dry weight).

Callus sampels were collected from all studied treatments of the micro elements compounds, Manganese sulfate (MnSO4.4H2O), Zinc sulfate Heptahydrate (ZnSO4.7H2O) and Cupric Sulfate (CuSO4.5H2O) for both Bartamuda and Sakkoty cv. for the following assay

1. Determination of free amino acids
Total amino nitrogen or free amino acids were determined according to Rosein [13]. For assay, one ml of sample was pipetted out into a series of test tubes, and then total volume made up to 4 ml with distilled water. One ml of ninhydrin reagent (4 %, 4 g ninhydrin was dissolved in 50 ml acetone and 50 ml acetate buffer) was added to each tube, mixed well, and the tubes were kept in a boiling water bath for 15 min. Then, the tubes were cooled and the volume was made up to 10 ml in measuring flask with ethanol 50 %. The pink color developed was measured using a

spectrophotometer at 570 nm DL-alanine. The concentration of total amino nitrogen as DL-alanine were calculated from the standard curve.

2. Extraction of Indoles and Phenols

One gram of fresh samples in three replicates were sectioned into minute pieces and extracted with 5 ml cold methanol 80 % and stored in cold condition for 24h. The combined extracts were collected and filtered. Then, the volume of sample was raised up to known volume with cold methanol.

A Determination of Total Indoles

The total indoles were determined in the methanolic extract using p-dimethyl amino benzaldehyde (PDAB) reagent, 1 g was dissolved in 50 ml HCl conc. and 50 ml ethanol 95 %) test according to Larsen et al., [14]. One ml of aliquot methanolic extract was pipetted into a test tube, then 4 ml of PDAB reagent was added and incubated at 30 – 40 °C for 1 h. The intensity of the resultant color was spectrophotometerically measured at 530 nm. A standard curve was established which refer to the relationship between different concentrations of IAA and their corresponding absorbance values.

B Determination of Total Phenols

Phenols determination was carried out according to Danial and George [15]. For estimation of total phenols, 1 ml of the methanol tissue extract was added to 0.5 ml of Folin-Ciocalteu's Phenol Reagent and shaken 3 min. Then, 1 ml saturated Na_2CO_3 (25 % w/v) plus 17.5 ml distilled water added. The mixtures were left for one hour at 30- 40 °C. Optical density of these samples was measured by a colorimeter using wavelength 730 nm. Concentrations of total phenols in different samples were calculated as mg phenol/100g FW. Amount of total phenolic compounds was calculated according to standard curve of pyrogalol (99.5 %).

Statistical analysis

The obtained data were subjected to analysis of variance. The mean values were compared using LSD test at the 5% level of probability. The data were tabulated and statistically factorial analysed according to the randomized complete block design with three replicates Snedecor & Cochran [16].

Results and Discussion

Effect of manganese sulfate ($MnSO_4.4H_2O$) concentration on some chemical component (total of amino acids, Indoles and Phenols) in embryogeinic callus stage of in vitro date palm (Sakkoty and Bartamuda cultivar)

Effect of manganese sulfate on total amino acids content (mg/g fresh weight)

Data in Table (1) clearly showed that no significant differences were found between the two cultivars under investigation (0.90, 0.90 mg/g fresh weight), was for Bartamuda and Sakkoty respectively. The manganese sulfate concentration 22.3mg/l was the most effective forming the highest significant value (1.73 mg/g fresh weight).Concerning the interaction between cultivars and manganese sulfate concentrations, the results illustrated that the highest significant value (1.75 mg/g fresh weight) was for Bartamuda cultivar embryogenic callus grown on medium contained 22.3mg/l manganese sulfate. The lowest value (0.29 mg/g fresh weight) was for Bartamuda cultivar embryogenic callus grown on medium contained 66.9 mg/l.

Table 1: Effect of manganese sulfate on total amino acids content (mg/g fresh weight).

Cultivar (A)	Manganese sulfate mg/ l (B)			
	22.3	44.6	66.9	Mean (A)
Bartamuda	1.75	0.67	0.29	0.90
Sakkoty	1.71	0.58	0.42	0.90
Mean (B)	1.73	0.62	0.35	
L.S.D 0.05: A=N.S, B=0.30, AB=0.43				

Effect of manganese sulfate on total indoles content (mg/g fresh weight)

Table 2: Effect of manganese sulfate on total indoles content (mg/g fresh weight).

Cultivar(A)	Manganese sulfate mg/ l (B)			
	22.3	44.6	66.9	Mean (A)
Bartamuda	0.33	0.16	0.15	0.21
Sakkoty	0.29	0.13	0.08	0.16
Mean (B)	0.31	0.15	0.11	
L.S.D 0.05: A=0.046, B=0.057, AB=0.081				

Data in Table (2) clearly showed that significant differences were observed between the two cultivars under investigation (0.21, 0.16 mg/g fresh weight, Bartamuda ,Sakkoty respectively), the manganese sulfate concentration 22.3 mg\l was the most effective, forming the highest significant value (0.31 mg/g fresh weight),concerning the interaction between cultivars and manganese sulfate concentrations, the results illustrated that the highest significant value (0.33 mg/g fresh weight) was recorded by Bartamuda cultivar embryogenic callus which was grown on medium contained 22.3mg/l manganese sulfate. The lowest value (0.08 mg/g fresh weight) was for Sakkoty cultivar embryogenic callus grown on medium contained 66.9 mg/l.

Effect of manganese sulfate on total phenols content (mg/g fresh weight)

Table 3: Effect of manganese sulfate on total phenols content (mg/g fresh weight).

Cultivar (A)	Manganese sulfate mg/ l (B)			
	22.3	44.6	66.9	Mean (A)
Bartamuda	0.48	0.80	1.17	0.82
Sakkoty	0.37	0.83	1.07	0.76
Mean (B)	0.42	0.82	1.12	
L.S.D 0.05, A=N.S, B=0.13, AB=0.19				

Data in Table (3) clearly showed that no significant differences were found between the two cultivars under investigation (0.82, 0.76 mg/g fresh weight, Bartamuda, Sakkoty respectively).The manganese sulfate concentration 66.9 mg/l was the most effective. The highest significant value (1.12 mg/g fresh weight), concerning the interaction between cultivars and manganese sulfate concentrations, the highest significant value (1.17 mg/g fresh weight) was for

Bartamuda cultivar embryogenic callus grown on medium contained 66.9 mg/l manganese sulfate. The lowest value (0.37 mg/g fresh weight) was for Sakkoty cultivar embryogenic callus grown on medium contained 22.3 mg/l.

Effect of Zinc sulfate Heptahydrate (ZnSO4.7H2O) concentration on some chemical component (total of amino acids, Indoles and Phenols) in embryogeinic callus stage of in vitro date palm (Sakkoty and Bartamuda cultivar)

Effect of zinc sulfate on the total amino acid content (mg/g fresh weight)

Table 4: Effect of zinc sulfate on total amino acid content (mg/g fresh weight).

Cultivar (A)	zinc sulfate mg/ l (B)			
	8.6	17.2	25.8	Mean (A)
Bartamuda	0.77	1.63	0.42	0.94
Sakkoty	0.57	1.64	0.36	0.86
Mean (B)	0.67	1.64	0.39	
L.S.D 0.05: A=N.S, B=0.30, AB=0.43				

Data in Table (4) showed that, no significant differences were noticed between the two cultivars under investigation (0.94, 0.86 mg/g fresh weight respectively) zinc sulfate concentration 17.2 mg/l was the most effective as it induced, the highest significant value was(1.64 mg/g fresh weight), concerning the interaction between cultivars and zinc sulfate concentrations, the highest significant value (1.64 mg/g fresh weight) was produced by for Sakkoty cultivar embryogenic callus grown on medium contained 17.2 mg/l zinc sulfate. The lowest value (0.36 mg/g fresh weight) was for Sakkoty cultivar embryogenic callus grown on medium contained 25.8 mg/l zinc sulfate.

Effect of zinc sulfate on the total indoles content (mg/g fresh weight)

Table 5: Effect of zinc sulfate on total indoles content (mg/g fresh weight).

Cultivar (A)	zinc sulfate mg/ l (B)			
	8.6	17.2	25.8	Mean (A)
Bartamuda	0.076	0.376	0.113	0.188
Sakkoty	0.090	0.400	0.133	0.207
Mean (B)	0.083	0.388	0.123	
L.S.D 0.05: A=N.S, B=0.070, AB=0.099				

Data in Table (5) showed that no significant differences were formed between the two cultivars under investigation (0.18, 0.20 mg/g fresh weight respectively), zinc sulfate concentration(17.2mg/l) was the most effective as it produced ,the highest significant value was(0.38 mg/g fresh weight), concerning the interaction between cultivars and zinc sulfate concentrations, the highest significant value (0.40 mg/g fresh weight) was for Sakkoty cultivar embryogenic callus grown on medium contained(17.2 mg/l) zinc sulfate. The lowest value (0.076 mg/g fresh weight) was for Bartamuda cultivar embryogenic callus grown on medium contained (8.6 mg/l) zinc sulfate.

Effect of zinc sulfate on the total phenols content (mg/g fresh weight)

Table 6: Effect of zinc sulfate on the total phenols content (mg/g fresh weight).

Cultivar (A)	zinc sulfate mg/ l (B)			
	8.6	17.2	25.8	Mean (A)
Bartamuda	0.81	0.31	1.20	0.78
Sakkoty	0.66	0.44	1.24	0.78
Mean (B)	0.73	0.38	1.22	
L.S.D 0.05: A=N.S, B=0.22, AB=0.31				

Data in Table (6) showed that no significant differences were formed between the two cultivars under investigation (0.78, 0.78 mg/g fresh weight respectively), zinc sulfate concentration 25.8mg/l was the most effective as it induced, the highest significant value was (1.22 mg/g fresh weight), concerning the interaction between cultivars and zinc sulfate concentrations, the highest significant value (1.24 mg/g fresh weight) was for Sakkoty cultivar embryogenic callus grown on medium contained 25.8 mg/l zinc sulfate. The lowest value (0.31 mg/g fresh weight) was for Bartamuda cultivar embryogenic callus grown on medium contained 17.2 mg/l zinc sulfate.

Effect of Cupric Sulfate (CuSO4.5H2O)   concentration on some chemical component (total of amino acids , Indoles and Phenols) in embryogeinic callus stage of in vitro date palm (Sakkoty and Bartamuda cultivar)

Effect of cupric sulfate on total amino acids content (mg/g fresh weight)

Table 7: Effect of cupric sulfate on total amino acids content (mg/g fresh weight).

Cultivar (A)	cupric sulfate mg/l (B)			
	0.025	0.050	0.075	Mean (A)
Bartamuda	1.36	0.61	0.30	0.75
Sakkoty	1.14	0.84	0.31	0.76
Mean (B)	1.25	0.72	0.31	
L.S.D 0.05: A=N.S, B=0.11, AB=0.16				

Data in Table (7) showed that no significant differences were found between the two cultivars under investigation (0.75, 0.76 mg/g fresh weight respectively),cupric sulfate concentration 0.025mg/l was the most effective as it induced, the highest significant value (1.25 mg/g fresh weight), concerning the interaction between cultivars and cupric sulfate concentrations, the highest significant value (1.36 mg/g fresh weight) was for Bartamuda cultivar embryogenic callus grown on medium contained 0.025 mg/l cupric sulfate. The lowest value (0.30 mg/g fresh weight) was for Bartamuda cultivar embryogenic callus grown on medium contained 0.075 mg/l cupric sulfate.

Effect of cupric sulfate on total indoles content (mg/g fresh weight)

Table 8: Effect of cupric sulfate on total indoles content (mg/g fresh weight).

Cultivar (A)	cupric sulfate mg/l (B)			
	0.025	**0.050**	**0.075**	**Mean (A)**
Bartamuda	0.40	0.16	0.09	0.22
Sakkoty	0.25	0.11	0.07	0.14
Mean (B)	0.32	0.14	0.08	
L.S.D 0.05: A=0.033, B=0.040, AB=0.057				

Data in Table (8) showed that, significant differences were found between the two cultivars under investigation (0.22, 0.14 mg/g fresh weight respectively),cupric sulfate concentration 0.025mg/l was the most effective as it produced the highest significant value (0.32 mg/g fresh weight), concerning the interaction between cultivars and cupric sulfate concentrations, the highest significant value (0.40 mg/g fresh weight) was for Bartamuda cultivar embryogenic callus grown on medium contained 0.025 mg/l cupric sulfate. The lowest value (0.07 mg/g fresh weight) was for Sakkoty cultivar embryogenic callus grown on medium contained 0.075 mg/l cupric sulfate.

Effect of cupric sulfate on total phenols content (mg/g fresh weight)

Table 9: Effect of cupric sulfate on total phenols content (mg/g fresh weight).

Cultivar (A)	cupric sulfate (CuSO4.5H2O)mg/l (B)			
	0.025	**0.050**	**0.075**	**Mean (A)**
Bartamuda	0.34	1.28	1.53	1.05
Sakkoty	0.47	1.56	1.83	1.28
Mean (B)	0.41	1.04	1.68	
L.S.D 0.05: A=0.18, B=0.23, AB=0.32				

Data in Table (9) clearly showed that, significant differences were found between the two cultivars under investigation (1.05, 1.28 mg/g fresh weight respectively), cupric sulfate concentration 0.075mg/l was the most effective as it produced the highest significant value was (1.68 mg/g fresh weight), concerning the interaction between cultivars and cupric sulfate concentrations, the highest significant value (1.83 mg/g fresh weight) was for Sakkoty cultivar embryogenic callus grown on medium contained 0.075 mg/l cupric sulfate. The lowest value (0.34 mg/g fresh weight) was recorded by Bartamuda cultivar embryogenic callus grown on medium contained 0.025 mg/l cupric sulfate.

The present results regarding the beneficial effect of manganese sulfate (MnSO4.4H2O) zinc sulfate (ZnSO4.7H2O) and cupric sulfate (CuSO4.5H2O) as microelements in congeniality with the findings of several investigators on some species. Plant-produced secondary compounds have been contributed into a wide range of commercial and manufactories applications. Obviously, in many cases, rigorously controlled plant *in vitro* cultures can generate the same valuable natural products [7]. There are many studies made on the method that could be used for the enhancement of the production of valuable secondary metabolites. Microelements are required in trace amounts (Manganese, iodine, copper, cobalt, boron, molybdenum, iron, and zinc) usually comprise the microelements for plant growth and development, and have many diverse roles The effects of the medium employed in various processes have been reported [8]. It has been reported that proper concentration microelements have been considered as nutrient factors or as abiotic elicitors, which trigger the formation of secondary metabolites [9]. Where, Metal ions cause stress at elevated concentrations and stress has been implicated in secondary metabolite production. Many studies were undertaken to assess the role of metal stress on growth and differentiation as well as on secondary metabolite production in plants [17]. Copper, for example, is essential for the function of many oxidases and oxygenases with a key role in secondary metabolism. Also, Copper deficiency strongly inhibits the activities of diamine oxidase, which is essential for the metabolism of the diamines putrescine and cadaverine of polyphenoloxidase, and of superoxide dismutase (Cu/ZnSOD) Micronutrients without redox functions are also directly or indirectly involved in plant secondary metabolism. Zinc is required for the activity of thousands of plant proteins. Manganese is essential for Mn-SOD activity. In the shikimate pathway, Mn stimulates the pre-chorismate step catalyzed by the metalloenzyme 3-deoxy-D-arabinoheptulosonate 7-phosphate synthase [9]. It could be suggested that the production of preformed phytochemicals can also be significantly enhanced by those, trace elements which are being widely used to stimulate the production of active principles from callus cultures.

Summary

Studies in this area could lead to the successful manipulation of secondary metabolism and could significantly increase the amounts of the compounds. It should be possible to achieve the synthesis of a wide range of compounds in date palm callus cultures.

References

[1] S. Gantait, M.M. El-Dawayati, J. Panigrahi, C. Labrooy, S.K. Verma, The retrospect and prospect of the applications of biotechnology in (Phoenix dactylifera L.), App. Microbial. Biotech. 102 (2018) 8229-8259. https://doi.org/10.1007/s00253-018-9232-x

[2] E. Ahmed, M. Arshad, M.Z. Khan, Secondary metabolites and their multidimensional prospective in plant life, J. Pharm. Phyto. 2 (2017) 205-14.

[3] R. Al-Alawi, J. Al-Mashiqri, J. Al-Nadabi, B. Al-Shihi, Y. Baq, Date palm tree (Phoenix dactylifera L.) natural products and therapeutic options, Front Plant Sci. 8 (2017) 1-12. https://doi.org/10.3389/fpls.2017.00845

[4] M.I. Dias, M.J. Sousa, R.C. Alves, I.C. Ferreira, Exploring plant tissue culture to improve the production of phenolic compounds: A review, Industrial Crops and Products, 1-82 (2016) 9-22. https://doi.org/10.1016/j.indcrop.2015.12.016

[5] I. Smetanska, Production of secondary metabolites using plant cell cultures, In Food biotech Springer, Berlin, Heidelberg, (2008) 187-228.

[6] W. Yue, Q.L. Ming, B. Lin, K. Rahman, C.J. Zheng, T. Han, L.P. Qin, Medicinal plant cell suspension cultures: pharmaceutical applications and high-yielding strategies for the desired

By-Products of Palm Trees and Their Applications   Materials Research Forum LLC
Materials Research Proceedings **11** (2019) 235-243   doi: https://doi.org/10.21741/9781644900178-19

secondary metabolites, Critical reviews in biotechnology 36-2 (2016) 215-32. https://doi.org/10.3109/07388551.2014.923986

[7] N.A. Fadzliana, S. Rogayah, N.A. Shaharuddin, O.A. Janna, Addition of L-Tyrosine to Improve Betalain Production in Red Pitaya Callus, Pertanika J. Tropical Agr. Sci. 40-4 (2017) 521-532.

[8] I. Smetanska, Production of secondary metabolites using plant cell cultures, In Food biotech Springer, Berlin, Heidelberg (2008) 187-228.

[9] C. Poschenrieder, J. Allué, R. Tolrà, M. Llugany, J. Barceló, Trace Elements and Plants Secondary Metabolism: Quality and Efficacy of Herbal Products. Trace Elements as Contaminants and Nutrients/ed. by MNV Prasad Published by John Wiley & Sons, Inc., Hoboken, New Jersey 25 (2008) 99-120. https://doi.org/10.1002/9780470370124.ch5

[10] Z.E., Zayed, Enhanced Indirect Somatic Embryogenesis from Shoot-Tip Explants of Date Palm by Gradual Reductions of 2, 4-D Concentration, In Date Palm Biotechnology Protocols, Humana Press, New York, 1 (2017) 77-88. https://doi.org/10.1007/978-1-4939-7156-5_7

[11] M.M. El-Dawayati, H.S. Ghazzawy, M. Munir, Somatic embryogenesis enhancement of date palm cultivar Sewi using different types of polyamines and glutamine amino acid concentration under in-vitro solid and liquid media conditions, Int J Biosci 12 (2018) 149–159. https://doi.org/10.12692/ijb/12.1.149-159

[12] T. Murashige, F. Skoog, A revised medium for rapid growth and bioassays with tobacco tissue cultures, Physiol. Plant 15 (1962) 473-497. https://doi.org/10.1111/j.1399-3054.1962.tb08052.x

[13] Rosein, H., A modified ninhydrin coloremetric analysis for amino acids, Archives of Biochemistry and Biophysics, 67 (1957) 10-15. https://doi.org/10.1016/0003-9861(57)90241-2

[14] P. Larsen, A. Harbo, S. Klungsour, T. Asheim, On the biogenesis of some indol compounds in Acetobacter xylinum. Physiologia Plantarum, 15 (1962) 552-655. https://doi.org/10.1111/j.1399-3054.1962.tb08058.x

[15] H.D. Danial, C.M. George, Peach seed dormancy in relation to endogenous inhibitors and applied growth substances, Journal of the American Society for Horticulture Science 17 (1972) 651- 654.

[16] G.W. Snedecor, W.G. Cochran, Statisical Methods, Oxford and J.B.H. Publishing Co., 6th edition, 1980, pp. 507.

[17] H. Gaosheng, J. Jingming, Production of Useful Secondary Metabolites Through Regulation of Biosynthetic Pathway in Cell and Tissue Suspension Culture of Medicinal Plants, Recent Advances in Plant in vitro Culture, Annarita Leva and Laura M. R. Rinaldi, IntechOpen, (2012). https://doi.org/10.5772/53038

By-Products of Palm Trees and Their Applications
Materials Research Proceedings 11 (2019) 244-252

Materials Research Forum LLC
doi: https://doi.org/10.21741/9781644900178-20

Effect of Vitamins (pyridoxine and nicotinic acid), Thiamine-Hcl and Myo-Inositol at Different Concentrations on Free Amino Acids and Indoles Content of Embryogeinic Callus of *in vitro* Date Oalm (Sakkoty and Bartamuda Cultivar)

Sherif F. El sharabasy[1,a*], Hussein A. Bosila[2], Abdel-Aal W. B.[1],
Bayome M. Mansour[2] and Abdel-Monem A. Bana[1]

[1]The Central Lab of Date Palm Researches and Development, ARC, Giza, Egypt

[2]Floriculture, Medicinal&Aromatic Plants. Dept., Fac. of Agric., Al-Azhar University, Cairo Egypt

[a]sharabasydates@yahoo.com

Keywords: vitamins, amino acid, indoles, embryogenic callus, tissue culture, date palm

Abstract. The potential of using tissue culture technique for the production of some bioactive compounds since it allows the manipulation of the biosynthetic routes to increase the production and accumulation of specific compounds. This study was conducted to investigate the effect of vitamins (pyridoxine and nicotinic acid), thiamine-Hcl at different cocentrations (0.5, 1.0 & 2.0 mg/l) and myo-inositol at different concentrations (25, 50, and 100mg/l) at different cocentrations supplemented in MS basal nutrient medium of embryogenic callus of date palm on the production of secondary metabolites of amino acids and indoles. Tow egyption cultivars (Sakkoty and Bartamuda cultivars) of date palm were used. Pyridoxine concentration at 0.5mg/l was the most effective concentration in the production of amino acids and indoles from embryonic callus of the tow studied cultivars of date palm. Nicotinic acid at 0.5mg/l showed also the best results of production of amino acids and indoles from embryogenic callus of two cultivars. Acording to thiamine at 2mg/l concentration was the most effective in inducing the highest significant value of amino acids and indoles from embryonic callus of two cultivars of date palm. Myo-inositol concentration at 25mg/l produced the highest significant value of amino acids and indoles.

Introduction

Many higher plants are major sources of natural product used as pharmaceuticals, agrochemicals, flavor and fragrance ingredients, food additives, and pesticides [1]. The search for new plant derived. In the search for alternatives to production of desirable medicinal compounds from plants, biotechnological approaches, specifically, plant tissue cultures, are found to have potential as a supplement to traditional agriculture in the industrial production of bioactive plant metabolites [2]. Date palm tree *Phoenix dactylifera* L. is a multipurpose tree from whole tree, the cultivation of this crop was distributed in North Africa and Middle East specially in Arabian Peninsula. Date palm tree can accumulate many chemicals in their tissues, as primary metabolites containing carbohydrates and proteins, and secondary metabolites which are produced from primary ones [3,4]. The yield of secondary compounds in plants cells can be enhanced by precursor feeding in culture medium it has been a normal and a popular approach to increase this bioactive compounds [5]. secondary metabolite formation has shown that the media components have an influence on metabolism [6]. Vitamins, myoinositol and thiamine-HCl are considered important copmonents which induce plant cell growth also thier role in stimulated the bioactve metabolites as precursors has been reported [6-9]. The aim of this work is to study the effect of some vitamins (Pyridoxine hydrochloride, Nicotinic acid, Thiamine hydrochloride, Myo- inositol) on (free amino acids content, total indols content) in embryogeinic callus stage of *in vitro* date palm (Sakkoty and Bartamuda cultivars).

Materials and Methods

Callus explants of two cultivars Bartamuda and Sakkoty were produced from indirect protocol of date palm micropropagation discribed by [10,11].

In this study received embryonic callus explants for both cultivars were cultured on basic nutrient medium for callus formation which composed of MS basal medium [12], supplemented 30 g/l sucrose and 3.0 g/l activated charcoal with 40 mg/l adenine – sulfate, 200 mg/l glutamine, 100 mg/l myo-inositol, 0.1 mg/l biotin, 170 mg/l NaH2PO4,0.1 mg/l thiamine HCl 0.5 mg/l pyridoxine, 0.5 mg/l nicotinic acid, 3.0 mg/l 2- isopentenyl adenine (2iP) + 10.0 mg/l 2,4 –D dichlorophenoxy acetic acid (2,4 – D).

Callus explants of tow cultivars Bartamuda and Sakkoty were produced from indirect protocol of date palm micropropagation discribed by [10,11].

In this study recived embryonic callus explants for both cultivars were cultured on basic nutrient medium for callus formation which composed of MS basal medium [12], supplemented 30 g/l sucrose and 3.0 g/l activated charcoal with 40 mg/L adenine – sulfate, 200 mg/l glutamine, 100 mg/l myo-inositol, 0.1 mg/l biotin, 170 mg/L NaH2PO4,0.1 mg/l thiamine HCL 0.5 mg/l pyridoxine,0.5 mg/l nicotinic acid, 3.0 mg/L 2- isopentenyl adenine (2iP) + 10.0 mg/l 2,4 –D dichlorophenoxy acetic acid (2,4 – D). The studied treatments were added as followed:-

1. Effect of vitamins

Effect of Pyridoxine hydrochloride concentration on secondary metabolites in embryogenic callus.

Pyridoxine hydrochloride concentration:

a) 0.5 mg/l   b) 1.0 mg/l   c) 2.0 mg/l.

Effect of Nicotinic acid concentration on secondary metabolites in embryogenic callus.

Nicotinic acid concentration:

a) 0.5 mg/l   b) 1.0 mg/l   c) 2.0 mg/l.

Effect of thiamine hydrochloride concentrations on secondary metabolites in embryogenic callus.

2. Effect ofThiamine-Hcl concentrations:

a) 0.5 mg/   b) 1.0 mg/l   c) 2.0 mg/l.

Effect of Myo- inositol concentrations

a) 25 mg/l   b) 50 mg/l c) 100 mg/l.

6.0 g/l agar were used to solidified culture medium which were distributed in culture jars (250 ml); each jar contained 25 ml of culture nutrient medium. Culture jars were immediately capped with polypropylene closure autoclaved at 121°C at 1.05 kg/cm^2 for 20 min. The cultured jars were incubated under total darkness at 27±1°C and data were recorded every (6 weeks) for three subcultures on total steroids (mg/g dry weight).

Callus sampels were collected from all studied treatments of the micro elements compounds, manganese sulfate (), zinc sulfate ($MnSO_4.4H_2O$) heptahydrate ($ZnSO_4.7H_2O$) and copper sulfate ($CuSO_4.5H_2O$) for both Bartamuda and Sakkoty cv. for the following assay.

1. Determination of free amino acids

Total amino nitrogen or free amino acids were determined according to Rosein [13]. For assay, one ml of sample was pipetted out into a series of test tubes, and then total volume made up to 4 ml with distilled water. One ml of ninhydrin reagent (4 %, 4 g ninhydrin was dissolved in 50 ml acetone and 50 ml acetate buffer) was added to each tube, mixed well, and the tubes were kept in a boiling water bath for 15 min. Then, the tubes were cooled and the volume was made up to 10 ml in measuring flask with ethanol 50 %. The pink color developed was measured using a spectrophotometer at 570 nm DL-alanine. The concentration of total amino nitrogen as DL-alanine were calculated from the standard curve.

2. Extraction of Indoles and Phenols

One gram of fresh samples in three replicates were sectioned into minute pieces and extracted with 5 ml cold methanol 80 % and stored in cold condition for 24h. The combined extracts were collected and filtered. Then, the volume of sample was raised up to known volume with cold methanol.

A-Determination of Total Indoles

The total indoles were determined in the methanolic extract using p-dimethyl amino benzaldehyde (PDAB) reagent, 1 g was dissolved in 50 ml HCl conc. and 50 ml ethanol 95 %) test according to Larsen et al. [14]. One ml of aliquot methanolic extract was pipetted into a test tube, then 4 ml of PDAB reagent was added and incubated at 30 – 40 °C for 1 h. The intensity of the resultant color was spectrophotometerically measured at 530 nm. A standard curve was established which refer to the relationship between different concentrations of IAA and their corresponding absorbance values.

B-Determination of Total Phenols

Phenols determination was carried out according to Danial and George [15]. For estimation of total phenols, 1 ml of the methanol tissue extract was added to 0.5 ml of Folin-Ciocalteu's Phenol Reagent and shaken 3 min. Then, 1 ml saturated Na_2CO_3 (25 % w/v) plus 17.5 ml distilled water added. The mixtures were left for one hour at 30- 40 °C. Optical density of these samples was measured by a colorimeter using wavelength 730 nm. Concentrations of total phenols in different samples were calculated as mg phenol/100g FW. Amount of total phenolic compounds was calculated according to standard curve of pyrogalol (99.5 %).

Statistical analysis

The obtained data were subjected to analysis of variance. The mean values were compared using LSD test at the 5% level of probability. The data were tabulated and statistically factorial analysed according to the randomized complete block design with three replicates Snedecor & Cochran [16].

Results and Discussion

1. Effect of Pyridoxine- HCl on total amino acids content (mg/g fresh weight).

Data in Table 1 clearly showed that no significant differences were found between the two cultivars under investigation (0.99, 0.94 mg/g fresh weight respectively),the pyridoxine concentration (0.5mg/l) was the most effective as it induced the highest significant value (1.65 mg/g fresh weight), concerning the interaction between cultivars and pyridoxine concentrations, these results illustrated that the highest significant value (1.65 mg/g fresh weight) was for Sakkoty and Bartamuda cultivars embryogenic callus grown on medium contained(0.5 mg/l) pyridoxine. The lowest value (0.47 mg/g fresh weight) was for Bartamuda cultivar embryogenic callus grown on medium contained (1.0 mg/l) pyridoxine.

Table 1: Effect of Pyridoxine -HCl on total amino acids content (mg/g fresh weight).

Cultivar (A)	Pyridoxine- HCL mg/l (B)			
	0.5	1.0	2.0	Mean (A)
Bartamuda	1.65	0.47	0.70	0.99
Sakkoty	1.65	0.54	0.77	0.94
Mean (B)	1.65	0.51	0.74	
L.S.D 0.05: A=N.S, B =0.19, AB=0.27				

2. Effect of Pyridoxine- HCl on total indoles content (mg/g fresh weight).

Table 2: Effect of Pyridoxine HCl on total indoles content (mg/g fresh weight).

Cultivar (A)	Pyridoxine HCL mg/l(B)			
	0.5	1.0	2.0	Mean (A)
Bartamuda	0.50	0.21	0.18	0.30
Sakkoty	0.59	0.30	0.15	0.35
Mean (B)	0.55	0.25	0.16	
L.S.D 0.05: A=0.03, B=0.041, AB=0.057				

Data in Table 2 clearly showed that, significant differences were found between the two cultivars under investigation (0.30, 0.35 mg/g fresh weight respectively), the pyridoxine concentration (0.5mg/l)was the most effective as it induced the highest significant value (0.55 mg/g fresh weight), concerning the interaction between cultivars and pyridoxine concentrations, the highest significant value(0.59 mg/g fresh weight) was for Sakkoty cultivar embryogenic callus grown on medium contained(0.5 mg/l) pyridoxine. The lowest value (0.15 mg/g fresh weight) was for Sakkoty cultivar embryogenic callus grown on medium contained 2.0 mg/l Pyridoxine.

3. Effect of Nicotinic acid on total amino acids content (mg/g fresh weight).

Table 3: Effect of Nicotinic acid on total amino acids content (mg/g fresh weight).

Cultivar (A)	Nicotinic acid mg/l (B)			
	0.5	1.0	2.0	Mean (A)
Bartamuda	0.72	0.49	0.22	0.48
Sakkoty	0.65	0.45	0.18	0.42
Mean (B)	0.68	0.47	0.20	
L.S.D 0.05: A=N.S, B=0.081, AB=0.115				

Data in Table 3 clearly showed that no significant differences were found between the two cultivars under investigation (0.48, 0.42 mg/g fresh weight respectively), the nicotinic acid concentration (0.5mg/l) was the most effective as it produced the highest significant value (0.68 mg/g fresh weight), concerning the interaction between cultivars and nicotinic acid concentrations, the highest significant value (0.72 mg/g fresh weight) was for Bartamuda cultivar embryogenic callus grown on medium contained(0.5 mg/l) nicotinic acid. The lowest value (0.18 mg/g fresh weight) was for Sakkoty cultivar embryogenic callus grown on medium contained (2.0 mg/l) nicotinic acid.

4. Effect of Nicotinic acid on total indoles content (mg/g fresh weight).

Table 4: Effect of Nicotinic acid on total indoles content (mg/g fresh weight).

Cultivar (A)	Nicotinic acid mg/l (B)			
	0.5	1.0	2.0	Mean (A)
Bartamuda	0.70	0.46	0.18	0.45
Sakkoty	0.71	0.42	0.12	0.42
Mean (B)	0.70	0.44	0.15	
L.S.D 0.05: A=N.S, B=0.070, AB =0.099				

Data in Table 4 showed that no significant differences were found between the two cultivars under investigation (0.45, 0.42 mg/g fresh weight respectively), the nicotinic acid concentration (0.5mg/l)was the most effective as it induced the highest significant value (0.70 mg/g fresh weight) then came (0.44 and 0.15 mg/g fresh weight respectively), concerning the interaction between cultivars and nicotinic acid concentrations, the highest significant value (0.71 mg/g fresh weight) was for Sakkoty cultivar embryogenic callus was grown on medium contained(0.5 mg/l) nicotinic acid. The lowest value (0.12 mg/g fresh weight) was for Sakkoty cultivar embryogenic callus grown on medium contained (2.0 mg/l) nicotinic acid.

5. Effect of thiamine-Hcl on total amino acids content (mg/g fresh weight).

Table 5: Effect of thiamine on total amino acids content (mg/g fresh weight).

Cultivar (A)	Thiamine HCl mg/l (B)			
	0.5	**1.0**	**2.0**	**Mean (A)**
Bartamuda	0.26	0.19	0.50	0.31
Sakkoty	0.13	0.19	0.42	0.24
Mean (B)	0.19	0.19	0.46	
L.S.D 0.05: A=0.046, B=0.057, AB=0.081				

Data in Table 5 showed that there are significant differences were found between the two cultivars under investigation (0.31, 0.24 mg/g fresh weight respectively), the thiamine concentration (2.0mg/l) was the most effective as it induced the highest significant value (0.46 mg/g fresh weight), concerning the interaction between cultivars and thiamine concentrations, the highest significant value(0.50 mg/g fresh weight) was for Bartamuda cultivar embryogenic callus grown on medium contained(2.0 mg/l) thiamine. The lowest value (0.13 mg/g fresh weight) was for Sakkoty cultivar embryogenic callus grown on medium contained (0.5 mg/l) thiamine.

6. Effect of Thiamine on total indoles content (mg/g fresh weight).

Table 6: Effect of thiamine on total indoles content (mg/g fresh weight).

Cultivar (A)	Thiamine HCl mg/l(B)			
	0.5	**1.0**	**2.0**	**Mean (A)**
Bartamuda	0.12	0.26	0.62	0.33
Sakkoty	0.14	0.34	0.77	0.42
Mean (B)	0.13	0.30	0.70	
L.S.D 0.05: A=0.033, B=0.040, AB = 0.575				

Data in Table 6 clearly showed that significant differences were found between the two cultivars under investigation (0.33, 0.42 mg/g fresh weight respectively), the thiamine concentration (2.0mg/l) was the most effective as it resulted in the highest significant value (0.70 mg/g fresh weight), concerning the interaction between cultivars and thiamine concentrations, the results illustrated that the highest significant value (0.77 mg/g fresh weight) was for Bartamuda cultivar embryogenic callus grown on medium contained(2.0 mg/l) thiamine. The lowest value (0.12 mg/g fresh weight) was for Bartamuda cultivar embryogenic callus grown on medium contained (0.5 mg/l) thiamine.

7. Effect of myo-inositol on total amino acids content (mg/g fresh weight).

Table. 7: Effect of myo-inositol on total amino acids content (mg/g fresh weight).

Cultivar (A)	Myo-inositol mg/l (B)			
	25	50	100	Mean (A)
Bartamuda	1.20	0.60	0.64	0.81
Sakkoty	1.00	0.61	0.60	0.73
Mean (B)	1.10	0.60	0.62	
L.S.D 0.05: A=N.S, B=0.14, AB=0.19				

Data in Table 7 showed that no significant differences were found between the two cultivars under investigation (0.81, 0.73 mg/g fresh weight respectively), the myo-inositol concentration (25mg/l) was the most effective as it induced the highest significant value (1.10 mg/g fresh weight), concerning the interaction between cultivars and myo-inositol concentrations, the highest significant value(1.20 mg/g fresh weight) was for Bartamuda cultivar embryogenic callus grown on medium contained (25 mg/l) myo-inositol. The lowest value (0.60 mg/g fresh weight) was for Sakkoty cultivar embryogenic callus grown on medium contained (100 mg/l) myo-inositol.

8. Effect of Myo-inositol on total indoles content (mg/g fresh weight).

Table. 8: Effect of myo-inositol on total indoles content (mg/g fresh weight).

Cultivar (A)	Myo-inositol mg/l (B)			
	25	50	100	Mean (A)
Bartamuda	0.61	0.23	0.27	0.37
Sakkoty	0.67	0.23	0.26	0.39
Mean (B)	0.64	0.23	0.26	
L.S.D 0.05: A=N.S, B=0.040, AB=0.057				

Data in Table 8 clearly showed that no significant differences were found between the two cultivars under investigation (0.37, 0.39 mg/g fresh weight respectively), the myo-inositol concentration (25mg/l) was the most effective as it produced the highest significant value (0.64 mg/g fresh weight), concerning the interaction between cultivars and myo-inositol concentrations, the highest significant value(0.67 mg/g fresh weight) was for Bartamuda cultivar embryogenic callus grown on medium contained(25 mg/l) myo-inositol. The lowest value (0.23 mg/g fresh weight) was for Bartamuda and Sakkoty cultivars the embryogenic callus grown on medium contained (50 mg/l) myo-inositol.

Secondary metabolite production can be induced by medium optimizations [17,18].Obviously, in many cases, rigorously controlled plant *in vitro* cultures can generate the same valuable natural products [5]. Vitamins are nitrogenous substances required in trace amounts to serve catalytic functions in enzyme systems. Plant cell grown in vitro are capable of synthesizing essential vitamins in suboptimal quantities; thus,culture media are often supplemented with vitamins to enhance growth. Various standard media formulations and modifications there of show wide differences in vitamin composition [5,19]. Thiamine is essential for many plant cells, it is also involved in cell biosynthesis and metabolism. Myo-inositol has been described as a natural constituent of plant which involved in cell membrane permeability. It stimulated the cell division when added at low concentrations to the culture medium [20-22]. On the light of our results these compounds additives have induced the content of free amino acids content and indole content in date palm callus dependent on the concentration.

Summary

Studies in this area could lead to the successful manipulation of secondary metabolism and could significantly increase the amounts of the compounds. It should be possible to achieve the synthesis of a wide range of compounds in date palm callus cultures.

References

[1] R. Muhaidat, M.A. Al-Qudah, O. Samir, J.H. Jacob,E. Hussein, I.N. Al-Tarawneh, E. Bsoul, S.T.Orabi, Phytochemical investigation and in vitro antibacterial activity of essential oils from Cleome droserifolia (Forssk.) Delile and C. trinervia Fresen.(Cleomaceae), South African J. Bot. 99 (2015) 21-8. https://doi.org/10.1016/j.sajb.2015.03.184

[2] M. Asif, Chemistry and antioxidant activity of plants containing some phenolic compounds,Chem. Int. 1 (2015), pp. 35-52.

[3] S. Gantait, M.M. El-Dawayati, J. Panigrahi, C. Labrooy, S.K. Verma, The retrospect and prospect of the applications of biotechnology in (Phoenix dactylifera L.), App. Microbial, Biotech. 102 (2018) 8229–8259. https://doi.org/10.1007/s00253-018-9232-x

[4] R. Al-Alawi, J. Al-Mashiqri, J. Al-Nadabi, B. Al-Shihi, Y. Baq, Date palm tree (Phoenix dactylifera L.) natural products and therapeutic options, Front Plant Sci 8 (2017) 1–12. https://doi.org/10.3389/fpls.2017.00845

[5] N.A Fadzliana, S. Rogayah, N.A. Shaharuddin, O.A. Janna, Addition of L-Tyrosine to Improve Betalain Production in Red Pitaya Callus, Pertanika J. Tropical Agr. Sci. 40-4 (2017) 521-532.

[6] A.P. Ling, S.L. Ong, H. Sobri, Strategies in enhancing secondary metabolites production in plant cell cultures, Med Aromat Plant Sci Biotechnol. 5 (2011) 94-101.

[7] A. Pérez, L. Nápoles, C. Carvajal, M. Hernandez, JC. Lorenzo, Effect of sucrose, inorganic salts, inositol, and thiamine on protease excretion during pineapple culture in temporary immersion bioreactors, In Vitro Cellular & Developmental Biology-Plant, 40-3 (2004) 311-316. https://doi.org/10.1079/ivp2004529

[8] A. Jacob, N. Malpathak, Manipulation of MS and B5 components for enhancement of growth and solasodine production in hairy root cultures of Solanum khasianum Clarke, Plant cell, tiss org cult. 80-3 (2005) 247-57. https://doi.org/10.1007/s11240-004-0740-2

[9] E.F. George, M.A. Hall, GJ De Klerk, The component of plant tissue culture media II. Organic additives, osmotic and pH effects and support system, In: Plant propagation by tissue culture the background (3rd Edn), Springer The Netherland 1 (2010) 115-174. https://doi.org/10.1007/978-1-4020-5005-3_4

[10] Z. E. Zayed, Enhanced Indirect Somatic Embryogenesis from Shoot-Tip Explants of Date Palm by Gradual Reductions of 2, 4-D Concentration, In Date Palm Biotechnology Protocols, Humana Press, New York 1 (2017) 77-88. https://doi.org/10.1007/978-1-4939-7156-5_7

[11] M.M. El-Dawayati, H.S. Ghazzawy, M. Munir, Somatic embryogenesis enhancement of date palm cultivar Sewi using different types of polyamines and glutamine amino acid concentration under in-vitro solid and liquid media conditions, Int J Biosci 12 (2018) 149-159. https://doi.org/10.12692/ijb/12.1.149-159

[12] T. Murashige, F. Skoog, A revised medium for rapid growth and bioassays with tobacco tissue cultures, Physiol. Plant 15 (1962) 473-497. https://doi.org/10.1111/j.1399-3054.1962.tb08052.x

[13] H. Rosein, A modified ninhydrin coloremetric analysis for amino acids. Archives of Biochemistry and Biophysics 67 (1957) 10-15. https://doi.org/10.1016/0003-9861(57)90241-2

[14] P. Larsen, A. Harbo, S. Klungsour, T. Asheim, On the biogenesis of some indol compounds in Acetobacter xylinum, Physiologia Plantarum, 15 (1962) 552 – 655. https://doi.org/10.1111/j.1399-3054.1962.tb08058.x

[15] H.D. Danial, C.M. George, Peach seed dormancy in relation to endogenous inhibitors and applied growth substances, Journal of the American Society for Horticultural Science, 17 (1972) 651- 654.

[16] G.W. Snedecor, W.G. Cochran, Statisical Methods, Oxford and J.B.H. Publishing Co., 6th edition, 1980, pp. 507.

[17] M.I. Dias, M.J. Sousa, R.C. Alves, Ferreira IC. Exploring plant tissue culture to improve the production of phenolic compounds: A review, Industrial Crops and Products, 82 (2016) 9-22. https://doi.org/10.1016/j.indcrop.2015.12.016

[18] I. Smetanska, Production of secondary metabolites using plant cell cultures, In Food biotech Springer, Berlin, Heidelberg (2008)187-228.

[19] C.L. Marbun, N. Toruan-Mathius, C. Utomo, T. Liwang, Micropropagation of embryogenic callus of oil palm (Elaeis guineensis Jacq.) using temporary immersion system, Procedia Chemistry, 2015. https://doi.org/10.1016/j.proche.2015.03.018

[20] Bettendorff L. Thiamine, Handbook of Vitamins, 5th Edition. 2014, pp. 267-323.

[21] T. Thorpe, S.E.A Yeung, GJ. de Klerek, A. Robert, E.F. George, The component of plant tissue culture media II. Organic additives, osmotic and pH effects and support system, In: Plant propagation by tissue culture the background (3red Edn), Springer The Netherland 1 (2010) 115-174. https://doi.org/10.1007/978-1-4020-5005-3_4

[22] A.P. Ling, S.L. Ong, H. Sobri, Strategies in enhancing secondary metabolites production in plant cell cultures, Med Aromat Plant Sci Biotechnol. 5 (2011) 94-101.

Materials Research Forum LLC
doi: https://doi.org/10.21741/9781644900178

Fiber, Paper, and Textile

By-Products of Palm Trees and Their Applications
Materials Research Proceedings 11 (2019) 255-261

Materials Research Forum LLC
doi: https://doi.org/10.21741/9781644900178-21

Investigations on the Effects of Cement Replacement and Calcium Chloride Addition on Selected Properties of Coconut Husk Fibre-Reinforced Roofing Tiles

Abel.O. Olorunnisola[1,a*] and Anthony .O. Adeniji[1,b]

[1]Department of Wood Products Engineering, University of Ibadan, Nigeria

[a]abelolorunnisola@yahoo.com, [b]toneeden@live.co.uk

Keywords: coconut husk, rice husk ash, calcium carbide waste, cement-bonded composite

Abstract. Provision of adequate and affordable housing is one of the continuing challenges posed by unprecedented urbanization in Nigeria and many other African countries. One of the solutions to this chronic problem is the development of non-conventional low cost building materials from recyclable agro-industrial wastes. This study was conducted to investigate the effects of $CaCl_2$ addition and partial replacement of cement with Rice Husk Ash (RHA) and calcium carbide waste (lime) on the density, water resistance and impact strength of cement-bonded composite roofing tiles reinforced with coconut husk (*Cocos nucifera*) fibres. Results indicated that $CaCl_2$ enhanced impact strength and dimensional stability of the composite samples, while RHA and lime lowered the impact strength of the roofing tiles.

Introduction

The need to improve housing supply in developing countries is great. So also are the needs to manage agro-industrial wastes in a sustainable manner and reduce the use of cement in building construction. Accumulation of unmanaged wastes results in environmental pollution. Recycling of such wastes, particularly agro-industrial wastes, as sustainable building construction materials appears to be viable solution not only to pollution problems but also to the problem of economic design of buildings.

The major types of roofing materials available in Nigeria are corrugated iron and aluminum sheets, slates and asbestos sheets. While corrugated iron sheets are prone to rusting and can be noisy when it is raining, asbestos roofing sheets are relatively expensive and have been outlawed in many countries due to the carcinogenic nature of asbestos fibres. Cement-bonded composites (CBCs) represent an important class of engineered construction materials in which some agro-industrial wastes could be used as partial replacement of cement, while others could serve as fibre reinforcement. Fibrous materials suitable for cement-bonded composite roofing and ceiling tile production in Nigeria include bamboo (*Bambusa vulgaris*), rattan cane, sugar cane bagasse (*Saccharum officinarum*), raffia palm (*Raphia africana*), luffa (*luffa cylindrica*), *Cissus populnea*, and coconut husk (*Cocos nucifera* Linn) among others [1-6].

There are about three million coconut palm trees producing approximately 70 million coconuts annually in Nigeria [7]. The average mature coconut weighs 680 g about 42% of which is made up of the husk [8]. The husk fibres, largely treated as waste, are a candidate material for CBC reinforcement. Potential agro-industrial waste products for partial replacement of cement in the country include welder's used carbide waste (lime) derived from ethyne (C_2H_2) gas, by the action of cold water on calcium carbide and plant ashes that have relatively high silica content and are therefore suitable as a pozzolana, including, RHA. It is generally believed that calcium

By-Products of Palm Trees and Their Applications   Materials Research Forum LLC
Materials Research Proceedings **11** (2019) 255-261   doi: https://doi.org/10.21741/9781644900178-21

carbide residue is rich in calcium hydroxide and behaves like hydrated lime. Hence, is has also been recommended as potential material for partial replacement of cement in concrete works [9].

The aim of this study was to evaluate the effects of $CaCl_2$ addition and partial replacement of cement on selected properties of coconut husk fibre-reinforced composite roofing tiles.

Methodology

Coconut fibres removed from the husk, were separated into individual strands and cut into 25 mm. Rice husk was air-dried for five days, charred and incinerated at 700^0C into white ash. Welder's used carbide waste (lime) obtained from a mechanical workshop was air-dried, pulverized and sieved. The fibre (2 %) was mixed with Portland cement, river sand, water and colouring material (Iron II Oxide), using a pre-determined water -cement ratio (control). For set I of the experimental samples, $CaCl_2$ was added at 2, 3 and 4% levels. Iron II Oxide was added at the rate of 2%. For sets II and III, cement was partially replaced with RHA and lime respectively at 5, 10 and 15%. All percentages were based on the mass of cement. Triplicate samples of 600 (L) x 300 (B) x 6 (T) mm corrugated roofing tiles were produced with each mixture, vibrated for 60 seconds at 50 Hz and cured for 28 days. The samples were tested for moisture content, density, impact energy, water absorption and thickness swelling using standard methods earlier reported [10,11]. Analysis of variance was conducted at 5% level of significance.

Results and Discussion

Density and Moisture Content of the Tiles

Samples of the red-coloured coconut husk fibre-reinforced composite roofing tiles produced are shown in Fig.1. The average density ranged between 1.3 and 1.6 g/cm^3 at an acceptable moisture content range of 2.5 – 5.5% (dry basis) Table 1. However, analysis of variance (Table 2) showed that neither the addition of $CaCl_2$ nor partial replacement of cement with RHA and lime had significant effect on density, though the densities of samples in which cement was partially replaced were generally lower in conformity with the findings from similar previous studies [10,11]. This is an indication that cement could be partially replaced to reduce the weight of the composite roofing tiles.

Fig. 1. Samples of coconut husk fibre-reinforced roofing tiles.

Table 1: Moisture Content and Density of the Composites.

Sample Composition	Average Moisture Content (%)	Average Density (g/cm^3)	Normalized Density[1]
5% RHA	3.8	1.5	0.94
10% RHA	3.4	1.3	0.81
15% RHA	3.7	1.4	0.87
5% Lime	4.1	1.5	0.94
10% Lime	4.3	1.6	1.0
15% Lime	4.0	1.4	0.87
2% $CaCl_2$	4.9	1.6	1.0
3% $CaCl_2$	2.5	1.7	1.1
4% $CaCl_2$	3.0	1.4	0.87
Control	5.5	1.6	1.0

[1]*The average specimen density divided by the average density of control specimen*

Table 2: Analysis of Variance of the Effect of Partial Replacement of Cement on Density.

Source of Variation	SS	df	MS	F	P-value	F crit
Between Groups	0.230427	6	0.038405	2.451591	0.078271	2.847726
Within Groups	0.219312	14	0.015665			
Total	0.449739	20				

Impact Resistance of the Roofing Tiles

The average impact resistance of the roofing tiles ranged between 0.31 and 0.53 J. These values are slightly higher than the range of values (0.19 – 0.22 J reported for banana fibre reinforced roofing tiles produced from similar cementitious admixtures [11]. Also, as in a similar study on rattan-cement composite roofing tiles [2], RHA had more negative effect on the impact resistance than waste carbide lime. However, neither the addition of $CaCl_2$ nor partial replacement of cement had significant effect on the impact resistance of the tiles, suggesting that the two substitution materials could be used without compromising the impact resistance of the composite tiles. A fairly strong positive correlation ($R^2 = 0.51$) was observed between impact strength and density of composite roofing tiles indicating density may be a good factor for determining their impact resistance.

Table 3: Impact Energy Resistance of the Composites.

Sample Composition	Average Impact Energy (J)
5% RHA	0.36
10% RHA	0.32
15% RHA	0.40
5% Lime	0.36

10% Lime	0.40
15% Lime	0.36
2% $CaCl_2$	0.45
3% $CaCl_2$	0.45
4% $CaCl_2$	0.49
Control	0.45

Water Absorption and Thickness Swelling

The average water absorption and corresponding thickness swelling values obtained are presented in Figure 2. The water absorption values ranged from 1.0 to 3.9%, while the thickness swelling values ranged from 0.9 to 2.0 % after 24 hour-immersion in cold water. These values compare favourably with established data on sorption properties of cement-bonded systems reinforced with organic fibres including fibre-reinforced composite roofing tiles produced with partial replacement of cement with lime and $CaCl_2$ [11-14]. While $CaCl_2$ at the three levels of application had no significant effect on the WA, the effect on TS was positive and significant. However, while partial replacement of cement with both RHA and lime led to significant increase in WA, the effect on TS was not significant.

Water absorption is a good indicator of the durability of the bioplastics material as the presence of water can cause cracking (associated with shrinkage and swelling phenomena), and premature biodegradation [15]. The relatively low water absorption values recorded is, therefore, desirable. The correspondingly low thickness swelling is also a good indicator of dimensional stability. As shown in Figures 3, 4 and 5, there were positive correlations between density and water absorption, density and thickness swelling, as well as water absorption and thickness swelling. Similar results were reported for cement-bonded rattan and coconut husk composites [15].

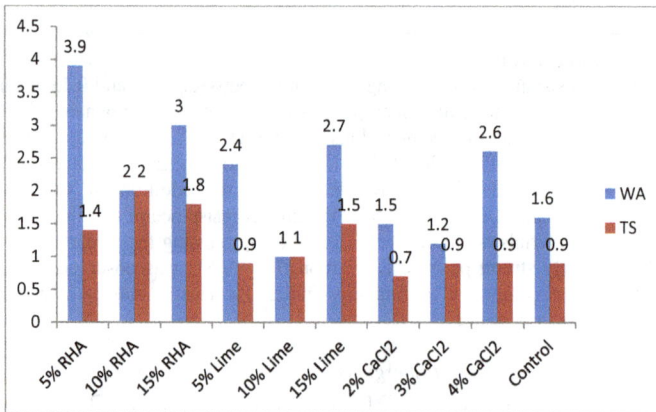

Fig.2. Water absorption by, and thickness swelling of the composite roofing tiles.

By-Products of Palm Trees and Their Applications   Materials Research Forum LLC
Materials Research Proceedings **11** (2019) 255-261   doi: https://doi.org/10.21741/9781644900178-21

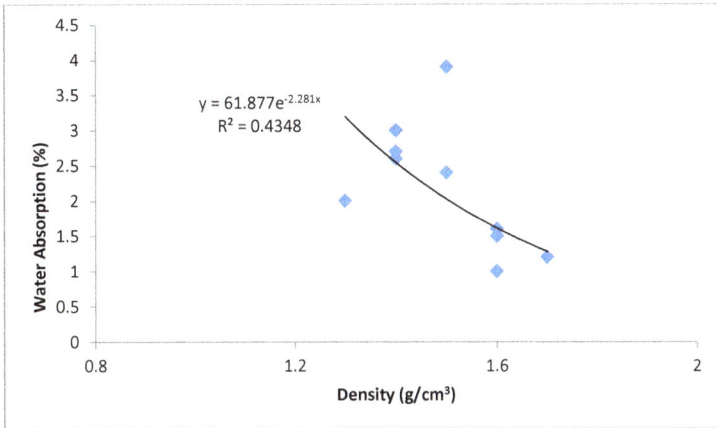

Fig. 3. Correlation between density and water absorption.

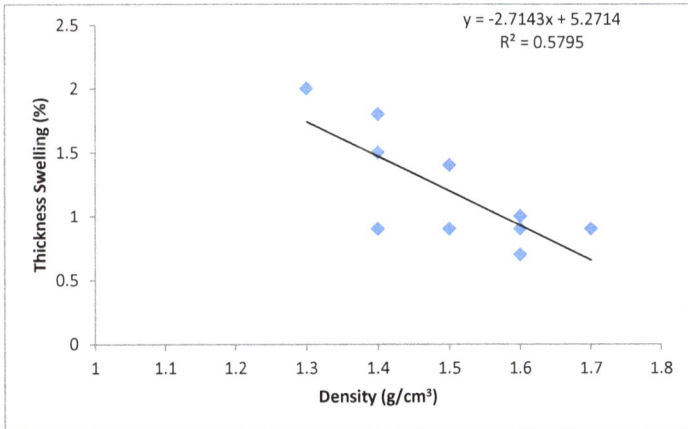

Fig.4. Correlation between density and thickness swelling.

By-Products of Palm Trees and Their Applications
Materials Research Proceedings 11 (2019) 255-261

Materials Research Forum LLC
doi: https://doi.org/10.21741/9781644900178-21

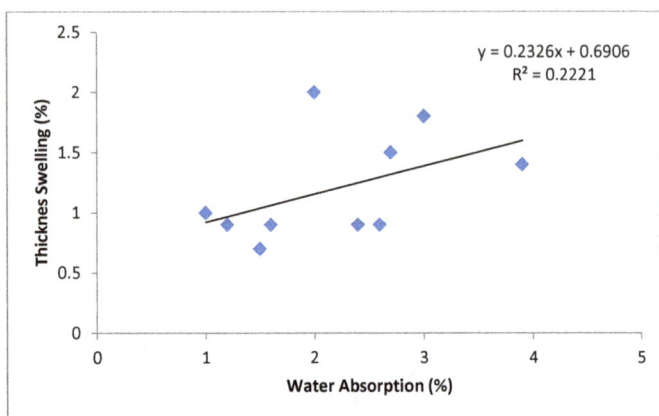

Fig.5. Correlation between Water absorption and thickness swelling.

Conclusion

This study showed that the use of $CaCl_2$ in the production of coconut husk fibre-reinforced roofing tiles would be beneficial in improving dimensional stability. Also, partial replacement of cement with RHA and lime up to 15% could help in reducing the weight (and perhaps cost of production) without having any deleterious effects on the water and impact energy absorption by the roofing tiles.

References

[1] A.O. Olorunnisola, Harnessing the forester's harvest for sustainable development, The Publishing House, University of Ibadan, Nigeria, 2013.

[2] A.O. Olorunnisola, A. Ogundipe, Impact and water resistance of rattan composites produced with rice husk ash and welder's carbide waste as partial replacement for portland cement, in F. Caldeira (ed.), Towards Forest Products and Processes with Lower Environmental Impact, University of Fernando Pessoa, Porto, Portugal (2014) 199-218.

[3] T.E. Omoniyi, Development and evaluation of roofing sheets from bagasse-cement composite, PhD Thesis, Department of Agricultural & Environmental Engineering, University of Ibadan, Nigeria, 2009.

[4] R.S. Odera, O.D. Onukwuli, E.C. Osoka, Tensile and compressive strength characteristics of raffia palm fibre-cement composites, Journal of Emerging Trends in Engineering and Applied Sciences 2-2 (2011) 231-234.

[5] M.O..Lazeez, Production and evaluation of cement-bonded composite using okra fibre as a potential reinforcement material, Project report, Department of Agricultural & Environmental Engineering, University of Ibadan, Nigeria, 2014

[6] K. Amoo, O.O. Adefisan, A.O. Olorunnisola, Development and evaluation of cement-bonded composite tiles reinforced with Cissus populnea fibres, International Journal of Composite Materials 64 (2016) 133-139.

[7] A.A. Badmus, Coconut processing technologies in Nigeria-A critical appraisal, A seminar paper, Department of Agricultural & Environmental Engineering, University of Ibadan, Nigeria, 2009.

[8] G.A. Badmus, N.A. Adeyemi, O.K. Owolarafe, Development of an improved manual dehusking lever, Nigerian Journal of Palms and Seeds, 16(2007) 36-47.

[9] WA Al-Khaja, Potential use of carbide lime waste as an alternative material to conventional hydrated lime of cement-lime mortars, Engineering Journal of Qatar University, 5(1992) 57-67.

[10] S.A. Nta, A.O. Olorunnisola, Experimental production and evaluation of cement bonded composite pipes for water conveyance, International Journal of Composite Materials 6-1 (2016) 9-14.

[11] A.O. Olorunnisola, F. Ope-Ogunseitan, Effects of lime and CaCl2 on impact strength and dimensional stability of banana fibre- reinforced composite roofing tiles, Proc. of the 15th Inorganic-Bonded Fiber Composites Conference, Fuzhou Empark Exhibition Grand Hotel, Fuzhou, China, (2016) 76- 84.

[12] A.O. Oyagade, Thickness swelling components and water absorption of cement-bonded particleboards made from gmelina wood, bagasse and coconut husk, Nigerian Journal of Forestry 30-1 (2000) 10-14.

[13] B. Ajayi, Preliminary investigation of cement-bonded particleboard from maize stalk residues, The Nigerian Journal of Forestry, 32-1 (2002) 33-37.

[14] A.O. Olorunnisola, O.O. Adefisan, Trial Production and Testing of Cement-bonded Particleboard from Rattan Furniture Waste, Wood & Fiber Science 3-1 (2002) 116-124.

[15] A.O. Olorunnisola, Compressive strength and water resistance behaviour of cement composites from rattan and coconut husk, Journal of Tropical Forest Resources 20-2 (2004) 1-13.

By-Products of Palm Trees and Their Applications
Materials Research Proceedings 11 (2019) 262-274

Materials Research Forum LLC
doi: https://doi.org/10.21741/9781644900178-22

Textile Palm Fibers from Amazon Biome

Lais Gonçalvez Andrade Pennas[1,a], Ivete Maria Cattani[1,b], Barbara Leonardi[1,c], Abdel-Fattah M. Seyam[2,d], Mohamad Midani[2,3,e], Amanda Sousa Monteiro[1,f], Julia Baruque-Ramos[1,g*]

[1]University of Sao Paulo, School of Arts, Sciences and Humanities, Av. Arlindo Bettio, 1000; 03828000, Sao Paulo, SP, Brazil

[2]North Carolina State University, College of Textiles, 1020 Main Campus Drive, Raleigh, NC 27606, USA

[3]German University in Cairo, College of Engineering and Materials Science, Al Tagamoa Al Khames, Cairo, Egypt

[a] laiis_s@hotmail.com, [b] ivetecattani@me.com, [c] leonardi.ba@gmail.com, [d] aseyam@ncsu.edu, [e] msmidani@ncsu.edu, [f] amandasousamont@gmail.com, [g] jbaruque@usp.br

Keywords: textile; palm fibers; Amazonia; Brazil; tucum; buriti; tururi; composite

Abstract. There are several species of Amazon palm trees from which can be obtained: food and oils (fruits and seeds), medicinal products, construction material (logs and leaves), handicraft, textiles, etc. Taking in account textile fibers, three palm origins stand out: tucum (*Astrocaryum chambira* Burret), buriti (*Mauritia flexuosa* Mart.) and tururi (*Manicaria saccifera* Gaertn.). Tucum fibers, obtained from grown leaves, are used in the manufacture of fabrics, handicrafts, nets, yarns and fishing nets. Buriti presents multiple uses, especially for handicraft products. A soft fiber ("linen") and another harder and rougher ("draff") are removed from the young leaves of the buriti palm, both being used. Tururi is the sac that wraps the fruits of the Ubuçu palm tree. The material is constantly used by the Amazonian riverside population and by artisans for handicrafts, fashion items and other products for tourism. In a joint project of the North Carolina State University (USA) and University of São Paulo (Brazil), multilayer composite materials were developed and characterized in 3D structure with quite promising results in terms of resistance and aesthetic finish similar to wood. Thus, the traditional and innovative uses of native vegetable fibers are ways of valuing the regional product and preserving their respective ecosystems.

Introduction

The Amazon biome comprises an area of 410 million hectares and is formed by three types of forests: dry land, wet land and flooded area. It encompasses extensive areas of "cerrados" (kind of savannas) and meadows. The Amazon biome develops around the Amazon basin and is present in eight countries of South America [1]. There are several species of palm trees from the Amazon biome, from which can be obtained: food or oils (fruits and seeds), biodiesel, medicinal and cosmetic uses, construction material (logs and leaves), handicraft material, including fibers for textile purposes, etc. Some examples are [2]: Açai (*Euterpe precatoria*); Cocao (*Attalea tessmanii*); Inaja (*Attalea maripa*); Jaci (*Attalea butyraceae*); Jarina (*Phytelephas macrocarpa*); Murmuru (*Astrocaryum murumuru*); Paxiubao (*Iriartea deltoidea*); Paxiubinha (*Socratea exorrhiza*); and Pataua (*Oenocarpus bataua*). The incentive for the employment of native vegetable fibers as an alternative textile material can increase local productivity and improving the income of the populations. Another point is that there is enormous creative potential. Aiming

By-Products of Palm Trees and Their Applications   Materials Research Forum LLC
Materials Research Proceedings 11 (2019) 262-274   doi: https://doi.org/10.21741/9781644900178-22

at technology, there is growing international interest in the use of these vegetable fibers, especially as non-conventional materials for the manufacture of composites instead of those made with wood or synthetic materials [3]. Taking in account the obtainment of textile fibers, three palm origins stand out: tucum buriti and tururi. A briefing enrolling the obtention of fibers, processing and manufacture of final products in shown in Figure 1.

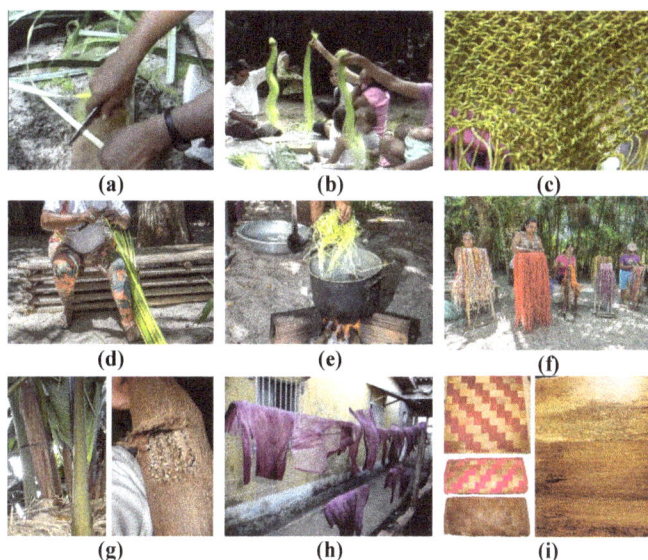

Figure 1. Tucum: (a) Obtention from grown palm leaves [4]; (b) Processing of obtained fibers [4]; (c) Macrame fabric [4] [5]. Buriti: (d) Obtention of "linen" and "draff" fibers from the young leaves [1] [6] [7] [8]; (e) Dyeing of fibers [1] [6] [7] [8]; (f) Working in manual looms [1] [6] [7] [8]. Tururi: (g) Sac covering the fruits in the palm and after collection [9] [10]; (h) Drying of tururi mats after dyeing [9] [10]; (g) Mat, bags and composite made from tururi [9] [10] [11].

In the present study the main physical-chemical characteristics of these palm fibrous material were compared and for tururi also the characteristics of composite structures were presented.

Material and methods
The fibers were taken respectively from: (i) Grown leaves of tucum palm trees (*Astrocaryum chambira* Burret), in the Community Ecological Village of Jurua, municipality of Ipixuna, Amazonas State, Brazil, GPS 07°03′04″S and 71°41′43″W; (ii) Yong leaves of buriti palm trees (*Mauritia flexuosa* Mart.), located in Marcelino Village, situated on the bank of "Preguiça" river, Barreirinhas city, Maranhao State, Brazil, GPS 02°45′18,8101″S and 42°49′04,2782″W; (iii) Sacs of tururi (*Manicaria saccifera* Gaertn), in the cities of Sao Sebastiao da Boa Vista and Muana, Para State, Brazil, central GPS positions respectively S -1°23 '53.4156"W - 49°38'14.9928" and S-1°20'40.3506 "W-49°17'45.3948". All locations are in comprised in Brazilian Amazon Forest biome and in all cases, the prospection radius was 5,000 m. It should

By-Products of Palm Trees and Their Applications   Materials Research Forum LLC
Materials Research Proceedings **11** (2019) 262-274   doi: https://doi.org/10.21741/9781644900178-22

be noted that obtaining these specimens does not require authorization from IBAMA (Brazilian Institute of Environment) or any other federal or state environmental agency, since the material is usually collected and marketed and its purchase and possession has no legal restriction in any of the Brazilian states. Other fibers employed in the present study, originated from the leaf of following plants, are: i) curaua (*Ananas erectifolius*) provided by the Federal University of Amazonas, produced in that Brazilian state; and ii) sisal (*Agave sisalana*) purchased in Sao Paulo city, while originating from Bahia state (both Brazilian regions).

The assays were performed at 20°C and 65% relative humidity (ABNT NBR ISO 139:2005) [12]. Cross microscopy structures were determined according to ABNT NBR 13538-1995 [13]. They were carried out on cross-sections of resin encapsulated fibers cut in rotational semi-automated microtome (Leica, RM 2245 model, Germany). The materials were analyzed in biological microscope (Leica, BME model, Germany) coupled to camera video digital imaging (Sony Color VideoCamera ESWAVEHAD, 55C-DC93-P model, China). All the obtained images were captured and processed by Video Analyzer 2000 system (Mesdan, Italy).

Tensile properties of the fibers (rupture load, elongation, tenacity and Young's modulus), from buriti and tucum fibers samples (obtained respectively from young and grown leaves) and fibers withdrawn from tururi sacs, were determined according to ASTM D 3 822-2001 [16] employing tester machine Instron (model 5569, Norwood, USA). Formerly, in order to determine tenacity (strength value shared by count number) fiber fineness (linear density or count number) was calculated in terms of TEX, defined as the weight in grams per 1,000 m of the fiber, by weighing a known length of the fiber. A gauge length of 25 mm, automatic pre-tension and crosshead speed of 50 mm/min and a cell of 1000 N were employed. The results are an average from at least twenty samples. The total length of the sample was approximately 100 mm, sufficient to allow the distance between the jaws of 25 mm.

The tenacity for fibers was determined by the presented in Equation 1 [14] [15]:

$$\gamma = \frac{F}{T_m} \qquad \begin{array}{l} \gamma = \text{tenacity (cN/tex);} \\ F = \text{breaking load (cN);} \\ T_m = \text{count number (tex).} \end{array} \qquad \text{(Eq.1)}$$

The Young's modulus (or textile initial modulus or module) of a fiber is determined by the slope of the tenacity-elongation curve in its initial linear part as presented in Equation 2 [15]:

$$\text{Young's modulus} = \frac{\gamma_1}{\varepsilon_1} \qquad \begin{array}{l} \gamma_1 = \text{Tenacity in the initial part of the tenacity-elongation curve (cN/tex)} \\ \varepsilon_1 = \text{elongation in the initial part of the tenacity-elongation curve (\%)} \end{array} \qquad \text{(Eq. 2)}$$

In addition, tururi fibrous material from sacs was tested in order to determine the values of tensile strength (testing the fibers withdrawn from these sacs and testing strips from the sacs that forms this material - *in natura* and after discoloration) and weight. Tensile properties of the surface fibrous material (rupture load, elongation, strength and Young's modulus), from samples obtained from cut strips of the fibrous material from the sacs, were determined according to ABNT NBR 13041:1993 [17] employing tester machine Instron (model 5569, Norwood, USA), employing using a crosshead speed of 100 mm/min and a cell of 1000 N. The results are an average from at least twenty samples. The samples were 20 mm wide and 300 mm length. Jaws

By-Products of Palm Trees and Their Applications   Materials Research Forum LLC
Materials Research Proceedings **11** (2019) 262-274   doi: https://doi.org/10.21741/9781644900178-22

of rubberized grips with dimensions of 3.8 x 5 cm were employed. The distance between the grips was 200 mm. The thickness of the samples was previously determined employing portable analogical thickness gauge (model 188F, Mesdan, Italy).

In order to define the weight of the surface of fibrous material, ABNT NBR 12984:2000 [18] standard was employed. A total of 20 samples of size 5.0 x 5.0 cm were tested. The samples were weighed on analytical balance (Sartorius, ED124S model, Germany). The weight calculation was calculated as g/m^2.

Regain determinations were performed according the method adapted from ISO/TR 6741-4: 1987 [19]. Percent Moisture Regain (or "Regain") is defined as the weight of water calculated as a percentage of dry weight. After acclimatization at 20°C and 65% relative humidity [13] [14] [15] the samples were weighed on analytical balance (Sartorius, ED124S model, Germany). The drying was performed in an oven with forced air circulation (Binder FD Model 115, Germany) at 70°C for 24 h or more until constant weight and the sample was again weighed. Twenty repetitions of each group were analyzed.

For tucum fiber, DSC, TGA and XRD tests were performed. DSC (Digital Scanning Calorimetry) and TGA (Thermogravimetry Analysis) tests were carried out in Thermogravimetric Analyzer Mettler Toledo (model TGA/DSC 2, Netherlands), temperature from 30 to 1000°C, 10°C/min, 50mL/min nitrogen atmosphere and 70µL alumina crucible. For lignocellulosic materials, the events (peaks characterized by inflection points) in the DTG (Derivative Thermogravimetry) curves can be associated to processes that occur to the different constituents of the analyzed material. Thus, in many cases, the approximate composition of the analyzed lignocellulosic material can be estimated by comparing the DTG and TGA curves and compare those results with those obtained through chemical determination [20].

In addition, the XRD spectra were obtained at room temperature (25°C) with a diffractometer Rigaku Miniflex 300 (Japan), Cu X-Ray tube, Kβ filter, 30 kV and 15 mA. Dispersion ranged from 4° to 80°, 2°/min continuous scan, 0.020° sampling width and 2θ/θ scan axis.

The tucum, buriti, tururi, sisal and curauá fibrous materials were analyzed by FTIR in equipment Thermo (model Avatar 370 FT-IR) employing cell of ATR / Germanium (Ge) (Nicolet, USA). The interval was from 4,000 to 700 cm^{-1}, performing 32 scans with 2 cm^{-1} resolution. The data acquisition was performed by OMNIC software, version 4.1, 2011[21].

Results and discussion
Cross-sectional microscopies
The cross-sectional microscopies of fibers are presented in Figure 2. They match with other ones of recognized textile employability.

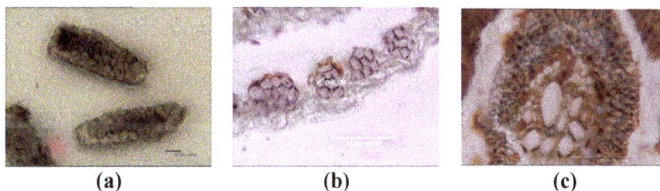

(a)   (b)   (c)

Figure 2. Cross microscopies: (a) Tucum: 1.280X magnification and 10 µm scale [original data]; (b) Buriti "linen": 1.280X magnification and 50 µm scale [1] [7] [22]; (c) Tururi: 640X magnification and 100 µm scale [3] [9].

By-Products of Palm Trees and Their Applications   Materials Research Forum LLC
Materials Research Proceedings **11** (2019) 262-274   doi: https://doi.org/10.21741/9781644900178-22

Despite of the similarities in the cross sections of studied species here and that ones from fibers of recognized textile employability, it cannot conclude, only through microscopic examination of their cross sections, the possible type of application for certain fiber. For this purpose, there is the necessity of a combined analysis of results from other physical and chemical tests. However, examination of cross sections by optical microscopy also is useful to evaluate the integrity of fiber cellular structure and the adequacy of procedures for processing of the fibers. The damage to cellular structures is visible such as deformation of cell shape [23]. The tucum cell diameter average value (from 10 determinations) is 5.5±1.5 µm (CV=27%). For buriti "linen" is 8.5 µm and for buriti "draff", 7.2 µm. The tururi cell diameter average value (from 10 determinations) is 8.7±5.1 µm (CV=58.6%). This value is compatible with the values of species of recognized textile employability. According to Reddy and Yang[25], the unit cell size ranges from 12.0 to 25.0 µm for cotton, 5.0 to 76.0 µm for flax and from 15.0 to 25.0 µm for jute.

Tensile and regain results

The values for tensile and regain results on fibers of buriti, tucum and withdrawn from tururi sacs are presented in Table 1.

Table 1 - Tensile and regain results on fibers of buriti, tucum and withdrawn from tururi sacs. Values expressed by average, standard coefficient and variation coefficient.

Fiber		Count Number* (tex)	Rupture Load (N)	Elongation (%)	Tenacity (cN/tex)	Young's Modulus (N/tex)	Regain (%)	References
Tucum		320±127 (39.5%)	119±49 (41.0%)	**6.6**±0.4 (5.8%)	37.4±5.6 (14.9%)	**8.3**±1.0 (12.5%)	**10.0**±0.3 (3.4%)	[original data]
Buriti**	linen	**223**±77.7 (34.8%)	64.1±27.4 (43.6%)	**8.3**±0.5 (6.8%)	28.4±5.5 (19.6%)	6.1±0.8 (13.1%)	**8.5**±0.3 (2.6%)	[1] [22]
	draff	**228**±134 (47.7%)	52.2±35.1 (67.2%)	**5.0**±0.8 (15.5%)	18.0±6.3 (34.9%)	6.0±1.6 (25.9%)	**9.0**±0.8 (8.4%)	
Tururi		**98.4**±15.2 (15.5%)	17.7±4.2 (23.5%)	**10.5**±2 (20%)	18.0±3.2 (18%)	3.4±0.5 (14.4%)	**12.0**±0.5 (4.3%)	[3] [9]

*For buriti and tururi the expressed values represent the count number of single fibers. For tucum the expressed value represents the count number of fiber bundles employed in tests, since the single fibers are very thin (near 2.4 tex and 90±12 cm natural fiber length). **From buriti young leave two different fibers are obtained: one more flexible (popularly called "buriti linen") employed in woven or knitted fabrics for fine handcrafts, and another one more rustic ("buriti draff") employed for confection of basketry, sets of placemats, etc.

Table 2 – Tenacity, elongation and Young's modulus values for species of recognized textile employability. Values adapted from the indicated references.

Fiber	Elongation (%)	Tenacity (cN/tex)	Young's modulus (N/tex)	Regain (%)	References
Sisal	2 – 3	35.3 – 44.1	12.4	11	[15] [24] [26] [28]
Curaua	4.5 – 6	135 - 326	30-80	9	[24] [26] [27]
Cotton	3 – 7	26.5 – 43.3	5.3–6.2	8.5	[15] [24] [25]
Hemp	1.8	51.2 – 60.0	19.4	8-12	[15] [24]
Jute	1.7 – 2.0	26.5 – 51.2	17.9	13.8	[15] [24] [25] [26]

A comparison between the determined values (Table 1) and the properties of other vegetal fibers (Table 2) was performed. The tenacity of tururi fiber is lower in relation to other analyzed fibers. It is comparable to the lower limit of this parameter for cotton and jute. The tenacity values for the other analyzed fibers are comparable to sisal (leaf fiber), cotton (seed fiber) and hemp and jute (stem fibers). All the values are inferior to the curauá (leaf fiber). However, it is remarkable that despite of curauá is a leaf fiber, their general employment is for manufacture of composites instead of textile purposes.

The obtained values of regain (Table 1) are consistent with other ones of recognized textile employability lignocellulosic fibers (Table 2).

Tensile tests and weight on tururi fibrous material strips

The test was performed with 20 samples (cut strips of the fibrous material from the sacs of tururi) with dimensions of 20 x 200 mm and an average thickness of 0.71 mm. The results of the tensile test are shown in first line of Table 3.

Table 3 – Tensile tests on cut strips of the tururi fibrous material. Values expressed by average, standard coefficient and variation coefficient.

Rupture Load (N)	Elongation (%)	Strength (MPa)	Young's modulus (MPa)	Weight (g/m^2)	References
213± 93 (43%)	5.9±1.0 (17%)	17.6±7.8 (44%)	552±288 (52%)	182±18 (10%)	[3] [9]
391	-	-	1,800-2,400	366-583	[29]
432	9.35	-	-	204.7	[30]
558.3	-	12.27	-	246.37	[31]

The values presented in Table 3 are similar (within the same order of magnitude). In the same way, for the weight values, but it is worthy of mention that for many tururi applications, it is stretched and the weight decreases in the proportion of its nonwoven structure opening.

There was no significant statistical variation between the tensile characteristics of the fibrous material in the natural condition (brown color) and after discoloration with hydrogen peroxide and solar illumination [3] [9].

By-Products of Palm Trees and Their Applications   Materials Research Forum LLC
Materials Research Proceedings **11** (2019) 262-274   doi: https://doi.org/10.21741/9781644900178-22

Evaluation of cellulose, hemicellulose and lignin contents and crystallinity index

The values of cellulose, hemicellulose and lignin contents and crystallinity index for all fibers, excepted tucum fiber, were obtained in literature as presented in Table 4.

Table 4. Estimation of the concentrations of holocellulose and lignin in the tucum fiber through the analysis of the TGA and DTG curves. Other values were obtained from literature.

Fiber	Holocellulose (wt%)		Lignin (wt%)	Pectin (wt%)	Waxes	Extractives (wt%)	Crystallinity Index* (%)	References
	Cellulose (wt%)	Hemicellulose (wt%)						
Tucum	68.4		21.7	-	-	-	80.25	[original data]
Buriti	65–71		21–27	-	-	5.4–6.0	71.2	[32]
Tururi	74.1	12	31.1	-	-	0.5	60.6	[29] [30]
Sisal	65–67	12	9.9	2-10	2-10	0.3-2	57.3	[32]
Curaua	71-74	9.9-21	7.5-11	-	0.79-0.9	2.5-2.8	43.5	[32]

*Calculated according empirical method proposed by Segal et al. in 1959 [33]

According the results shown in Table 4 the obtained values are close. However, Segal's method [33] must be used with care, since it cannot able to reflect the real degree of biomass crystallinity, but instead provide a parameter for comparison. Even little variations in their compositions or cellulose crystallinity are expressive in order to determine differences in mechanical and thermal properties, being that higher tensile strength and higher thermal stability are obtained for fibers that contain more crystalline cellulose [32].

FTIR

The Fourier transform infrared spectroscopy (FTIR) from 4,000 to 500 cm^{-1} in equipment Thermo (model Avatar 370 FT-IR) employing cell of ATR / Germanium (Ge) (Nicolet, USA) comparing the spectra of three palm fibers from: (i) grown leaves of tucum (*Astrocaryum chambira* Burret); (ii) young leaves of buriti (*Mauritia flexuosa* Mart.); and (iii) sacs of tururi fibrous material (*Manicaria saccifera* Gaertn.); and two leaf fibers from: (i) curaua (*Ananas erectifolius*) and (ii) sisal (*Agave sisalana*) - is shown in Figure 3.

Figure 3. FTIR from 4,000 to 500 cm⁻¹ in equipment Thermo (model Avatar 370 FT-IR) employing cell of ATR / Germanium (Ge) (Nicolet, USA) - Transmittance from 96.4 to 99.8%. From top to down: tucum (dark blue line), buriti (orange line), tururi (red line), sisal (light blue line) and curauá (red line). The overlap of FTIR spectra of 'linen' and 'draff' of buriti indicate a strong similarity between their chemical characteristics.

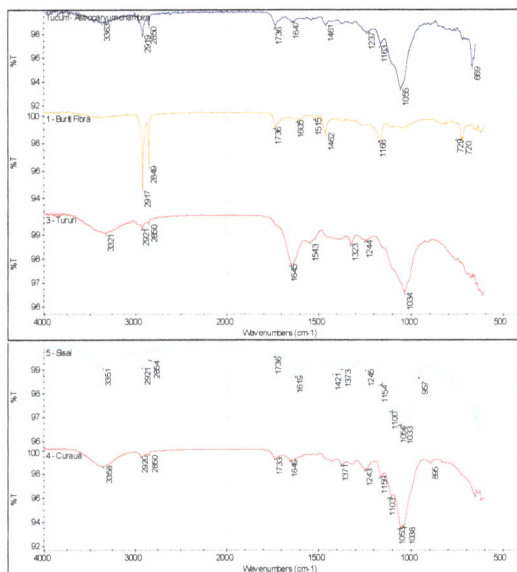

Taking in account the informations available by Stuart [34], an interpretation for the assignments of each fiber correspondent peak is presented in Table 5.

Analyzing the findings in Figure 3 and Table 5, it is possible to notice the great similarity in two wavenumber regions, near 3000 cm⁻¹ and near the 3500 cm⁻¹, indicating the presence of CH and OH respectively [35]. These are the major bands observed in the graphic, which was expected, since all analyzed materials have vegetal origin. It is still possible to compare the bands obtained around 1000 cm⁻¹, in the region so called "fingerprint" of FTIR spectrum [34].

The characteristic absorption bands (cm⁻¹) for cotton are: 3450-3250; 2900; 1630; 1430; 1370; 1100-970; 550. These bands have similarity with the fibers analyzed in present study. Near the 1750 cm⁻¹ band there is an area which indicates the presence of carbonyl group (C=O). The angular deformation between 3339 and 3564 cm⁻¹ indicates the presence of hydroxyl groups which, in the cellulose chain are able to interact with each other, forming hydrogen bonds of two types: intramolecular (between the hydroxyl groups of the same chain), which are responsible for the stiffness of the chains, and intermolecular (between the hydroxyl groups of adjacent chains) are responsible for the formation of the supramolecular structure [36].

Thus, by FTIR analysis, considering the similarities and differences between the spectra (Figure 3 and Table 5), denoting different compositions and/or molecular structures and the presence of cellulose, hemicellulose and lignin in the fibers analyzed in present study.

Table 5. Infrared bands determined for tucum (Astrocaryum chambira), buriti (Mauritia flexuosa), tururi (Manicaria saccifera), sisal (Agave sisalana) and curaua (Ananas erectifolius) and their respective assignments.

Wavenumber (cm^{-1})					Assignment[***]
Tucum[*]	Buriti[*]	Tururi[**]	Sisal[**]	Curaua[**]	
3363	-	-	3351	3358	3700-3200 O-H stretching
-	-	3321	-	-	3700-3200 O-H stretching
2919	2917	2921	2921	2920	---
2850	2849	2850	2854	2850	2900-2700 C-H aldehyde stretching
1736	1736	-	1736	1733	1740-1720 C=O aliphatic aldehyde stretching
1647	-	1645	-	1649	1680-1600 C=C stretching
-	1605	-	-	-	1680-1600 C=C stretching
-	-	-	1619	-	1680-1600 C=C stretching
-	1515	-	-	-	---
,-	-	1543	-	-	---
1461	1462	-	-	-	---
-	-	-	1421	-	---
-	-	-	1373	1371	---
-	-	1323	-	-	---
1237	-	1244	1245	1243	1300-1100 C-O stretching
1163	1168	-	1154	1159	1300-1100 C-O stretching
-	-	-	1100	1103	1100 C-O-C stretching
1055	-	-	1056	1053	---
-	-	1034	1033	1038	---
-	-	-	957	-	1000-600 =C-H out-of-plane bending
-	-	-	-	895	1000-600 =C-H out-of-plane bending
-	729	-	-	-	1000-600 =C-H out-of-plane bending
-	720	-	-	-	1003-600 =C-H out-of-plane bending
669	-	-	-	-	1004-600 =C-H out-of-plane bending

*original data; **[3] [9]; ***Assignments according the interpretation of informations by Stuart [34].

Tururi composite development

In a joint project of the North Carolina State University (USA) and University of Sao Paulo (Brazil), multilayer composite materials were developed and characterized in 3D structure with quite promising results in terms of resistance and aesthetic finish similar to wood. The goal of this research was to develop and characterize multilayer 3D green composites from Tururi fibrous material and identify applications based on their performance. A total of 12 composite samples were fabricated using Vacuum Assisted Resin Transfer Molding Technique (VARTM) to study the effect of the structural parameters, namely, number of Tururi fibrous layers, fiber orientation, and fiber volume fraction on the tensile and impact behavior of the final composites. It was found that increasing the % stretch of the Tururi sac, and using an angle-ply stacking arrangement significantly reduced the anisotropy of the produced composite, and resulted in a quasiisotropic material. In the 0°/90° arrangement, the tensile breaking load for 0% stretch was the same in the sac and cross directions, whereas in the 100% stretch it was random. Moreover, the breaking load of 100% stretch was generally higher than the 0% stretch. Additionally, when the breaking load was normalized by the preform areal density, it was found that composites with higher number of layers have lower normalized breaking load. Finally, increasing the stretch improved the resin penetration and increased the normalized breaking load. In the 0°/45°/-45°/90° arrangement, the tensile breaking load for 100% stretch was the same in the sac and cross directions, whereas in the 0% stretch it was different. Moreover, the breaking load of 100% stretch was generally higher than the 0% stretch. Additionally, when the breaking load was normalized by the preform areal density, it was found that the number of layers and % stretch have negligible effect on the normalized breaking load. It was found that a proper stacking sequence, can produce composites from tururi fibers with quasi-isotropic tensile behavior, and with the proper combination of number of layers, and stretch %, the tensile properties of the produced composite can be optimized [11].

Furthermore, the impact properties of these composites were characterized in light of structural variables, namely, number of layers, fiber orientation, and % stretch. It was found that the number of layers had the most significant effect on the impact resistance of the tururi composites, followed by the % stretch, whereas the preform orientation had a slight significant effect on the normalized impact energy; moreover, it significantly affected the failure mechanism. Thus, in order to design composites from tururi fibers with high impact resistance, the first factor to be considered is the number of layers which should not exceed a certain threshold after which the resin penetration becomes impaired. The second factor is the % stretch, which should be minimized to maintain higher fiber volume fraction. Finally, the fiber orientation should be random to improve the impact resistance, failure mechanism, and the damage tolerance of the structure [37].

The results are promising and indicate that tururi fibers are good candidate for the reinforcement of polymer composites. Natural fiber composites from tururi fibers exhibit a natural wood grain appearance; therefore, it can be used as a wood alternative in applications like floor laminates, counter tops, and indoor and outdoor furniture [11] [37].

Conclusion
The analyzed palm fibers (tucum, buriti and tururi) have employment potential in different kind of products, handcrafts or composites, which could generate articles such as utensils, furniture, flooring or construction. On the other hand, it is very important to preserve the palm species present in Amazon Forest biome, from which they are extracted in a sustainable way from local communities. Thus, more studies are necessary in order to know about the availability of this material thinking in an industrial scale. On the other hand, the work of local communities and craft cooperatives, which employ this material, could be stimulated respecting social and cultural aspects. The employment of natural fibers, in a sustainable context, bring implications of great importance to society, such us the environmental care, incoming generation and the possibility of qualifying the products produced by these communities adding technical and design attributes.

References

[1] I. M. Cattani, Buriti fiber (Mauritia flexuosa Mart.): registration in local community (Barreirinhas-MA, Brazil), physicochemical characterization and study of impregnation with resins, MSc thesis, University of Sao Paulo (Brazil), 2016.
https://doi.org/10.11606/d.100.2016.tde-29092016-100227

[2] V. Froes et al., Especies botanicas do artesanato, in: V. Froes (Ed.), Linha do Tucum: Artesanato da Amazonia, Instituto de Estudos da Cultura Amazonica, Rio de Janeiro, 2010, pp. 72-105. https://doi.org/10.18542/amazonica.v6i2.1875

[3] A. S. Monteiro et al., Tururi palm fibrous material (Manicaria saccifera Gaertn.) characterization. Green Materials 3-4 (2016) 120-131. https://doi.org/10.1680/jgrma.15.00024

[4] N. Nunes, Documentary "Linha do Tucum: a linha da lealdade". Direction Noilton Nunes. Production Imagine Filmes. 2009. 50 min. <http://youtu.be/RpbJH8lgzJ0>

[5] L. A. G. Pennas, J. Baruque-Ramos, Tucum fiber: reflections about Amazonian biodiversity, traditional knowledge and sustainable fashion, in: 4th International Fashion and Design Congress (CIMODE 2018), CRC Press/Balkema, Boca Raton, 2018, in press.

[6] M. A. Goulart, Buriti Photography Catalog. State of Maranhao (Brazil), 2014.

[7] I. M. Cattani, J. Baruque-Ramos, Brazilian Buriti Palm Fiber (Mauritia flexuosa Mart.), in: R. Fangueiro, S. Rana (Eds.), Natural Fibres: Advances in Science and Technology Towards Industrial Applications. Springer Netherlands, Heidelberg, 2016, pp. 89-98. https://doi.org/10.1007/978-94-017-7515-1_7

[8] I. M. Cattani, Video "Fibra de Buriti: da folha ao produto - Artesanato e Design". 2016. 3:57 min. <https://www.youtube.com/watch?v=ZAIgaxmmzvA>

[9] A. S. Monteiro, Tururi (Manicaria saccifera Gaertn.): textile characterization, processes and handicraft techniques in Amazon local community (PA - Brazil), MSc thesis, University of Sao Paulo (Brazil), 2016. https://doi.org/10.11606/d.100.2016.tde-04082016-144047

[10] A. S. Monteiro, J. Baruque-Ramos, Amazonian Tururi Palm Fiber Material (Manicaria saccifera Gaertn.) in: R. Fangueiro, S. Rana (Eds.), Natural Fibres: Advances in Science and Technology Towards Industrial Applications. Springer Netherlands, Heidelberg, 2016, pp. 127-137. https://doi.org/10.1007/978-94-017-7515-1_10

[11] A. F. Seyam et al., Effect of structural parameters on the tensile properties of multilayer 3D composites from Tururi palm tree (Manicaria saccifera Gaertn) fibrous material. Composites Part B: Engineering 111 (2017) 17-26. https://doi.org/10.1016/j.compositesb.2016.11.040

[12] ABNT ISO 139:2005: Texteis – Atmosferas normais de condicionamento de ensaios ("Textiles - Standard atmospheres for conditioning tests").

[13] ABNT NBR 13 538-1995: Material textil - Analise qualitativa ("Textile material - Qualitative analysis").

[14] B.P. Saville, Physical testing of textiles, The Textile Institute Woodhead Publishing Ltd., Cambridge, 2007.

[15] E. R. Kaswell, Wellington Sears Handbook of Industrial Textiles, Massachusetts Institute of Technology (MIT) and Wellington Sears Company, Cambridge, 1963. https://doi.org/10.1201/9780203733905

[16] ASTM D 3 822-2001: Standard test method for tensile properties of single textile fibers.

[17] ABNT NBR 13041:1993: Naotecido - Determinaçao da resistencia a traçao e alongamento - Metodo de ensaio ("Nonwoven - Determination of tensile strength and elongation" - Test method").

[18] ABNT NBR 12984:2000: Naotecido - Determinaçao da massa por unidade de area - Metodo de ensaio ("Nonwoven - Determination of mass per area unit - Test method").

[19] ISO/TR 6741-4 -1987: Textiles - Fibres and yarns - Determination of commercial mass of consignments - Part 4: Values used for the commercial allowances and the commercial moisture regains. https://doi.org/10.3403/bsiso6741

[20] J. Bouchard et al., Quantification of residual polymeric families present in thermo-mechanical and chemically pretreated lignocellulosics via thermal analysis. Biomass 9(3) (1986) 161–171.

[21] Nicolet FT-IR User's Guide. https://instrumentalanalysis.community.uaf.edu/files/2013/01/FT-IR_manual.pdf

[22] I. M. Cattani, J. Baruque-Ramos, Buriti palm fiber (Mauritia flexuosa Mart.): characterization and studies for its application in design products, Key Engineering Materials 668 (2016) 63-74. https://doi.org/10.4028/www.scientific.net/kem.668.63

[23] P. Ganan et al., I. Biological natural retting for determining the hierarchical structuration of banana fibers. Macromolecular Science 4 (2004) 978-983. https://doi.org/10.1002/mabi.200400041

[24] R. S. Blackburn, Biodegradable and sustainable fibres. CRC Press and Woodhead Publishing Ltd., Cambridge, 2005.

[25] N. Reddy, Y. Yang, Properties and potential applications of natural cellulose fibers from cornhusks. Green Chemistry 7 (2005) 190-195. https://doi.org/10.1039/b415102j

[26] K. G. Satyanarayana et al., Studies on lignocellulosic fibers of Brazil. Part I: Source, production, morphology, properties and applications. Composites Part A: Applied Science and Manufacturing 38(7) (2007) 1694-1709. https://doi.org/10.1016/j.compositesa.2007.02.006

[27] M. A. S. Spinace et al., Characterization of lignocellulosic curaua fibres." Carbohydrate Polymers 77(1) (2009) 47-53. https://doi.org/10.1016/j.carbpol.2008.12.005

[28] N. Chand et al., SEM and strength characteristics of acetylated sisal fibre. Journal of materials science letters 8(11) (1989) 1307-1309. https://doi.org/10.1007/bf00721503

[29] A. Porras et al., Characterization of a novel natural cellulose fabric from Manicaria saccifera palm as possible reinforcement of composite materials. Composites Part B: Engineering 74 (2015) 66-73. https://doi.org/10.1016/j.compositesb.2014.12.033

[30] A. K. F. Oliveira, J. R. M. D'Almeida, Characterization of ubuçu (Manicaria saccifera) natural fiber mat. Polymers from Renewable Resources 5(1) (2014) 13. https://doi.org/10.1177/204124791400500102

[31] C. Duarte, Fabrication and characterization of polyester resin composite reinforced with fabric of tururi fiber extracted from palm ubuçu - Manicaria saccifera. MSc thesis, Federal University of Para (Brazil), 2011.

<http://repositorio.ufpa.br/jspui/bitstream/2011/5104/1/Dissertacao_FabricacaoCaracterizacaoM aterial.pdf>

[32] M. Poletto et al., Native cellulose: structure, characterization and thermal properties. Materials 7(9) (2014) 6105-6119.

[33] L. Segal et al., An empirical method for estimating the degree of crystallinity of native cellulose using the X-ray diffractometer. Textile Research Journal 29(10) (1959) 786-794. https://doi.org/10.1177/004051755902901003

[34] B. H. Stuart, Infrared Spectroscopy, Kirk-Othmer Encyclopedia of Chemical Technology, John Wiley & Sons, Inc., New York, 2005. https://doi.org/10.1002/0471238961

[35] D. Ray, B. K. Sarkar, Characterization of alkali-treated jute fibers for physical and mechanical properties, Journal of Applied Polymer Science 80(7) (2001) 1013–1020. https://doi.org/10.1002/app.1184

[36] M. M. Houck, Identification of Textile Fibers, Woodhead Publishing, Cambridge, 2009.

[37] M. Midani et al., Effect of structural parameters on the impact properties of multilayer composites from tururi palm (Manicaria saccifera Gaertn.) fibrous material, Journal of Natural Fibers (2018) 1-14. https://doi.org/10.1080/15440478.2018.1491369

By-Products of Palm Trees and Their Applications   Materials Research Forum LLC
Materials Research Proceedings 11 (2019) 275-285   doi: https://doi.org/10.21741/9781644900178-23

Adsorption of Methylene Blue onto Chemically Prepared Activated Carbon from Date Palm Pits: Kinetics and Thermodynamics

A. M. Youssef [2], H. EL-Didamony[3], M. Sobhy[1,a*], S. F. EL- Sharabasy[1,b]

[1]Central Laboratory of Date Palm Research and Development, Agricultural Research Center, Egypt

[2]Department of Chemistry, Faculty of Science, Mansoura University, Mansoura, Egypt

[3]Department of Chemistry, Faculty of Science, Zagazig University, Zagazig, Egypt

E-mail: mahasobhy1000@yhaoo.com

Keywords: date palm pits activated carbon, methylene blue, kinetics, thermodynamics

Abstract. Three activated carbons were prepared using phosphoric acid (P) as an activating agent from date palm pits (DPP) as a precursor via thermal pretreatment producing (CP212, CP214 and CP124) samples, where the ratio of raw material to phosphoric acid is (2:1and 1:2) respectively at curing time two days for first sample and four days for the second and third sample, the activating temperature was 550°C, the precursor was washed with distilled water, dried, crushed, and then sieved. In order to study the effect of phosphoric acid modification, the characteristics of the activated carbon produced were determined before and after acid modification and subsequently compared. These characteristics include surface morphology, surface area, average pore diameter and pore volume. Characterization results showed that modification of date palm pits with phosphoric acid enhanced the surface area of the activated carbon from 427.8 to620.3 m^2/g. The average pore diameter was also enhanced from 1.14 to 1.82 nm. SEM analysis confirmed the improvement in surface area and pore development resulting from the phosphoric acid modification.

Introduction

MB is synthetic thiazine dye of an amorphous nature with a molecular formula $C_{16}H_{18}ClN_3S.xH_2O$. It is also called basic blue, tetra methylthionine chloride and colour index (Cl) number 52012. The molecular weight of MB is 320 and its maximum wave length 662nm. It is dark green powder, with a characteristic deep blue colour in aqueous solution where it dissociates into an MB cation and a chloride anion dye, Methylene blue is a common dye mostly used by industries involve in textile, rubber, paper, plastics, leather, pharmaceutical cosmetics, and food industries. Effluents discharged from such industries contain residues of dyes. Consequently, the presence of very low concentrations in effluent is highly visible [1, 2]. Discharge of colored waste water without proper treatment can results in numerous problems such as chemical oxygen demand (COD) by the water body, and an increase in toxicity. Organic dyes are harmful to human beings the need to remove color from waste water become environmentally important. It is rather difficult to treat dye effluents because of their synthetic origins and mainly aromatic structures, which are biologically non-degradable. Moreover, their degradation products may be mutagenic and carcinogenic [3,4]. Many dyes may cause allergic dermatitis, skin irritation, and dyes function of kidney, liver, and brain, reproductive and central nervous system [5]. It is estimated that 10–15%of the dyes are lost in the effluent during the dyeing processes. Activated carbon has been extensively used in wastewater treatment, chemical

By-Products of Palm Trees and Their Applications   Materials Research Forum LLC
Materials Research Proceedings 11 (2019) 275-285   doi: https://doi.org/10.21741/9781644900178-23

recovery and catalytic support industries primarily due to large surface area and presence of different pore sizes [6]. In most reported cases chemical activation is preferred as it, getting better porous, more surface area and high yielding activated carbons [7]. Date palm pits are considered one of the most useful and an abundant renewable agricultural waste, in 2006 world production of dates, was about 7 million tons. About 14% of the fruit is waste material, in the form of seeds. Date stones represent about 10% of the date weight. The chemical composition of date stone consists of hemicelluloses (23%), lignin (15%), and cellulose (57%) [8]. Date palm pits is not consumable by humans in any form; it has a high content of crude fiber (around 19%) that may cause digestibility problems in ruminant animals as well The main objective of this research is to prepare phosphoric acid (H_3PO_4)-activated carbons from date palm pits.

The following methods were used to characterize the prepared activated carbons: nitrogen adsorption at -196°C, scanning electron microscopy for analysis of the surface chemistry, pHpzc, surface pH analysis, and measurement of MB adsorption capacity at different temperatures. Special attention was paid to kinetic studies.

Materials and Methods

1. Materials

All the primary chemicals used in this study were of analytical grade. Methylene blue with 99.99% purity was obtained from Sigma-Aldrich Company and phosphoric acid. Distilled water was utilized throughout the experiments for solution preparation and glassware cleaning.

2. Preparation of activated carbons

In our study, we used (Sawi date palm pits) and the material was collected from pastry factory in (Shubra Al Khaimah). The date palm pits were washed with hot deionized water to remove dust and other impurities, and dried at105°C.Raw materials were ground into fine particles and sieved to a particle size of (2mm). Phosphoric acid activated carbon samples were prepared by soaking date palm pits in 50 wt. % analytical grade phosphoric acid at room temperature at different ratios of raw material to phosphoric acid (CP212, CP214 and CP124) for curing time two days for the first sample and four days for the second and third sample .The slurry was occasionally stirred, the solid was then separated and heated gradually in absence of air in a stainless steel reactor at a rate of heating = 10 °C /min up to 550 °C and then maintained at that temperature for 3 h. After cooling, the activated carbon mass was washed with distilled water till washing solution attained a pH value of 6.0 then the washed material was dried at 110 °C for 24 h and stored in clean dry bottles [9].

3. Characterization methods

3.1. Determination of % ash contents, weight loss on drying, pH of the supernatant and pHpzc

The ash content percentage fixed carbons were carried out on each sample of the activated carbon and the precursor. For each sample a crucible was placed in furnace at 650 °C for 1 h, cooled down in a desiccator, and the weight of ignited crucible was recorded, 2.0 g sample of activated carbon was placed in a crucible and transferred into a muffle furnace at 650 °C for six hours. After heating, the crucible was allowed to cool to room temperature in desiccator. The percentage ash content was determined using Eq. (1) [10, 11].

Where, W_f is the weight of sample pulse crucible after firing, W_c is the weight of empty crucible and W_s is the weight of sample before firing.

The weight loss during drying was determined for the date palm pits and activated sample by weighting 0.5 g of the sample and heating for 24 h in oven at 110 C° until the weight of the sample became constant using Eq. (2).

Where, W_b and W_a is the weight of sample before and after drying respectively

pH of the supernatant was determined by adding 0.5 gm of adsorbent to 25 mL of deionized water and the mixture was shaking for 48 h, the supernatant was then filtered to remove solid adsorbent and the pH of the supernatant was measured using Jenway pH-meter. pHpzc of activated carbon was also measured as 50 mL of 0.01 M NaCl solutions was put into several closed Erlenmeyer flasks. The pH in each flask was adjusted to a value between 2 and 6 by adding HCl (0.01 M) or NaOH (0.01 M) solutions. We take 0.15 g from the sample and added them to each flask, the flasks were agitated for 48 h, and the final pH was then measured. The pHpzc is the point where pH final - pH initial = zero [12].

3.2. Scanning Electron Microscopy (SEM) analysis

The surface morphology was obtained using Scanning Electron Microscope (Quanta 250 FEG) working at a high voltage of 15 kV. The samples were coated with gold by a gold sputtering device for clear visibility and conductivity. Thereafter, the coated sample was placed on the sample holder for analysis [13, 14].

3.3.Surface area and pore size analysis

The surface area and average pore size of samples were determined using a gas sorption analyzer. The adsorption-desorption isotherm of nitrogen was determined at its boiling point - 196 °C by ASAP 2020 instrument. The samples were degassed under vacuum at 350 °C for 3h prior to measurement. The nitrogen adsorption-desorption data were recorded at liquid nitrogen temperature of -196 °C. The adsorption equilibrium time was set at 60 S. The Brunauer, Emmett and Teller (BET) method was used to calculate the surface area, using the data obtained from the N_2 adsorption isotherm within the P/P^o range of 0.05 - 0.30, where P is the equilibrium pressure and P^o is the saturation pressure.

3.4. Adsorption equilibrium studies

Batch adsorption experiments were undertaken in a series of Erlenmeyer flasks containing 0.25g of ACS and equal volumes of methylene blue solutions at varying concentration (50-500 mg /L). The flasks were shaken in a thermostatic shaker at the desired temperature with shaking speed of 150 rpm for 24 h. The equilibrium concentrations of methylene blue in the supernatant were analyzed spectrophotometrically using (HPST uv/vis spectrophotometer) at λ_{max} 662 nm. The effects contact time were studied with respectto the adsorption of MB byCP212, CP214 and CP124.The effect of temperature was studied by conducting a batch adsorption experiment at 25, 35, and 45°Cfor CP124 as a selected sample.

Results and Discussion

1. Characterization of precursor and activated carbon samples.

Table 1: Ash content weight loss drying, pH of supernatant and pHpzc of samples.

sample	Percentage ash (%)	Weight loss drying (%)	pH of supernatant	pHpzc
DPP	0.77	13.24	5.95	6.32
CP212	5.170	4.59	2.35	3.20
CP214	4.765	4.960	2.65	3.5
CP124	3.214	5.155	3.00	4.3

Table 1 shows that ash content for DPP is 0.77%it is very low due to the higher amount of cellulose and lignin [15], while for modified AC it is in the range of 3.214–5.170 %. We can say that activating agent has the ability to dissolve apart of ash content; Moisture content is one of the factors that will influence adsorption capacity of activated carbon (AC). Moisture content in case of DPP was (13.24 %) which is higher compared with that for activated samples and this may be related to the decomposition of polar cellulose and hemi cellulose during chemical activation at higher temperature in absence of air, while for sample activated by H_3PO_4 the weight loss during drying increase with increasing the percent of activating agent and curing time, This is a logical result of the higher surface area, The pH value of activated carbon sample is mainly based on the inorganic ingredients in the source material or added during manufacture. In this study, acidic pH values were obtained. This is due to phosphorus-containing compounds such as polyphosphates, which may form during impregnation [16].

2. Scanning electron microscopy (SEM)

Scanning electron microscopy images were taken to observe the surface topography of the sample. Basically, the pore structure of activated carbon was observed. The micrographs of activated carbon are shown in (Fig.1) which showed that chemically activated carbon sample have high porosity compared to date palm pits (DPP). This was because during carbonization process of raw material with phosphoric acid at high temperature (550°C), most volatile matter was lost and thus created a system with advanced pore structure. Impregnation with phosphoric acid followed by the second carbonization dehydrated the cellulose material, resulting in weakening of the precursor structure and creation of pores. During chemical activation process, phosphoric acid was responsible for decomposition of organic material to release volatile matter and development of micro porous structure which could increase the adsorption capacity [17] Besides, we found that for the sample activated with phosphoric acid by increasing the percent of activating agent it minimizing the formation of tars and other liquids which could clog up the pores and inhibit the development of pores as clear in sampleCP124 [18]. The porosity created in the carbon structure was also resulted from phosphoric acid removal from carbon structure by intense washing

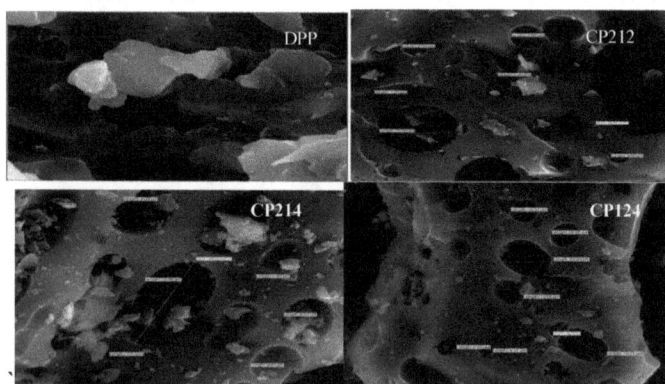

Fig. 1: SEM for (DDP) and activated carbons CP212, CP214 and CP124.

Materials Research Forum LLC
doi: https://doi.org/10.21741/9781644900178-23

3. Surface area and pore structure.

The BET surface area (S_{BET}), external and mesopore surface area ($S_{ext+meso}$) and the mean pore radius (r^-) of the activated carbons were determined by application of Brunauer-Emmett-Teller (BET) analysis software with the instrument. The BET surface area was determined by means of the standard BET equation applied in the relative pressure range from 0.06 to 0.30 [19]. The data are summarized in Table 2.

Table 2: Textural properties of the investigated carbons as determined from nitrogen adsorption at -196°C

Adsorbents	S_{BET} (m²/g)	C_{BET}	$S_{ext+meso}$ (m²/g)	V_μ (cm³/g)	V_P (cm³/g)	r^- (nm)
DPP	0	0	0	0	0	0
CP212	427.8	-45.05	112.3	0.178	0.243	1.14
CP214	470.6	-43.74	110.9	0.201	0.266	1.13
CP124	620.3	-187	330.7	0.172	0.567	1.82

We can conclude that: (i) the surface area for the date palm pits (DPP) is zero due to the absence of the activating agent. (ii) Activation with H_3PO_4 raise surface area, total pore volume, micropores volume and pore radius. The last results based on that activation of biomass based materials are dehydrating agents. These dehydrating agents penetrate deep into the biomass structure and cause the organic molecules to disintegrate into smaller molecules. After the release of these smaller molecules, tiny pores were created. Besides helping in the development of new pores or expansion of the pore, it also affects in enhancing the surface area. Normally, it was observed that micropores and mesopore formation results into a larger surface area of the activated carbons.(Fig.2) presents the nitrogen adsorption/desorption isotherms at -196 °C for the investigated activated carbon sample. It is clear that all above sample give hysteresis loop, this hysteresis loop is usually attributed to thermodynamic or network effects or combination of two effects [20].

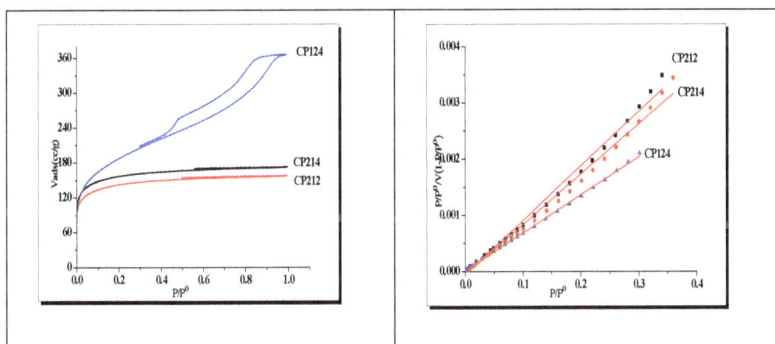

Fig. 2: Nitrogen adsorption isotherms and Linear BET plots of activated carbon sample at -196 °C

By-Products of Palm Trees and Their Applications   Materials Research Forum LLC
Materials Research Proceedings **11** (2019) 275-285   doi: https://doi.org/10.21741/9781644900178-23

4. Adsorption of Methylene blue.
4.1 Effect of contact time on adsorption.
The adsorptions of the methylene blue by activated carbon (CP214 andCP124) were studied at various time intervals (0.25- 25 h). (Fig .3) showed that the adsorption of methylene blue is fast in the initial stage by passing the time it will decrease till a constant value at about 15 h (equilibrium time). The fast adsorption at the initial stage occur due to the presence of high uncovered surface area and active sites on the adsorbent [21,22]. The slow adsorption of methylene blue by passing the time due to the small availability of uncovered surface area and fewer remaining active sites, resulting in lengthy time for adsorption to reach equilibrium of methylene blue [23].

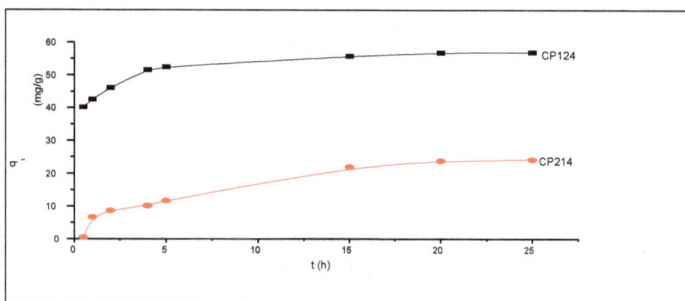

Fig. 3: Effect of contact time on removal of methylene blue onto CP214 and CP124

A kinetic study considers as method to give us information about the adsorption mechanism in our studies pseudo-first and pseudo-second order models were applied for CP214and CP124. The linearized form of the PFO can be expressed as

$$\log(q_e - q_t) = \log q_e - \frac{K_1}{2}.303\, t \qquad (3)$$

Where q_e is the equilibrium adsorbed amount (mg/g), q_t is the amount adsorbed in time t (mg/g), K_1 is the pseudo-first-order rate constant (g/mg. min) and t is the time in minute .The linearized form of the PSO can be expressed as

$$\frac{t}{:q_t} = \frac{1}{K_2 q_e} + \frac{1}{q_e^{\boxminus}}\, t \qquad (4)$$

Where K_2 is the pseudo second order rate constant (g/mg. min)
(Fig. 4) depicts the application of linear pseudo-first-order and linear pseudo second order kinetics model. Upon analysis of Table 3 one can concluded that the adsorption follow pseudo– second order kinetics model based on: (i) Correlation coefficient R^2 for PSO ranged between 0.9904 and 0.9943 indicating the good applicability of PSO linearized form. On the other hand, R^2 for PFO is very low (ranged between 0.89107and 0.67129). (ii) Calculated q_e (mg/ g) from PSO model are closer to q_m (mg/g) calculated from Langmuir adsorption model while that calculated from PFO is very high. (iii) K_2 ranged between 0.0159 and 0.0513 indicating the

By-Products of Palm Trees and Their Applications   Materials Research Forum LLC
Materials Research Proceedings **11** (2019) 275-285   doi: https://doi.org/10.21741/9781644900178-23

higher rate of adsorption especially in case of CP124 which exhibit a sharp increase in adsorption at beginning time of adsorption as shown in Fig (3).

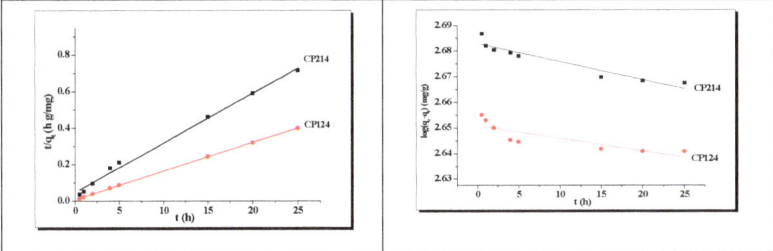

Fig. 4: Pseudo–first order and Pseudo–second order kinetic model for adsorption of MB onto activated carbon CP214 and CP124 samples.

Table 3: Parameter of pseudo-first-order and pseudo-second-order kinetics model for adsorption of methylene blue onto activated carbon samples

Pseudo–second order kinetic model			Pseudo–first order kinetic model			q_m (mg/g)	Sample
R^2	K_2 (g/mg h)	q_e (mg/g)	R^2	K_1 (h^{-1})	q_e (mg/g)		
0.9904	0.0159	36.764	0.89107	-16.28×10^{-4}	480.83	20	CP214
0.9943	0.0513	63.694	0.67129	-11.37×10^{-4}	477.50	81.36	CP124

4.2. Adsorption isotherms.

The purpose of carrying out the batch adsorption isotherm is to determine the best isotherm model for methylene blue adsorption by activated carbons and to determine the maximum adsorption capacity (q_{max}). In this study we used Langmuir models [24] which describe the monolayer adsorption of the adsorbate onto the sorbent. The linear equation for this model is given by

$$\frac{C_e}{q_e} = \frac{1}{b\,q_m} + \frac{C_e}{q_e} \tag{5}$$

Where q_e is the amount adsorbed at equilibrium time (mg/g), C_e is the equilibrium concentration of methylene blue (mg / L), q_m is the maximum adsorption capacity (mg / g), and b is known by Langmuir constant (L/mg) and it is related to the heat of adsorption. If we plot C_e/q_e versus C_e for methylene blue activated carbons, it will give a straight line with slope = $1/q_m$ and an intercept = $1/ bq_m$, as shown in (Fig.5A).One of the essential characterizes of Langmuir isotherm can be expressed by a separation factor, R_L: which is defined as showed in equation (6).

$$R_L = \frac{1}{1 + b\,C_0} \tag{6}$$

Here, C_o is the highest initial solute concentration. The R_L value implies whether the adsorption is unfavorable ($R_L>1$), linear ($R_L=1$), favorable ($0< R_L<1$), or irreversible (($R_L= 0$).

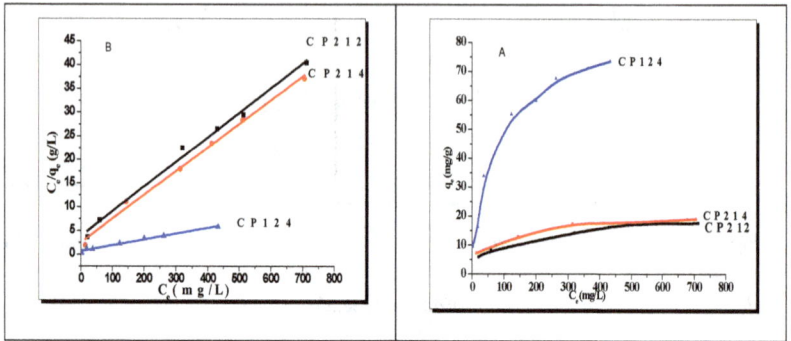

Fig. 5: *Adsorption isotherms (A) and linear Langmuir equation of methylene blue onto activated carbon samples (B).*

Table 4: *Langmuir parameter for adsorption of MB onto investigated carbon samples.*

Langmuir parameters				Sample
R_L	R^2	b (L /mg)	q_{max} (mg/g)	
0.0847	0.9909	0.0135	19.249	CP212
0.0585	0.9933	0.0201	20	CP214
0.0617	0.9808	0.019	81.366	CP124

From (Fig 5A) we observed that a very steep initial raise at the beginning of adsorption, this may be due to the fact that as initial concentration increases there are adsorption sites available for adsorption of methylene blue, thus enhancing the % uptake [25]. Upon inspection of Table 4, (i) Correlation coefficient (R^2) ranged between 0.9808 and 0.9909 indicating the accepted applicability of Langmuir models as shown in (Fig 5B). (ii) R_L value lies between 0 and 1 indicating that adsorption of methylene blue on activated carbon was favorable .It is clear that the R_L values decrease with the increase in initial dye concentration, indicating that the adsorption was favorable at higher MB concentration. (iii) CP212 and CP214 approximately have the same value of q_m due to there is no observable difference in surface area or surface chemistry, while by doubling amount of H_3PO_4 activating agent as in CP124 sample there is observable increase in q_m by about 4 times based on increasing surface area by about 150 m^2/g.

4.4. Effect of temperature and thermodynamic study

The effect of temperature on the removal of MB in aqueous was studied by varying the temperature between 25, 35 and 45 °C. The data presented in Fig 6 and it showed that adsorption of MB on CP124, activated carbon sample increase with raising temperature from 25 to 45 °C, due to the enlargement of pore size or creation of some new active sites on the surface of adsorbent due to rupture of bond. These also lead to enhancement the transformation of MB from the bulk of solution to the surface

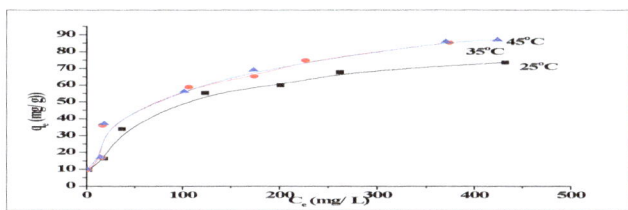

Fig. 6: Adsorption of methylene blue on CP124 activated carbon sample at 25, 35 and 45 °C.

Thermodynamic parameters were calculated for CP124 sample, to explain the nature of adsorption process on the surface and within the pores of activated carbons. The distribution coefficient, Entropy change, and Gibbs free energy change of the adsorption process is related to the equilibrium constant by the classic Van't Hoff equation

$$K_d = \frac{C_s}{C_e}$$ (10)

Where K_d is the distribution coefficient for the adsorption, C_s is the surface concentration of MB and C_e the equilibrium concentration. According to thermodynamics, the Gibbs free energy change is also related to the entropy change and heat of adsorption at constant temperature by the following equation:

$$\Delta G° = \Delta H° - T\Delta S°$$ (11)
$$\Delta G° = - RT \ln K_d$$ (12)

Where, R is the universal gas constant (8.314×10^{-3} kJ K^{-1}mol^{-1}), T is temperature in K. $\Delta G°$ is the free energy change (kJ mol^{-1}), $\Delta H°$ is the change in enthalpy (kJ mol^{-1}), $\Delta S°$ is the entropy change (kJ mol^{-1} K^{-1}), Thus $\Delta H°$ and $\Delta S°$ can be determined by the slope and intercept. The value of $\Delta S°$, $\Delta H°$ and $\Delta G°$, are showed in Table 5.The negative values of free energy change mean spontaneous process with a high affinity of the adsorbate to the surface of adsorbent. The positive value of enthalpy change and standard entropy indicated spontaneous and endothermic nature of adsorption process.

Table 5: Thermodynamics parameter for adsorption of MB onto CP124

ΔG (kJ mol^{-1})			ΔH° (kJ mol^{-1})	ΔS° (kJ mol^{-1} K^{-1})	Sample
318 K	308 K	298 K			
-5.056	-4.246	-3.436	20.702	0.081	CP124

Conclusions

Three activated carbons were prepared from (DPP) using activating agents: H$_3$PO$_4$, ash content, nitrogen adsorption, SEM and pH$_{PZC}$. S$_{BET}$ ranged between 427.8 and 620.3 m^2/g. The optimum condition for removal of MB was confirmed at pH7, equilibrium time 24 h and increased with temperature. Maximum adsorption capacities for the three adsorbents were found to be 19.2, 20,

By-Products of Palm Trees and Their Applications   Materials Research Forum LLC
Materials Research Proceedings 11 (2019) 275-285   doi: https://doi.org/10.21741/9781644900178-23

and 80 mg/g for CP212, CP214 and CP124, respectively at 25 °C. The adsorption of MB on the prepared activated carbons follow pseudo-second order kinetic model with correlation coefficients reached to 0.9933. Thermodynamic measurements showed spontaneous and endothermic nature of adsorption process.

References

[1] M. Alkan, O. Demirbas, S. Celikc, M. Dogan, Sorption of acid red 57 from aqueous solutions ontosepiolite, J Hazard. Mater. 116 (2004) 135–145.
https://doi.org/10.1016/j.jhazmat.2004.08.003

[2] K. Turhan, S.A. Ozturkcan, Decolorization and Degradation of Reactive Dye in Aqueous Solution by Ozonation in a Semi-batch Bubble Column Reactor, Water, Air, Soil Pollut. 224 (2012) 1353. https://doi.org/10.1007/s11270-012-1353-8

[3] T. Liu et al., Biointerfaces Adsorption of methylene blue from aqueous solution by graphene, Colloids Surfaces Biointerfaces 90 (2012) 197–203.
https://doi.org/10.1016/j.colsurfb.2011.10.019

[4] P. Saha, R. Das Mishra, R. Husk, Adsorption of safranin onto chemically modified rice husk in a upward flow packed bed reacto: artificial neural network modeling, Biotechnol. Adv. 44, (2012) 7579–7583.

[5] T.A. Khan, S. Sharma I. Ali, Adsorption of Rhodamine B dye from aqueous solution onto acidactivated mango (Magnifera indica) leaf powder: Equilibrium, kinetic and thermodynamic studies, J.Toxicol. Environ. Heal. Sci. 3, (2011) 286–297

[6] F. Rodriguez-Reinoso, M. Molina-Sabio, Activated carbons from lignocellulosic materials by chemical and/or physical activation: an overview, Carbon 30 (1992) 1111-1118.
https://doi.org/10.1016/0008-6223(92)90143-k

[7] C.A. Toles, W.E. Marshall, L. HWartelle, A. McAloon, A Steam-orcarbon dioxide-activated carbons from almond shells: physical, chemical and adsorptive properties and estimated cost of production. Bioresour Technol 75-3 (2000) 197-203. https://doi.org/10.1016/s0960-8524(00)00058-4

[8] N. M. Haimour, S. Emeish, Utilization of date stones for production of activated carbon using phosphoric acid, Waste Manag 26 (2006) 651-660.
https://doi.org/10.1016/j.wasman.2005.08.004

[9] A. M. Youssef, M. N Alaya, N. Nawar, Adsorption properties of activated carbon obtained from polymer wastes, Adsorb. Sci. Technol 11 (1994) 225.
https://doi.org/10.1177/026361749401100405

[10] E.O. Odebunmi, O.F. Okeola, Preparation and characterization of activated carbon from Waste Material, J. Chem. Soc. Nigeria 26-2 (2001) 149- 155.

[11] F.A. Adekola, H.I. Adekoge, Adsorption of Blue Dye on activated carbon from rice husk, coconut shell and coconut coirpith, Ife journal of Science, Nigeria 7(1) (2005) 151 – 157.
https://doi.org/10.4314/ijs.v7i1.32169

By-Products of Palm Trees and Their Applications   Materials Research Forum LLC
Materials Research Proceedings **11** (2019) 275-285   doi: https://doi.org/10.21741/9781644900178-23

[12] A.F. Hassan, A.M. Youssef, P. Priecel, Removal of deltamethrin insecticide over highly porous activated carbon prepared from pistachio Nutshells, Carbon Lett 14 (2013) 234-242. https://doi.org/10.5714/cl.2013.14.4.234

[13] W.J. Liu, F.X. Zeng, H. Jiang, X. S. Zhang, Preparation of high adsorption capacity bio-chars from waste biomass" Bioresource technology 102(17) (2011) 8247-8252. https://doi.org/10.1016/j.biortech.2011.06.014

[14] L. Wang, X. Wang, B. Zou, X. Ma, Y. Qu, C. Rong, Y. Li, Y. Su, Z. Wang, Preparation of carbon black from rice husk by hydrolysis, carbonization and pyrolysis" Bioresource technology 102(17) (2011) 8220-8224. https://doi.org/10.1016/j.biortech.2011.05.079

[15] M.M. Yeganeh, T. Kaghazchi, M. Soleimani, Effect of raw materials on properties of activated carbons, Chem Eng Technol 29 (2006) 1247. https://doi.org/10.1002/ceat.200500298

[16] A.M Puziy, O.I. Poddubnaya, A. Martinez-Alonso, F.Suarez-Garcia, J.M.D Tascon, Synthetic carbons activated with phosphoric acid I. Surface chemistry and ion binding properties, Carbon 40 (2002) 1493–1505. https://doi.org/10.1016/s0008-6223(01)00317-7

[17] R. Malik, D. S. Ramteke, S. R. Wate, Adsorption of malachite green on groundnut shell waste based powdered activated carbon, Waste management 27-9 (2007) 1129–1138. https://doi.org/10.1016/j.wasman.2006.06.009

[18] W.B. Wan Nik, M.M. Rahman, A.M. Yusof, F.N. Ani, C.N. Che Adnan, Production of activated carbon from palm oil shell waste and its adsorption characteristics, in Proceedings of the 1st International Conference on Natural Resources, Engineering and Technology (2006) 646–654.

[19] M. Aroua, S. Leong, L. Teo, W. Daud, Real time determination of kinetics of adsorption of lead (II) onto shell based activated carbon using ion selection electrode" Bioresour. Technol. 99 (2008) 5786-5792. https://doi.org/10.1016/j.biortech.2007.10.010

[20] F. Rouquerol, J. Rouquerol, K.S.W. Sing, Adsorption by powders and porous solid, Academic press, San Diego, 1999. https://doi.org/10.1016/b978-012598920-6/50006-3

[21] Y. Wu, S. Zhang, X. Guo, H. Huang, Adsorption of chromium (III) on lignin, Bioresour Technol 99 (2008) 7709.

[22] M.K. Aroua, S.P.P. Leong, L.Y. Teo, C.Y. Yin, W.M.A.W. Daud, Real time determination of kinetics of adsorption of lead (II) onto palm shell-based activated carbon using ion selective electrode Bioresour Technol. 99 (2008) 5786. https://doi.org/10.1016/j.biortech.2007.10.010

[23] Y. Li, Q. Du, X. Wang, P. Zhang, D. Wang, Z. Wang, Y. Xia, Removal of lead from aqueous solution by activated carbon prepared from Enteromorphaprolifera by zinc chloride activation, J Hazard Mater 183 (2010) 583-589. https://doi.org/10.1016/j.jhazmat.2010.07.063

[24] I. Langumir, The adsorption of gases on plane surface of glass, mica and platinum, J. Am. Chem. Soc 40 (1918) 1361-1403. https://doi.org/10.1021/ja02242a004

[25] R. Aravindhan, N.N. Fathima, J.R. Rao, B.U. Nair, Equilibrium and thermodynamic studies on the removal of basic black dye using calcium alginate beads, Colloids Surf A 299 (2007) 232. https://doi.org/10.1016/j.colsurfa.2006.11.045

By-Products of Palm Trees and Their Applications   Materials Research Forum LLC
Materials Research Proceedings **11** (2019) 286-292   doi: https://doi.org/10.21741/9781644900178-24

Chemical Composition and Pulping of Tunisian Almond and Fig Stems – A Comparison with Tunisian Date Palm Rachis

Ibtissem Moussa[1,2,a,*], Ramzi Khiari[1,3,4,b], Ali Moussa[2,c], and
Mohamed Farouk Mhenni[1,d]

[1] Research Unity of Applied Chemistry & Environment, Faculty of Sciences, The university of Monastir, Tunisia

[2] National Engineering School of Monastir, The university of Monastir, Tunisia.

[3] CNRS, LGP2, The university of Grenoble Alpes, France

[4] Department of Textile, Higher Institute of Technological Studies of Ksar Hellal, The university of Monastir, Tunisia

[a]moussa.ibtissem@hotmail.fr, [b]khiari_ramzi2000@yahoo.fr, [c]ali.moussa76@yahoo.fr, [d]farouk.mhenni@gmail.com

Keywords: date palm rachis, almond stems, fig stems, cellulose fiber, characterization

Abstract. In the present paper, the main objectives of this paper are the characterization of two Tunisian cellulosic by-products (almond and fig stems). The first part of this work is devoted to the determination of the chemical composition. Their chemical composition was showed that they present amounts of holocellulose, lignin and cellulose similar to those encountered in Tunisian date palm rachis. In the second part of this paper, the ensuing pulps of almond and fig stems are characterized in terms of degree of polymerization, fiber length and fiber width. These properties were compared to the properties of a Tunisian date palm rachis that was fully characterized by Khiari et al. (2010). Finally, the results of this work clearly demonstrated that almond and fig stems can be considered as a possible alternative source of fibers for cellulose derivatives and/or as lignocellulosic fibers for fiber-reinforced composite materials or papermaking application.

Introduction

The increase in fibers demand will be met by increased forestry production, which will give growth to global deforestation, with unsafe results to the environment. However, owing to the increasing fiber concerns and the potential increases in wood expenses, non-wood materials like annual plants have received more attention to produce pulp and/or paper and/or cellulose derivatives and/or composites. Lignocellulosic wastes are used as animal feed or burn in the soil or left to decompose. The utilization of these cheap and widely disposable wastes does not resolve just the environment pollution but also makes additional value.

Pulping methods have been modified these days by addition of certain chemicals, e.g. anthraquinone, to the basic pulping chemicals (soda and kraft) [1–4]. The choice of pulping chemicals is informed by literature indicating that addition of anthraquinone to soda pulping would have better advantages [5,6]. Adding anthraquinone as catalyst in sodium hydroxide system increases the pulp yields, decreases the kappa numbers, and improves the strength properties. It also increases lignin removal by promoting cleavage of inter-unit bonds in the lignin molecules that are not cleaved in the absence of anthraquinone. It also helps minimize recondensation of lignin reactions by reacting with the carbohydrates to increase lignin removal during pulping process and produced cellulose with high yield.

By-Products of Palm Trees and Their Applications
Materials Research Proceedings 11 (2019) 286-292

Materials Research Forum LLC
doi: https://doi.org/10.21741/9781644900178-24

The valorization of isolated fibers from Tunisian date palm rachis have been investigated by Khiari et al. (2010) for the making of paper, green composites and cellulose derivatives [7–9]. Date palm waste is widely available agricultural crops in Tunisia, which has more than four million dates palm trees occupying 33 thousand hectares. In our paper, two lignocellulosic materials were studied, largely disposable in Tunisia, as a source of cellulosic fibers, namely: almond and fig stems.

According to the Food and Agricultural Organization (FAO), Tunisia is ranked as the 8th producing country of almond with about 3.8% of the total world production [10]. The main production space is located in coastal areas (Bizerte, Cap Bon, Mahdia, Sfax, Zarzis, Djerba, Kerkenah, etc.) and in some mountainous regions (Gafsa, Beni Khedache, etc.). Fig (Ficus carica L.) is amongst the coventional Mediterranean species. Figs have recently attracted a great deal of attention and therefore are widespread across the world. The entire world produces over 1 000 000 tons of figs yearly, of which 82% are manufactured in Mediterranean countries [11]. In Tunisia, the production is all about 29 000 tons; it represents 3% of overall international production [11]. Almond and fig stems are by-products which have no proven uses. They are incinerated or dumped. This paper aims at deep understanding of the chemical and structural composition of these wastes and its cellulosic fibers. These properties were compared to the properties of a Tunisian date palm rachis that was fully characterized by Khiari et al. (2010).

Materials and Methods
Materials
The almond (*Prunus amygdalus* L.) and fig (*Ficus carica* L.) (Fig. 1) used during this work were cultivated in Monastir (city on the central coast of Tunisia).

Fig. 1. A) Almond tree (Prunus amygdalus L.) and B) Fig tree (Ficus carica L.).

Chemical Composition
The chemical composition of the raw materials studied here was established. Thus, the contents of Klason lignin, holocellulose, as well as α-cellulose were assessed by using different standards or methods, as summarized in Table 2 [1,4]. The amounts of lignin and α-cellulose were established by using the following respective TAPPI methods: T222 om-02; T203 cm-99. Finally, the holocellulose content was determined according to the method described by Wise (1946) [12]. The ashes content was determined, according to the standard procedure TAPPI T211 om-02, by calcinations of the materials at $525 \pm 25°C$ for at least 4 hours.

The evaluation of extractives was carried out in different liquids according to common standards, namely, cold and hot water solubility (T207 cm-99), 1% sodium hydroxide solution solubility (T212 om-02), and ethanol–toluene solubility (T204 cm-97).

Ethanol–toluene extractives: This method describes a procedure for determining the amount of solvent-soluble, non-volatile material in wood and pulp.

Cold and hot water extractives: The cold-water procedure removes a part of extraneous components, such as inorganic compounds, tannins, gums, sugars, and coloring matter present in wood and pulp. The hot-water procedure removes, in addition, starches.

1% sodium hydroxide solution extractives: Hot alkali solution extracts low-molecular-weight carbohydrates consisting mainly of hemicellulose and degraded cellulose in our agricultural wastes. The solubility could indicate the degree of a fungus decay or of degradation by heat, light, oxidation, etc. As agricultural waste decays or degrades, the percentage of the alkali-soluble material increases. The solubility of our agricultural wastes indicates also an extent of cellulose degradation during pulping and bleaching processes.

Ion Chromatography

The ion chromatography is an analytical technique which able to separate and quantify quantitatively only five monosaccharides (Glucose, Xylose, Galactose, Arabinose and Mannose). It is a method which gave the best recovery of sugars with minimum hydrolysis of sucrose.

The analytical hydrolysis procedure uses a two-step acid hydrolysis [13]. Primary hydrolysis of 350 mg sample was performed with 3 mL 72% (w/w) H_2SO_4 for an hour at 30°C. Hydrolysates were diluted to 4% (w/w) H_2SO_4 with distilled water. A secondary hydrolysis performed for 60 minutes at 120°C in autoclave (1 bar) to fractionate the biomass into forms which are more easily quantified. Fucose was added as an internal standard. The hydrolysates were diluted with H_2O. Following filtration, 10 µL samples of hydrolysates were injected directly onto the chromatographic system without any additional treatment.

Sugar contents of hydrolysates were determined by CI. The chromatographic system consisted of a 738-autosampler (Alcott Chromatography, Norcross, GA), a GPM-1 or a GP40 gradient high-pressure pump (Dionex Corp., Sunnyvale, CA), and a pulsed amperometric detector (PAD) (Dionex).

Extraction of Cellulose

The extraction of cellulose was prepared based on our previous studies [1,4]. The preparation of extracted bleached cellulose was conducted in two steps: pulping and bleaching. In our case, the operation of delignification (pulping) consisted of extracting and isolating the cellulosic fibers by adopting a chemical soda-anthraquinone process. The delignification stage of the almond and fig stems were performed according to experimental conditions described in a previous publication [1,4]. The obtained pulp was separated from black liquor and washed several times, until a neutral pH was attained. The bleaching step was performed according to experimental conditions described in a previous publication [1,4] using 100 mL of sodium hypochlorite solution (30%, v/v) (12 % of active chlorine) in an alkaline basic medium pH (pH varied between 9 and 11) for 180 min at 45°C. Finally, the bleached fibers were extensively washed with water until their pH was neutral, then purified by an anti-chlorine treatment and air dried before further use.

Morfi

Morphological properties of the fibres were studied by a morfi (LB-01) analyzer developed by Techpap – France [14]. The main fiber parameters were assessed by image analysis of a diluted suspension flowing in a transparent flat channel observed by a CCD video-camera. The average weighted length and the average width were measured and evaluated.

Carbanilation Reaction of Cellulose

Cellulose (15 mg) was place in test tube equipped with micro stir bars and dried overnight under vacuum at 40°C. Following the addition of anhydrous pyridine (4 mL) and phenyl isocyanate (0.5 mL), the test tube was permitted to stir for 48 hours at 70°C. Then, methanol (1 mL) was added to quench the phenyl isocyanate. Next, the contents of test tube was put into 7:3 (v:v) methanol:water (100 mL) to precipitate the derivatized cellulose. Finally, the solid was filtrated, washed with the methanol: water solution followed by deionized water and dried overnight under vacuum at 40°C.

Molecular weight distribution of cellulose

The derivatized cellulose was dissolved in tetrahydrofuran (THF) (1 mg.mL^{-1}), filtered through a 0.45 mm filter and placed in a 2 mL auto-sampler vial. The molecular weight distributions of the cellulose tricarbanilate samples were analyzed on an Agilent GPC security 1200 system. Molecular weight was calculated by the software relative to the polystyrene calibration curve. Weight average degree of polymerization (DP$_w$) was obtained by dividing M$_w$ by the molecular weight of the tri-carbanilated cellulose repeat unit (519 g.mol^{-1}).

Results and Discussion
Chemical Composition

The ash contents of almond and fig stems were rather high (3-5%) as indicated in (Table 1), but typical for tropical non-woody plants. The hot water extractives from fig and almond stems were (12.7 and 16.7%), cold water (9 and 12%), ethanol–toluene extractives (4 and 7%), and 1% NaOH (21 and 29%) were rather high due to the presence of many soluble polysaccharides and phenolic compounds.

On the other hand they are an indication of easy access and degradation of the cell wall materials by weak alkali. The cellulose content of fig stems was 47% which meant good pulp yields at suitable alkali utilization. The lignin content of fig stems was 19% which was relatively moderate. This should result in moderate cooking chemical charges and a short cooking cycle. Cellulose to lignin ratio was higher than 2 and predicted normal pulping with alkaline methods. High cellulose content and low content of lignin as well as organic and inorganic substances indicate the suitability of biomass for the production of cellulose fibres.

In paper production the primary role of hemicelluloses is to imbibe water and thus to contribute to fiber swelling. This leads to internal lubrication of the fiber and improves its flexibility and ease of beating. The swelling pressure contributes to loosening of the structure and fibrillation. The hemicelluloses being amorphous and adhesive in nature tend to hornify as the fiber shrinks and dries. Thus hemicelluloses serve as a matrix binding substance between fibers in a pulp.

The chemical composition of date palm rachis is summarized in Table 1, which shows that the amounts of extractives in water (cold, hot) are in the range of 5-8%, which is slightly lower when compared with almond and fig stems. Whereas ethanol-toluene extractives, 1% NaOH extractives, and ashes are comparable to those of our annual agricultural crops, holocellulose was found to be relatively high (around 74%).

The Tunisian almond and fig stems wastes are characterized by large amounts of cellulose. The cellulose content of fig stems (47%) was higher than date palm rachis.

The Klason lignin of date palm rachis is in the range of 27%, which is higher than fig stems and lower than almond stems. Moreover, rachis has slightly lower lignin content than other woody biomass like eucalyptus (29–32%) and pine (28%) [15].

Table 1. Characterization of almond and fig stems

Components	Fig stems	Almond stems	Date palm rachis [8]
Cold water extractives (%)	9.24	12.17	5
Hot water extractives (%)	12.70	16.70	8.1
Ethanol–toluene extractives (%)	4.18	7.02	6.3
1% NaOH extractives (%)	21.57	29.01	20.8
Ash (%)	5.10	3.39	5
Klason lignin (%)	19.64	34.35	27.2
Holocellulose (%)	60.11	50.66	74.8
Cellulose (%)	47.06	31.41	45

Their carbohydrate composition, are given in Table 2. As seen, there was significant difference in the carbohydrate composition. Almond and fig stems consisted mainly of glucan, xylan and galactan. Glucomannan and mannan were present in relatively minor quantities. Glucose and xylose are the main components of fig and almond stems, while galactose and arabinose were present in relatively minor quantities.

Table 2. Carbohydrate composition (mg/g dry weight) of almond and fig stems

	Almond stems	Fig stems
Glucan	342.777	511.458
Xylan	94.458	96.036
Galactan	60.631	26.333
Mannan	0.000	10.519
Glucomannan	0.153	0.107

Glucan and xylan contents were 20.62 and 10.53% (based on dry matter) in leaflets, while they were 38.34 and 20.07% in rachis [16]. Carbohydrate contents (glucan and xylan) of rachis are comparably high with conventional lignocellulosic biomass around the world like corn stover (30–38% and 20–25%), wheat straw (34–40% and 21–26%), and sugarcane bagasse (32–43% and 22–25%) [17].

Given the high carbohydrates and low lignin content, date palm rachis seems to be a potential source of woody biomass for biorefinery or it can be considered as a possible alternative source of fibers for cellulose derivatives and/or as lignocellulosic fibers for fiber-reinforced composite materials or papermaking application. Conversely, leaflets have signifcantly lower carbohydrates content and higher lignin content than conventional lignocellulosic biomass, making date palm leaflets a potential biomass candidate for lignin production.

Morphological Investigation

As shown in Table 3, based on morfi analyse, it can be noticed that fibers extracted from almond and fig stems present the same length, but the diameter of the fiber of almond stems is lower than the fig stems. The degree of polymerization (DP) of fibers extracted from almond and fig stems were determined by using gel permeation chromatography (GPC) technique. The relation in DP and the strength of fibers is rather obvious i.e. higher the DP higher the tensile strength [18–21].

By-Products of Palm Trees and Their Applications
Materials Research Proceedings **11** (2019) 286-292

Materials Research Forum LLC
doi: https://doi.org/10.21741/9781644900178-24

The DP of the fiber extracted from date palm rachis is lower than fibers extracted from almond and fig stems.

The principal factors controlling the strength of paper are fibre length and strength [22]. The fibers from the two stems studied (almond and fig stems) were in the range of hardwood fibers, with short fiber length, as shown in Table 3, with more or less moderate walls and fiber width. The fibres from the date palm rachis were also in the range of hardwood fibres with short fibre length (0.89 mm) and width (22.3 μm).

Table 3. Morpholigical of fibers

	M_n (Daltons)	DP_n	M_w (Daltons)	DP_w	M_w/M_n	Fiber length (mm)	Fiber width (μm)
Almond stems	643 436	1021	958 721	1521	1.49	0.518	19.6
Fig stems	589 934	1113	985 190	1563	1.67	0.516	22.4
Date palm rachis [8]	-	-	-	1203	-	0.890	22.3

Conclusion

The chemical composition of almond and fig stems was showed that they present amounts of holocellulose, lignin and cellulose similar to those encountered in date palm rachis. Given the high carbohydrates and low lignin content, almond stems especially the fig stems seem to be potential sources of woody biomass for biorefnery or they can be considered as alternative sources of fibers for cellulose derivatives or fiber-reinforced composite materials or papermaking application. The performance of the fiber extracted from almond stems is better than fibers extracted from fig stems and date palm rachis.

References

[1] I. Moussa, R. Khiari, A. Moussa, R.E. Abouzeid, M.F. Mhenni, F. Malek, Variation of Chemical and Morphological Properties of Different Parts of Prunus Amygdalus L and Their Effects on Pulping, Egypt. J. Chem. 62 (2019) 343–356. https://doi.org/10.21608/ejchem.2018.4827.1429

[2] Z. Marrakchi, R. Khiari, H. Oueslati, E. Mauret, F. Mhenni, Pulping and papermaking properties of Tunisian Alfa stems (Stipa tenacissima)—Effects of refining process, Ind. Crops Prod. 34 (2011) 1572–1582. https://doi.org/10.1016/j.indcrop.2011.05.022

[3] A. El Gendy, R. Khiari, F. Bettaieb, N. Marlin, A. Dufresne, Preparation and application of chemically modified kaolin as fillers in Egyptian kraft bagasse pulp, Appl. Clay Sci. 101 (2014) 626–631. https://doi.org/10.1016/j.clay.2014.09.032

[4] I. Moussa, R. Khiari, A. Moussa, M.F. Mhenni, M. Naceur Belgacem, Physico-chemical characterization of polysaccharides and extraction of cellulose from annual agricultural wastes, Cellul. Chem. Technol. 52 (2018) 841–851.

[5] P.W. Hart, A.W. Rudie, Anthraquinone - A Review of the Rise and Fall of a Pulping Catalyst, Tappi J. 13 (2014) 23-31.

[6] A. Rodríguez, R. Sánchez, M.E. Eugenio, R. Yáñez, L. Jiménez, Soda-anthraquinone pulping of residues from oil palm industry, Cellul. Chem. Technol. 44 (2010) 239–248.

[7] K. Ramzi, M. Nizar, M. Farouk, B.M. Naceur, M. Evelyne, Sodium carboxylmethylate cellulose from date palm rachis as a sizing agent for cotton yarn, Fibers Polym. 12 (2011) 587–593. https://doi.org/10.1007/s12221-011-0587-1

[8] R. Khiari, M.F. Mhenni, M.N. Belgacem, E. Mauret, Chemical composition and pulping of date palm rachis and Posidonia oceanica – A comparison with other wood and non-wood fibre sources, Bioresour. Technol. 101 (2010) 775–780. https://doi.org/10.1016/j.biortech.2009.08.079

[9] R. Khiari, E. Mauret, M.N. Belgacem, F. Mhenni, Tunisian date palm rachis used as an alternative source of fibres for papermaking applications, Bioresources 6 (2011) 265–281.

[10] Z. Hu, M.P. Srinivasan, Y. Ni, Preparation of Mesoporous High-Surface-Area Activated Carbon, Adv. Mater. 12 (2000) 62–65. https://doi.org/10.1002/(sici)1521-4095(200001)12:1%3C62::aid-adma62%3E3.0.co;2-b

[11] Information on: http://www.fao.org/faostat/en/#home.

[12] L. Wise, Chlorite holocellulose, its fractionation and bearing on summative wood analysis and on studies on the hemicelluloses, Paper Trade J. 122 (1946) 35-43.

[13] J.F. Saeman, W.E. Moore, R.L. Mitchell, M.A. Millett, Techniques for the determination of pulp constituents by quantitiative paper chromatography, Tappi J. 37 (1954) 336–343.

[14] R. Passas, M. Lecourt, P. Nougier, W. Minko, B. Khelifi, Effets de la remise en suspension des pâtes sur leur caractérisation morphologique, ATIP. Assoc. Tech. 58 (2002) 6–13.

[15] A.J. Ragauskas, G.T. Beckham, M.J. Biddy, R. Chandra, F. Chen, M.F. Davis, B.H. Davison, R.A. Dixon, P. Gilna, M. Keller, P. Langan, A.K. Naskar, J.N. Saddler, T.J. Tschaplinski, G.A. Tuskan, C.E. Wyman, Lignin Valorization: Improving Lignin Processing in the Biorefinery, Science 80 (2014) 1246843–1246843. https://doi.org/10.1126/science.1246843

[16] C. Fang, J.E. Schmidt, I. Cybulska, G.P. Brudecki, C.G. Frankær, M.H. Thomsen, Hydrothermal Pretreatment of Date Palm (Phoenix dactylifera L) Leaflets and Rachis to Enhance Enzymatic Digestibility and Bioethanol Potential, Biomed Res. Int. 2015 (2015) 1–13. https://doi.org/10.1155/2015/216454

[17] T.S. Khan, U. Mubeen, Wheat Straw: A Pragmatic Overview, Curr. Res. J. Biol. Sci. 4 (2012) 673–675.

[18] N.R.J. Hyness, N.J. Vignesh, P. Senthamaraikannan, S.S. Saravanakumar, M.R. Sanjay, Characterization of New Natural Cellulosic Fiber from Heteropogon Contortus Plant, J. Nat. Fibers 15 (2018) 146–153. https://doi.org/10.1080/15440478.2017.1321516

[19] S.S. Saravanakumar, A. Kumaravel, T. Nagarajan, P. Sudhakar, R. Baskaran, Characterization of a novel natural cellulosic fiber from Prosopis juliflora bark, Carbohydr. Polym. 92 (2013) 1928–1933. https://doi.org/10.1016/j.carbpol.2012.11.064

[20] Y. Hu, O. Hamed, R. Salghi, N. Abidi, S. Jodeh, R. Hattb, Extraction and characterization of cellulose from agricultural waste argan press cake, Cellul. Chem. Technol. 51 (2017) 263–272.

[21] W. Wang, R.C. Sabo, M.D. Mozuch, P. Kersten, J.Y. Zhu, Y. Jin, Physical and Mechanical Properties of Cellulose Nanofibril Films from Bleached Eucalyptus Pulp by Endoglucanase Treatment and Microfluidization, J. Polym. Environ. 23 (2015) 551-558. https://doi.org/10.1007/s10924-015-0726-7

[22] R. Liu, H. Yu, Y. Huang, Structure and morphology of cellulose in wheat straw, Cellulose 12 (2005) 25–34. https://doi.org/10.1023/b:cell.0000049346.28276.95

Food Applications

By-Products of Palm Trees and Their Applications
Materials Research Proceedings 11 (2019) 295-301

Materials Research Forum LLC
doi: https://doi.org/10.21741/9781644900178-25

New Technologies for Value Added Products from Coconut Residue

Navin K. Rastogi

Department of Food Engineering, CSIR-Central Food Technological Research Institute (CFTRI), Mysore 570 020, India

nkrastogi@cftri.res.in

Keywords: coconut palm, wet processing, by-products utilization

Abstract. This paper deals with the technologies developed in the field of coconut research at CFTRI in the last three decades including process for desiccated coconut, technology development for the production of spray dried coconut milk powder, wet processing of coconut, vinegar generation from coconut water, virgin coconut oil, tender coconut based beverage, coconut spread etc. CFTRI is in forefront in developing technologies for coconut-based products. Some of these technologies have been successfully transferred and most of the produce is being exported. Our current research efforts are focused on production of low fat dietary fiber from coconut residue after the milk extraction, concentration of coconut water by membrane processing, preservation of coconut water by emerging technologies will also be discussed.

Introduction

The word 'Coco' is derived from Spanish word 'Macoco', which refers to three holes on coconut that resemble the face of an ape. The coconut is mainly produced in southern states of India such as Kerla, Karnataka, Andhra Pradesh, Maharashtra etc. The tree of coconut is called as *kalptaru*, because all the parts of it are useful in one form or the other. Specially, the kernel of matured coconut is most valuable and is used for edible purpose as such or in dehydrated form. The dried kernel known as 'copra' and is the richest source of vegetable oil and the coconut oil cake is a valuable feed for livestock and a source of protein. The coconut shell is mainly used as a fuel, for making decorative items, shell powder, shell charcoal and biodegradable containers etc. The husk yields fiber, which is converted into coir and its products. The coir pith obtained during the defibring process is used as an ideal soil conditioner. The coconut water is one of the valuable by-products of the coconut processing industries, which can be subjected to fermentation to produce vinegar. The economy of the coconut-processing sector is mainly dependent on the copra and coconut oil, and on desiccated coconut to a lesser extent. About 60% of the total coconut production is used for edible purpose, 3.5% as tender coconut, 35% as milling copra for oil extraction and balance is processed into products like desiccated coconut. Coconut oil contributes about 6% of the total edible oil demand.

In order to develop the diversified products from coconut and to improve the economy of this sector Coconut Development Board, India has taken a welcome step in sponsoring research projects at different institutions for the development of technologies in this regard. CSIR-CFTRI is in forefront in developing technologies pertaining to diversified products from coconut.

Coconut Related Technologies Developed at CFTRI

A. Desiccated Coconut Powder

The process includes removal of shell and paring, disintegration of white endosperm, final drying in the drier and then packaging. On an average 1000 nuts give 110 kg of DCP.

By-Products of Palm Trees and Their Applications   Materials Research Forum LLC
Materials Research Proceedings 11 (2019) 295-301   doi: https://doi.org/10.21741/9781644900178-25

Several units located in India produce about 15,000 tons annually. All are in the small-scale sector with capacities ranging from 0.5-1.0 ton per day. DCP is mainly used in biscuits, sweets, bakery products and other food preparations.

B. Spray Dried Coconut Milk Powder

Coconut milk is a product of the region of tropical climate and is in great demand in the international culinary. It is a white milky product extracted from the endosperm of coconut and constitutes into an emulsion stabilized by proteins and probably, by some ions found in oil-water interface. Coconut milk is an important dietary in coconut producing countries. It is valued mainly for its characteristic nutty flavour and also for its nutritional values. It is an ingredient for many fish, shellfish, meat, poultry, vegetable dishes, confectioneries, sweets, serbhats, beverages and other type of preparations. Under ambient conditions coconut milk shows poor stability and the emulsion separates into two distinct phases: a heavy aqueous phase and a lighter creamy phase. Coconut milk extracted from freshly grated coconut meat undergoes very fast progressive deterioration at room temperature due to its high content of fat, moisture and other organic components, which quickly deteriorate upon exposure to microorganism, light, oxygen and high temperature.

Based on our previous experience and published reports, it was opined that dehydration is the best feasible method for the preservation of coconut milk. Dehydrated coconut milk powder which retains the natural flavour and texture of coconut milk, yet has good keeping quality, would lead to greater convenience and increased consumption. Individual and institutional user would be relieved from the task of extracting milk. Further, it offers additional advantages such as less storage space and extended shelf life. Very little information is available in the literature regarding the dehydration of whole coconut milk, though lot of information is available on dehydration of skim coconut milk powder. And out of available information most of it either remained proprietary or available in languages other than English or in the form of patents. The only report available is work of Hassan [1].

The process for the production of whole coconut milk powder involves various unit operations such as size reduction, extraction of milk, stabilization of emulsion, homogenization and spray drying. The coconut milk contains large amount of fat, which poses difficulty in achieving its stabilization during spray drying. A process was developed to stabilize coconut milk, which enables spray drying [2]. Recently, technology for the production of whole coconut milk powder has been transferred to an Indian industry. The process was developed for detachment of coconut kernel from shell [3]. The white endosperm after removal of shell and paring was passed through rotary wedge cutter having a sieve plate (3 mm. hole) through which shredded meat is forced out. This moist coconut grating was expressed in a screw press to extract coconut milk. Coconut milk, thus obtained, was homogenised in two stage of high and low pressures, respectively, over a period of 30 minutes. Then coconut milk was formulated by addition of certain ingredients and chemicals etc. This formulated coconut milk was pasteurised at 60-70°C for 5 minutes and homogenised again. Finally, the milk was spray dried at temperature 100-150°C at a feed flow rate of 120 ml/minute and packed [4]. The flow sheet of the process is given in Figure 1.

Apart from the processing technology, quality and yield of the extract (coconut milk) is affected by several factors such as varietal differences, coconut maturity, meat particle size, processing temperature and extraction pressure. It is observed that maturity of coconut has considerable effect on the yield of coconut milk. It was observed that the yield was less if the coconuts were over or under matured. When the meat particle size was less it was observed that

By-Products of Palm Trees and Their Applications   Materials Research Forum LLC
Materials Research Proceedings **11** (2019) 295-301   doi: https://doi.org/10.21741/9781644900178-25

yield of coconut milk was higher because of the effective rupture of cells. Increased pressure in the screw press resulted in increased yield of milk.

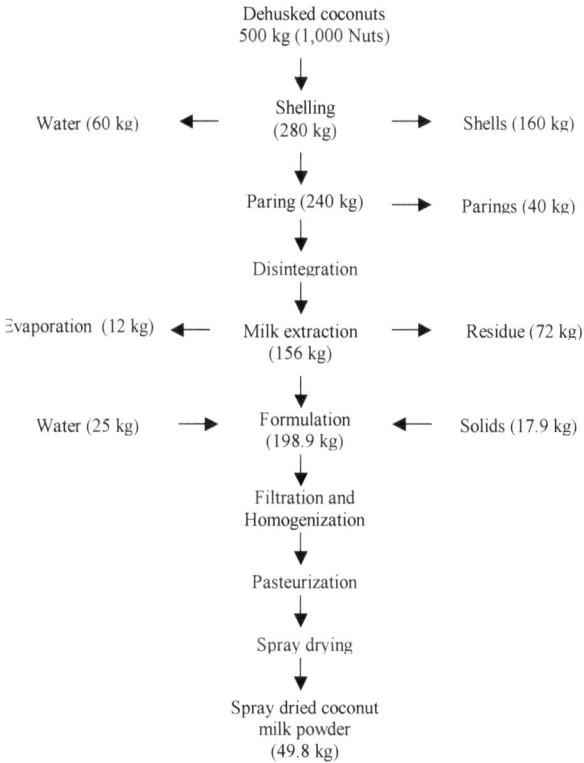

Fig. 1: Flow sheet for the production of spray dried coconut milk powder.

C. Virgin Coconut Oil by Wet Processing

The white endosperm portion of coconut is disintegrated and squeezed in screw press to recover coconut milk, which is filtered, and cream is separated by centrifugation. The cream is stirred vigorously to get the virgin coconut oil (VCO) by a process called phase inversion. The oil thus obtained is very clear, nutritious and has a longer shelf life. The residual coconut cake can be dried and sold as medium/low fat desiccated coconut, which may find application in bakery and formulation of low calorie foods. The skim milk obtained from centrifugation can be concentrated and spray dried. The value-added by-products render the whole process quite economical one. Virgin coconut oil (VCO) is prepared from fresh mature coconuts by wet processing without any heat treatment [5]. VCO is colourless and having an intense coconut aroma. It is rich in lauric fatty acid, which is a proven antiviral and anti-bacterial agent. The high-grade VCO has a long shelf life due to presence of natural anti-oxidents in coconut oil. Coconut milk is extracted from deshelled, pared and disintegrated coconuts. Further, coconut

Materials Research Forum LLC

doi: https://doi.org/10.21741/9781644900178-25

cream is separated, which is subjected to tempering, conditioning and separation techniques to separate VCO. Material balance for VCO production is presented in Figure 2. Coconut oil is a product of the region of tropical climate and is in great demand in the cosmetic industry and international culinary. It is valued mainly for its characteristic nutty flavour and also for its nutritional values. The novelty of our invention lies in the manner in which the VCO is obtained without heat treatment of coconut milk or fermentation of coconut cream, thereby keeping its characteristic flavour and nutrients intact.

Ultrafiltration in combination with spray drying was also explored as a method for the production of coconut whey protein powder. The coconut whey was centrifuged to remove fat and then it was subjected to ultrafiltration using membranes of MWCO of 5, 10, 30 and 50 kDa. The retentate and permeate were collected. It was found that MWCO of 5 kDa gave maximum retention of proteins in the retentate (96%). The ultrafiltration was performed in the pressure range of 2 to 10 bar. The separation process occurring across a membrane discriminates solute molecules on the basis of their sizes. The retentate collected was then spray dried to get coconut protein powder.

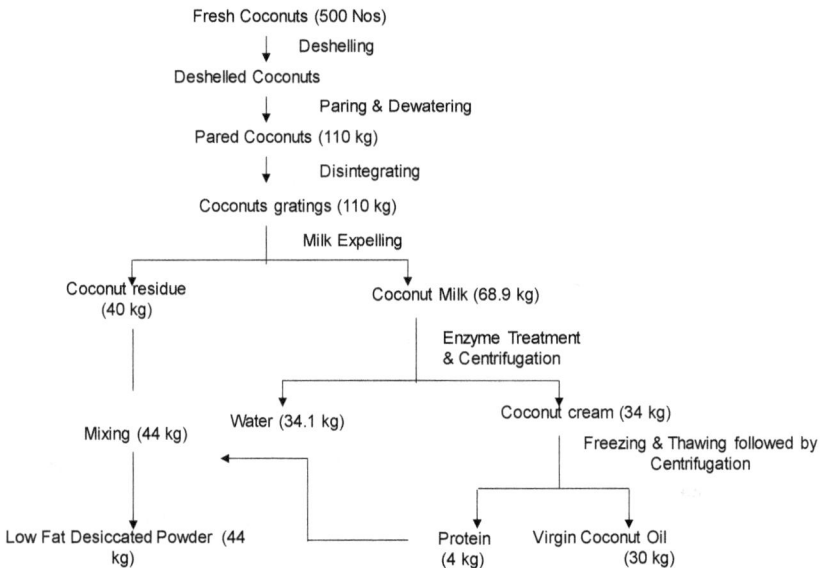

Fig. 2: Material balance for VCO production.

D. Mature Coconut Water

Coconut water (yield~28%) is a very important by-product from coconut processing industries, because it contains about 2.5-3.0% sugar, which can be utilised as a fermentation substrate for the production of vinegar. Vinegar is the product of two stage fermentation process. In the first stage, sugars are converted into ethyl alcohol by the action of yeast (*Saccharomyces cerevisiae*) in a anaerobic fermentation process. In the second stage of fermentation process, bacteria (Acetobector) oxides ethyl alcohol to acetic acid [6]. It is an aerobic fermentation of exothermic nature. During the fermentation process series of reaction takes place. Since minimum requirement of total fermentable sugars in the substrate should be in the range 8-10%, it is supplemented from the external source. The cost of sugar (or jaggery) may be compensated with the recovery of value added by-products, besides solving the pollution problems and amount incurred in its disposal.

The vinegar generator is designed to provide the maximum surface exposure for a volume of fermented coconut water in order to supply enough air for the acetic bacteria to efficiently and quickly oxidises the alcohol to acetic acid. The generator assembly usually comprises a feed trough, and acetifier and receiving trough. In essence, it is counter current gas absorber wherein the acetic bacteria causes the oxidation of ethyl alcohol to acetic acid. The feed is uniformly sprayed over the surface of an inert porous packing medium (corn cobs) at the surface of which the oxidation takes place. The stock which drains off from the packing by gravity into the base of the generator is run out and pumped back into the feed vat from which it is recycled until acetification is complete. When the vinegar is reached its maximum strength (4.0%), it is aged before bottling. In certain coconut growing countries like Philippines, coconut water is used to produce a sweet dessert dish called Nata-de-coco which is also popular as a component of fruit salad, ice-creams, fruit salads and fruit cocktails.

The use of coconut water adds favorably to the economics of the existing coconut industries. Mature coconut water procured from desiccated coconut industry and was filtered and centrifuged to remove suspended solids and fats. Mature coconut water was passed through cationic, anionic and mixed bed resins to remove saltiness caused by the minerals in the final product. The mature coconut water was then subjected to thin film evaporator to achieve 47°B. Further, coconut spread was prepared by partial replacement of sugar with concentrate from mature coconut water along with addition of other ingredients such as citric acid, pectin and benzoic acid followed by thermal treatment. Addition of coconut dietary fiber, which can be evenly suspended in the spread, provides a characteristic coconut flavor, texture and taste. Partial or optimized level of replacement of sugar (50%) with mature coconut water concentrate yielded a very highly acceptable product having characteristic taste and flavor [7].

E. Dietary Fiber

Dietary fiber refers to the plant substances including plant cell wall (cellulose, hemi cellulose, pectin and lignin) as well as intracellular polysaccharides such as gums and mucilage that are not digested by human digestive enzymes [8]. The main components of dietary fiber are cellulose, hemi cellulose, starch, pectin substance (polygalacturonic acid components) and lignin [9]. Amongst these, only cellulose and a portion of retrograded starch (called resistance starch) are insoluble in water, while the other are soluble. The non-starch polysaccharides act as bulking agent or roughage in the food. Dietary fiber is considered as a physiologically inert material although the bulking and laxative properties of many fiber sources have long been appreciated [10]. It has been shown to play an important role in the prevention of the risk of carcinogenesis, atheroscler-osis and in the control and proper management of diabetes mellitus [11]. During the wet processing of coconut, fresh coconuts, after shelling and paring, were disintegrated and ex-pressed to extract coconut milk, which was either used for the preparation of virgin coconut oil or spray dried coconut milk powder. The grinding of coconut residue after the fat extraction led to the rupture of the honey comb physical structure (matrix) resulting in a flat ribbon type structure, thereby providing an increase in surface area for water and fat absorption, which can be utilized as dietary fiber [12, 13]. The grinding has to be done in such a way that the resultant product size must be ~550 micrometer. During solvent extraction the fat content of the product should be reduced to less than 2% and product has to be dried completely. The presence of higher content of fat and moisture resulted in inadequate grinding, which led to the reduction in water as well as fat absorption properties. A very fine or very coarse product will lead to decrease in water as well as fat absorption properties. Hydration properties of coconut dietary fiber were compared with other commercially available dietary fibers Figure 3. Except for apple fiber (5.43 g/g) and citrus fiber (10.66 g/g), the water retention capacity of coconut dietary fiber (5.4 g/g) was higher compare to all other the samples. Water holding capacity of coconut fiber (7.1 g/g) was also more than that of the other samples. Coconut fiber showed highest swelling capacity (20 ml/g) as compared to any other fiber studied. This showed that coconut fiber has the maximum capacity to swell when compared to other fibers, which is the most desirable parameter for physical functioning of dietary fiber [12, 13].

F. Tender Coconut Water
Coconut beverage was produced from the solid (white meat) as well as liquid (coconut water) endosperm from tender coconut having good shelf life, sensory properties and characteristics coconut flavour. It is of white color and viscous in nature. The tender coconut beverage can be packaged, distributed and sold commercially due to the presence of natural electrolytes, refreshing and fresh taste of coconut. The tender coconut water and thin solid endosperm from the tender coconut and was homogenized in the ratio 4:1, the homogenized mixture was formulated with sugar and xanthan. The formulated beverage was filtered with cheesecloth, heated up to 90°C and filled in pre-sterilized glass bottles. The bottles were subjected to hot water to expel the air present in the headspace. Further, the bottles were sealed and autoclaved. Figure 4 represents the flowchart for the production of tender coconut beverage

The product was found to have characteristic taste and aroma of tender coconut and no preservative was used in this process. The novelty of the present invention lies in the way of selecting the processing steps, conditions/ parameters and components to obtain a value added product, which is not hitherto available, without losing the characteristic flavour of tender coconut.

Tender coconut water concentrate is a rich source of proteins and micro-nutrients, besides, preserving the flavor, color and nutrition. The process includes initial concentration of coconut water by reverse osmosis and further concentration was achieved by osmotic membrane distillation [14]. This novel approach involves no heat treatment, thus retaining the flavors of tender coconut largely.

Conclusion
Other technologies available are canned coconut chunk in brine, canning of coconut cream and process for canning of tender coconut water, coconut beverage from tender coconut and coconut spread. Of course, the coconut research may take vibrant trend with the advent of latest technologies such as membrane processing and other techniques. These things are only possible when regulatory and governing agencies come forward to sponsor further research projects in this potential field. It would be very encouraging if the food industries come closer to R&D institutions and work together for development of state-of-the-art technologies and can finance part of the research.

Acknowledgement
The author wishes to gratefully acknowledge the constant encouragement of Dr. KSMS Raghavarao, Director, CFTRI. This work was carried out under the project funded by Coconut Development Board (CDB), Kochi, India.

References

[1] M.A. Hassan, Spray drying of coconut milk Pertanika, 8 (1985) 127-130.

[2] N.K. Rastogi et al., A process for the preparation of stable whole coconut milk, Indian Patent number 184681 (1995).

[3] N.K. Rastogi, K.S.M.S. Raghavarao, S.G. Jayaprakashan, A process for detachment of coconut kernel from its shell. Indian Patent Appl. 2638/DEL/96 (1996).

[4] N.K. Rastogi, K.S.M.S. Raghavarao, Production of dehydrated coconut milk powder, In the National Seminar on Processing and Marketing of Coconut (SPAMCO II), Bangalore (1992).

[5] N.K. Rastogi et al., A process for the production of virgin coconut oil, Indian Patent Appl. number 443/DEL/2009 (2009).

[6] C.T. Dwarakanath, Vinegar Fermentation with Special Emphasis on Possibilities of Coconut water, In the National Seminar on Processing and Marketing of Coconut (SPAMCO II), Bangalore (1992).

[7] N.K. Rastogi, K.S.M.S Raghavarao, M. Prakash, A process for the production of coconut spread based on mature coconut- water concentrate and coconut dietary fiber, Indian Patent Application number 0287/DEL/2009 (2009).

[8] G.A. Spiller, CRC Hand book of: Dietary fiber in human nutrition. CRC Press, New York, 2000, pp. 9-10.

[9] M.S. Wolthuis, H.F.F. Albers, J.G.C. Jeveren, J.W. Jong, J.G.A.J. Hautvast, R.J.J Hermus, M.B. Katan, W.G. Brydon, M.A. Eastwood, The American Journal of Clinical Nutrition 33 (1980) 1745-1756. https://doi.org/10.1093/ajcn/33.8.1745

[10] K.L. Roehrig, Food Hydrocolloids 2 (1988) 1-18.

[11] T.P. Trinidad, D. Valdez, A.C. Mallillin, F.C. Askali, A.S. Maglaya, M.T. Chua, J.C. Castillo, A.S. Loyala, D.B. Masa, Indian Coconut Journal 7 (2001) 45-50. https://doi.org/10.1016/j.ifset.2004.04.003

[12] S.N. Raghavendra, N.K. Rastogi, K.S.M.S. Raghavarao, R.N. Tharanathan, Dietary fiber from coconut residue: effect of different treatments and particle size on the hydration properties, European Food Research and Technology 218-6 (2004) 563-567. https://doi.org/10.1007/s00217-004-0889-2

[13] S.N. Raghavendra, S.R. Ramchandra Swamy, N.K. Rastogi, K.S.M.S Raghavarao, Sourav Kumar, R.N. Tharanathan, Grinding characteristics and hydration properties of dietary fiber from coconut residue, Journal of Food Engineering 72 (2006) 281-286. https://doi.org/10.1016/j.jfoodeng.2004.12.008

[14] S.N. Raghavendra, N.K. Rastogi, K.S.M.S. Raghavarao, M. Prakash, A process for the preparation of tender coconut beverage, Indian Patent Application number 283/DEL/2009 (2009).

By-Products of Palm Trees and Their Applications   Materials Research Forum LLC
Materials Research Proceedings 11 (2019) 302-312   doi: https://doi.org/10.21741/9781644900178-26

Production of Single Cell Protein from Date Waste

Mohamed Al-Farsi[1*], Alaa Al Bakir[2], Hassan Al Marzouqi[3]
and Rejoo Thomas[2]

[1]Food & Water Lab. Center, Ministry of Regional Municipalities & Water Resources, Muscat, Oman

[2]R&D Section, Al Foah Company, Al Ain, UAE

[3]R&D Department, Abu Dhabi Food Control Authority, AD, UAE

Keywords: single cell protein; date's waste; fungal strains; protein, amino acids

Abstract. This study aimed to utilize the waste of date's industry to produce single cell protein. Five fungal strains were evaluated and the production conditions were optimized. A. oryzae was selected as the optimum strain due to its vigorous growth and high protein production. Ammonium sulfate at 0.8% was the best source of nitrogen for the selected strain, pH at 5.5 and the medium ratio of 75 g in 250 ml flask were the best for growth. The single cell protein produced has a good source of nutrition, as the ratio of essential to the total amino acids was 46%. These results benefit establishing large-scale production to produce single cell protein from date's waste which creates a source of income to this sector and prevent pollution from such waste.

Abbreviations

Single cell protein (SCP); American Type Culture Collection (ATCC); Potato dextrose agar (PDA); Peptone yeast extract glucose agar (PYG); Dates waste agar (DWA).

Introduction

Single cell protein (SCP) is dried cell of microorganisms, which used as protein supplement in human foods and animal feeds. The SCP is cheap and competes well with other source of protein and may provide good nutritive value. Besides high protein content (60-82%), SCP contains fat, carbohydrates, vitamins and minerals [1,2]. SCP also rich in essential amino acids like lysine and methionine which are limiting in most plant and animal foods [3]. With increase in population and worldwide protein shortage, the use of SCP as a food and feed is more needed [4]. A number of agricultural and agro-industrial waste products have been used for production of SCP, including orange waste, mango waste, cotton stalks, kinnow-mandarin waste, barley straw, corn cops, rice straw, corn straw, onion juice and sugar cane bagasse [5], cassava starch [6], wheat straw [7], banana waste [8], capsicum powder [9] and coconut water [10].

Date syrup production end with waste consist of date fiber and seed. According to Al-Farsi et al [11], the production of date syrup will end with 59% syrup, 23% press cake and 12% seed. Therefore, for instance Al Baraka Dates Company in Dubai, UAE produced 4000 tons of date syrup in the year 2016 [12], this production will end with 1560 kg of press cake, which can be used for SCP production. Al Farsi et al [11] reported the composition of the syrup waste for three varieties, their protein ranged between 3.6-5.2%, fat between 1.4-2.2% and carbohydrates between 81.9-83.3%. The usage of such wastes as a sole carbon and nitrogen source for production of SCP by microorganisms could be simply attributed to their presence in nature on large scale and their cheap cost. Also, utilization of such waste prevents pollution problems and sanitary hazard as well as creating another source of income to this sector.

Different type of microorganisms can be used for SCP production, such as algal, bacteria, fungi and yeast. The microorganisms used for SCP should be low in nutritional requirements, rapid growth rate, stability during growth, non-pathogenic, low nucleic acid content, non-toxic and good digestibility [13]. While the substrate should be non-toxic, abundant, non-exotic, cheap and able to support rapid growth of organisms resulting in high quality of biomass [14]. During microorganism process for conversion of lignocellulosic wastes into feed, at least one of the three objectives must be reached: 1- An increase in the protein level 2- An increase in digestibility 3- An increase in the essential amino acids [15].

In this work five different fugal strains were evaluated for SCP production from date's syrup waste and the medium condition of the selected strain was optimized for maximum production. This study could be the first research on utilizing dates waste to produce SCP, as no other studies found in this area.

Materials and Methods
Dates waste
The dates waste used in this study was procured from Al Foah Company, Al Ain, UAE. This is a by-product of date's syrup production consists of date fiber and seed, which packed in polyethylene bags and stored in -30°C until used.

Fungal Strains
Five fungal strains were obtained from American Type Culture Collection (ATCC, Manassas, Virginia, USA) based on their characteristics as SCP producers and capability of utilizing lignocellulosic by-products. The strains were *Trichoderma reesei* (ATCC 13631), *Fusarium venenatum* (ATCC 20334), *Thermomyces lanuginosus* (ATCC 34626), *Aspergillus oryzae* (ATCC 14895) and *Fusarium graminearum* (ATCC 20333). All cultures were processed according to the specific directions of ATCC.

Culture Media
Two types of media were used for maintenance and routine subculture including:
1- standard ATCC recommended media for freeze dried culture which were as follows:
   - potato dextrose agar (PDA) for *Trichoderma reesei and Fusarium graminearum.*
   - malt agar medium (ATCC 323) for *Fusarium venenatum*
   - malt extract agar medium (ATCC 325) for *Aspergillus oryzae*
   - peptone yeast extract glucose agar (PYG, ATCC 663) medium for *Thermomyces lanuginosus*
2- Dates waste agar (DWA) used as production medium for fungal strains. The medium contains 25% date waste, 2% agar, 0.3% ammonium sulfate and 72.7% water. The pH was kept at the natural date fibers at 5.3.

Strain Selection
The five fungal strains were grown on DWA after autoclaving the medium in 250 ml flasks at 121 °C for 15 min (duplicates). The inoculation was carried out by spore suspension with absorbance of 1.00 at 600 nm and surface fermentation was conducted under static condition at 25 °C, except *Thermomyces lanuginosus* , which was incubated at 45 °C for 120 hrs. The biomass and growth medium were dried at 70 °C until constant weight, ground to fine powder and analyzed for protein content. The cultures were stored at - 36 °C until processed for culturing.

Determination of Optimal Production Conditions
- Optimal nitrogen source: Ammonium chloride, ammonium sulfate and urea were used at 0.2% level in the growth medium DWA with a control containing no added nitrogen.
- Optimal concentration of nitrogen source: The selected nitrogen source was added to the growth medium DWA at 0, 0.2, 0.4, 0.6, 0.8 and 1.0%.
- Optimal medium weight to flask volume: The production medium DWA was dispensed in 250 ml flasks at 45, 60, 75, and 90 g/ each flask prior to sterilization and inoculation with the selected strain spore suspension, incubated and processed as described above.
- Optimal Initial pH: The pH of the DWA medium was adjusted by 1N HCL at 4.0, 4.5, 5.0, 5.5, 6.0 and 6.5 prior to sterilization.

Protein Determination
Protein content of SCP was determined by analyzing the powdered samples for nitrogen content in CHNSO analyzer (HEKAtech GmbH, Wegberg, Germany) and the obtained values were converted to protein by the factor 6.25.

Determination of amino acids
Amino acid content of dates fiber and SCP were determined according to the Official European Union regulation No 152/2009 [16]. This method determines free and total amino acids (peptide bound and free) using amino acid analyzer (Shimadzu, Japan).

Statistical analysis
Results were expressed as mean ± standard deviation on a wet weight basis. Statistical significance (t-test: two-sample equal variance, using two-tailed distribution) was determined using the Microsoft Excel Statistical Data Analysis. Differences at $p < 0.05$ were considered to be not significant.

Results and discussion
Strain selection
The DWA medium which used as a medium for fungal growth was simple to prepare and has clear economic potential in large scale part of the project. No hydrolyses treatment been used for this media, as the date fiber macerated from dates syrup processing. The presence of sugar in date fibers along with complex lignocelluloses components has induction effect on production of essential carbohydrate hydrolyses. Figure 1 present growth of the five fungal strains and figure 2 shows the protein content produced by these fungal strains. It's clearly shows the vigorous growth of *A. oryzae* compares to other fungi, also the protein content produced by *A. oryzae* was significantly higher than others. Based on the obtained results (vigorous growth and high protein) *A. oryzae* was selected as the optimum strain for DWA and used for optimal production condition. Jin et al [17] found *A. oryzae* as the best option to produce SCP from starch waste water. Also, Ahmadi et al [18] reported 57% protein content produced by *A. oryzae* from rice bran.

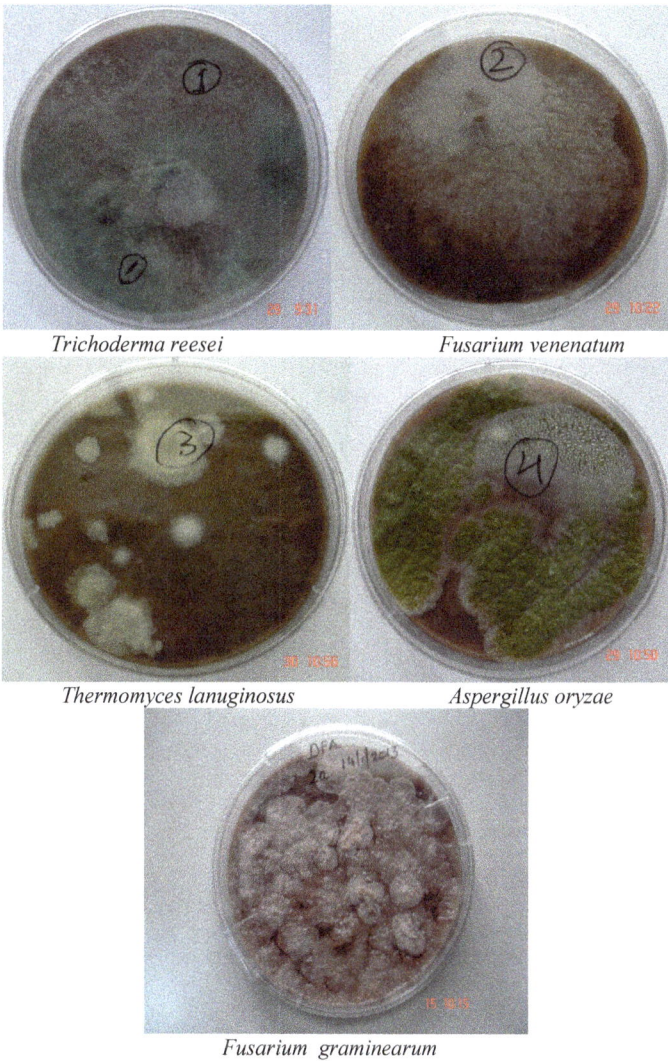

| *Trichoderma reesei* | *Fusarium venenatum* |
| *Thermomyces lanuginosus* | *Aspergillus oryzae* |

Fusarium graminearum

Fig. 1. Growth of the five fungal strains.

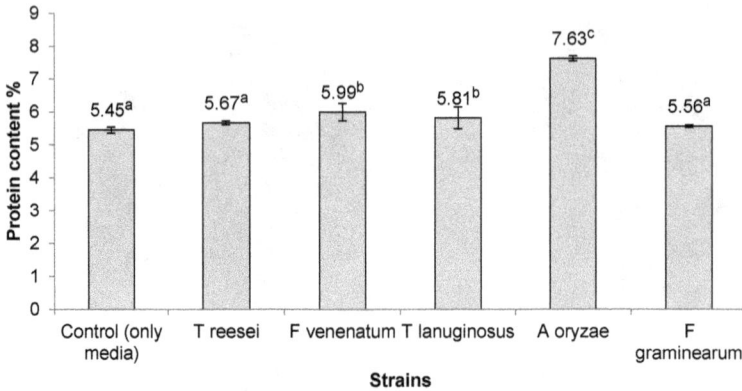

Fig. 2. Protein content (%) produced by different fungal strains. The values followed by the same letters are not significantly different (p < 0.05).

Nitrogen source

In order to select the optimum nitrogen source for maximum fungal growth, three nitrogen sources, ammonium chloride, ammonium sulfate and urea were used. Figure 3 present the effect of nitrogen source on protein content produced by *A. oryzae* after 5 days growth. The result shows that ammonium sulfate was the best source as the protein content produced was significantly highest (13.8%). Rao et al [19] also found ammonium sulfate is the best nitrogen source for fungi strain *Penicillium janthinellum* to produce SCP from bagasse. The possible reason may be that ammonium sulfate has some additional growth factors such as some amino acids, mineral, certain vitamins, which gave better growth results compare to other nitrogen sources [13]. This source is rather inexpensive and easy to mix with the media. Therefore, ammonium sulfate was selected as the optimal nitrogen source for *A. oryzae* growth on DWA medium. However, Mondal et al [3] found inorganic nitrogen supplementation had suppressive effect by decreasing SCP produced by yeast.

Concentration of nitrogen source

The selected ammonium sulfate was used in different concentration to determine the optimum concentration for *A. oryzae* growth on DWA medium. Figure 4 shows the effect of difference concentration of ammonium sulfate on protein content produced by *A. oryzae*. The protein content increased with increase of ammonium sulfate to reach to maximum content when using 1% ammonium sulfate. However, the difference between 0.8 and 1.0% concentration was insignificant, therefore, 0.8% ammonium sulfate was selected as the optimum concentration for growth medium. This result supported by Ahangi et al [18,20], they found that lower glucose level and higher nitrogen level in fungi medium resulted higher protein production.

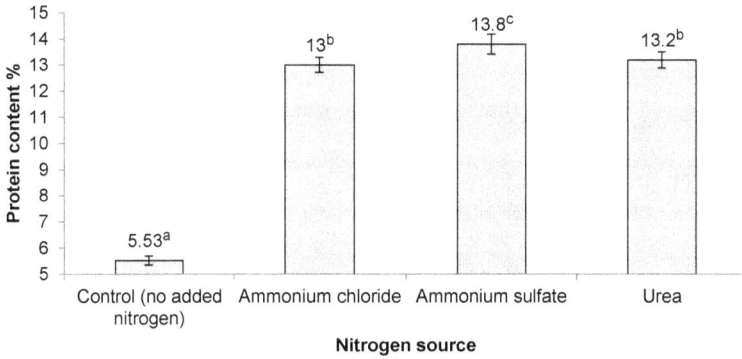

Fig. 3. *Effect of nitrogen source on protein content (%) produced by Aspergillus oryzae in dates waste agar (DWA). The values followed by the same letters are not significantly different (p < 0.05).*

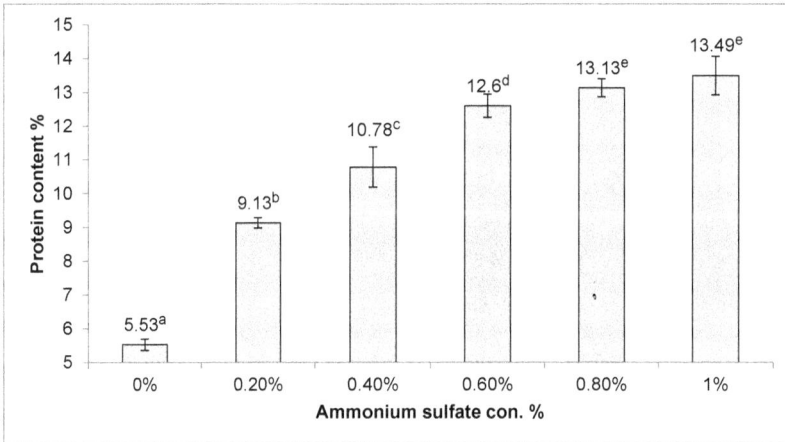

Fig. 4. *Effect of ammonium sulfate concentrations (%) on protein content (%) produced by Aspergillus oryzae in dates waste agar (DWA). The values followed by the same letters are not significantly different (p < 0.05).*

Medium weight / flask volume ratios

Figure 5 presents the effect of various medium weight/flask volume ratio on protein content produced by *A. oryzae*. Protein content obtained from different substrate levels revealed that 75g in 250 ml flask was the best ratio for growth. Therefore, the ratio of 75 g/250 ml was selected as the optimal ratio. This factor is important in estimating the required media quantities according to the fermentation vessel volume. The substrate cost is the largest single cost factor in SCP production, thus, simplifying the manufacture and purification of substrate can reduce cost production.

Medium pH

The initial pH of the DWA medium was sensitive to the growth of *A. oryzae*. Different initial pH values were used to check the optimum pH value for maximum yield of the biomass. The results of present study showed in Figure 6 that yield of biomass increased from pH 4 and optimum production was observed at 5.5 yielding 16.25 % of crude protein. This pH 5.5 is close range of natural date fibers pH (5.3). Further increase in initial medium pH leads to decline in protein production. These results supported by several studies; Ravinder et al [19-21] found the optimal growth of *A. oryzae* in rice bran was in pH range of 5-7. Jin et al [16,17] found pH 4.5-5.5 was the optimum pH for *A. oryzae* in starch waste water. Also, Yousufi [20-22] reported pH 5.0 as the optimum pH for production of SCP from soymilk using *A. oryzae*.

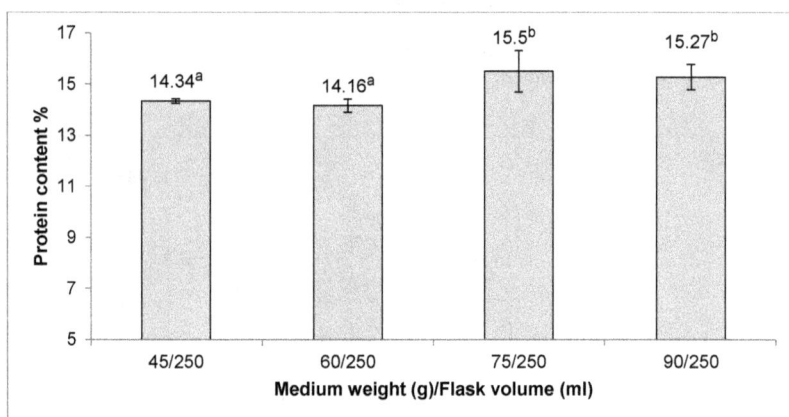

Fig. 5. Effect of various medium weight / flask volume ratios on protein content (%) produced by Aspergillus oryzae in dates waste agar (DWA). The values followed by the same letters are not significantly different (p < 0.05).

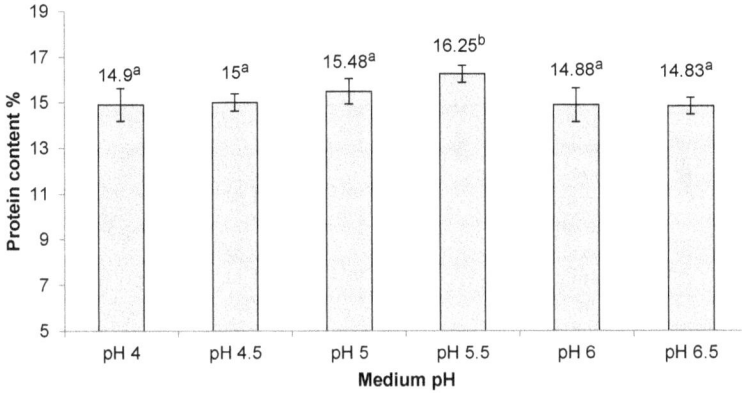

Fig. 6. Effect of medium pH values on protein content (%) produced by Aspergillus oryzae in dates waste agar (DWA). The values followed by the same letters are not significantly different (p < 0.05).

Amino acid content

The nutritious value and potency of SCP from any source is based on its composition and should be analyzed for the properties of their components such amino acid profile before the final product is used as food or feed supplementation. Table 1 present the content of amino acids in dates waste and SCP produced by this study. The total content of amino acids in dates waste was 6.02 % and 10.14 % in SCP. Glutamic, aspartic and proline acids were the major amino acids in SCP, and their value are 1.11, 1.01 and 0.88 % respectively. The results indicated that the ratio of essential amino acids to total amino acids of the SCP produced was 46%, which is a good source of nutrition. Khanifar et al [21-23] reported higher ratio of essential amino acids produced by white rot fungi from wheat straw, which was 65.6%. The nutrition benefits of SCP depend on culture condition, pre-treatment of substrates, nutrient supplementation, types of fermentation processes and strain [22-24].

Table 1. Amino acids content (%) of dates fiber and single cell protein (SCP) from dates waste

Amino acids %	Date fiber	SCP
Lysine *	0.28 [a]	0.48 [b]
Methionine *	0.14 [a]	0.22 [b]
Cystine	0.11 [a]	0.19 [b]
Aspartic acid	0.64 [a]	1.01 [b]
Threonine *	0.31 [a]	0.52 [b]
Serine	0.35 [a]	0.58 [b]
Glutamic acid	0.74 [a]	1.11 [b]
Proline	0.30 [a]	0.88 [b]
Glycine	0.42 [a]	0.63 [b]
Alanine	0.43 [a]	0.68 [b]
Valine *	0.37 [a]	0.57 [b]
Isoleucine *	0.27 [a]	0.44 [b]
Leucine *	0.53 [a]	0.85 [b]
Tyrosine	0.22 [a]	0.22 [a]
Phenylalanine *	0.30 [a]	0.72 [b]
Histidine *	0.16 [a]	0.28 [b]
Arginine *	0.32	0.57 [b]
Tryptophan	0.13 [a]	0.19 [b]
Total	**6.02 [a]**	**10.14 [b]**

* The essential amino acids. Values followed by the same letter, within a row, are not significantly different ($p < 0.05$).

Conclusions
Optimization of SCP production from dates waste using fungal strains been conducted. From the five fungal strains used, *A. oryzae* was the best with the optimized condition of; ammonium sulfate at 0.8% as nitrogen source, 75g of medium in 250ml flask and at pH of 5.5. The SCP produced was a good source of nutrition as the ratio of essential amino acids to the total amino acids reach 46%. This preliminary study provides the optimum conditions for producing SCP from date's waste, which is useful for large scale industry.

References

[1] M.J. Asad , M. Asghan, M. Yaqub, K. Shahzad, Production of single cell protein delignified corn cob by Arachniotus species, Pak. J. of Agric. Sci. 37 (2000) 3-4.

[2] P. Jamel, M.Z. Alam, N. Umi, Media optimization for bio proteins production from cheaper carbon source, J. of Engi. Sci. and Techno. 3-2 (2008) 124-130.

[3] A. Mondal, S. Sengupta, J. Bhowal, D. Bhattacharya, Utilization of fruit wastes in producing single cell protein, International Journal of Science Environment & Technology, 1-5 (2012) 430-438.

[4] A.T. Nasseri S. Rasoul-Amini, M.H. Morowvat, Ghasemi, Single cell protein: Production and process, Ame. J. Food Technol. 6 (2011) 103-116. https://doi.org/10.3923/ajft.2011.103.116

[5] N.M. Nigam, Cultivation of Candida langeronii in sugarcane bagasse hemi cellulose hydrolysate for the production of single cell protein, W.J.Microbiol and biotechnol. 16 (2000) 367- 372.

[6] H. Tipparat, A.H. Kittikun, Optimization of single cell protein production from cassava starch using Schwanniomyces castellii, W.J. Microbiol. & Biotechnol. 11 (1995) 607-609. https://doi.org/10.1007/bf00360999

[7] S.A.A. Abou Hamed, Bioconversion of wheat straw by yeast into single cell protein, Egypt, J. Microbiol. 28-1 (1993) 1-9.

[8] P.M.A. Saquido, V.A. Cayabyab, F.R. Vyenco, Bioconversion of banana waste into single cell protein, J. Applied Microbiol. & Biotechnol. 5-3 (1981) 321-326.

[9] G. Zhao, W. Zhang, G. Zhang, Production of single cell protein using waste capsicum powder produced during capsanthin extraction, Lett Appl Microbiol. 50 (2010) 187-91. https://doi.org/10.1111/j.1472-765x.2009.02773.x

[10] M.E. Smith, A.T. Bull, Protein and other compositional analysis of Saccharomyces fragilis grown on coconut water waste, J. Applied Bacteriol. 41 (1976) 97-107.

[11] M. Al-Farsi, C. Alasalvar, M. Al-Abid, K. Al-Shoaily, M. Al-Amry and F. Al-Rawahy, Compositional and functional characteristics of dates, syrups and their by-products, Food Chem. 104(2007) 943-947. https://doi.org/10.1016/j.foodchem.2006.12.051

[12] Al Barakah Date Company, Al Barakah Dates (2016): www.albarakahdatesfactory.com, accessed 11th March 2019.

[13] M.A. Shahzad, M. Rajok, Single cell protein production from Aspergillus terreus and its evaluation in broiler chicks, Int. J. Biosci. Biochem. Bioinform. 1 (2011) 137-141. https://doi.org/10.7763/ijbbb.2011.v1.25

[14] D. Dhanasekaran, S. Lawanya, S. Saha, N. Thajuddin and A. Panneerselvam, Production of single cell protein from pineapple waste using yeast, Innovative Romanian Food Biotechnol. 8 (2011) 26-32.

[15] D.N. Kamara, F. Zadrazil, Microbiological improvement of lignocellulosic in animal feed production: Rrview. Elsevier, Essex, UK (1988) 56-63.

[16] EU., The official European union regulation no. 152/2009. Official Journal of the European Union (2009) 23-31.

[17] B. Jin, H.J. Van Leeuwen, B. Patel, Q. Yu, Utilization of starch processing wastewater for production of microbial biomass protein and fungal alph-amylase by Aspergillus Oryzae. Bioresource Technology 66 (1998) 201-206. https://doi.org/10.1016/s0960-8524(98)00060-1

[18] R. Ravinder, L. Venkateshwar, P. Ravindra , Studies on Aspergillus oryzae Mutants for the Production of Single Cell Proteins from Deoiled Rice Bran. Food Technol. Biotechnol. 41-3 (2003) 243-246.

[19] M. Rao, A. Varma, S. Deshmukh, Production of single cell protein, essential amino acids and xylanase by Penicillium janthinellum. BioResource 5 (2010) 2470-2477.

[20] Z. Ahangi, S.A. Shojaosadati, H. Nikoopour, Study of mycoprotein production using Fusarium oxysporum PTCC 5115 and reduction of its RNA content, Pakistan. J. Nutr.7 (2008) 240-243. https://doi.org/10.3923/pjn.2008.240.243

[21] R. Ravinder, L. Venkateshwar Rao, P. Ravindra, Production of SCP from de-oiled rice bran, Food Technol. Biotechnol. 41 (2003) 243–246.

[22] M.K. Yousufi, Impact of pH on the single cell protein produced on okara-wheat grit substrates using Rhizopus oligosporus and Aspergillus Oryzae, J. Envir. Sci. Tox. Food Tech 1-2 (2012) 32-35. https://doi.org/10.9790/2402-0123235

[23] J. Khanifar, H. Ghoorchian, A.R. Ahmadi, R. Hajihosaini, Comparison of essential and non essential amino acids in the single cell protein of white rot fungi from wheat straw, Afri. J Agri. Rese. 6-17 (2012) 3994-3999.

[24] Anupama, P. Ravindra, Value-added food: Single Cell Protein. Biotechnol. Advances 18 (2000) 459-479. https://doi.org/10.1016/s0734-9750(00)00045-8

Materials Research Forum LLC
doi: https://doi.org/10.21741/9781644900178

Design and Architecture

By-Products of Palm Trees and Their Applications   Materials Research Forum LLC
Materials Research Proceedings **11** (2019) 315-324   doi: https://doi.org/10.21741/9781644900178-27

Design for Enhancing Material Appreciation: An Application on the Palm Tree Midribs

Alaa El Anssary[1,a *], Nariman G. Lotfi[1,b]

[1] German University in Cairo, Egypt

[a]alaa.elanssary@guc.edu.eg, [b]nariman.gamal@guc.edu.eg

Keywords: palm tree midribs, material appreciation, emotional design, user perception

Abstract. Despite academic and professional efforts to extend the uses of palm midribs in production, there is lack of user appreciation or interest in the end product. Most studies focus on the palm technical characteristics to compete with standard wood, disregarding emotional factors that are essential in product promotion. Recent applications eliminate material naturalness, creating artificial substitutes and confusing users who prefer material genuineness. The study aims to analyze perceptions that could lead to the integration of palm midribs into local products by demonstrating visual and tactile attributes. The research highlights designer involvement in product development. Due to the material nature, engineers and designers must collaborate to develop products technically and emotionally. Two experimental design methods were conducted with participants from each group concerned with developing the material. Participants evaluated the material surface from their perspectives in terms of visual and composition. Results revealed that diverse feelings of tactile attributes create sentiments and intellectual curiosities, evoking value and appreciation for the material. Although both groups perceive and communicate differently, they have similar objectives: successful material implementation. In order to enhance designer-engineer cooperation, a multidisciplinary platform was developed enabling material features to be integrated in product development. By focusing on palm by-product naturalness designers and engineers can create meaningful and delightful user experiences.

Introduction

The byproducts from date palm tree (*Phoenix dactylifera*) are natural materials available vastly in Egypt. Despite that they have been around for centuries, researchers are only starting to approach their full potential as materials for manufacturing. As imported wood prices are increasing, Egyptian manufacturers are eager to find a substitute material with similar performance and characteristics and lesser cost. One particular part of the date palm tree, the midrib, is a renewable material source, allowing its use in sustainable methods and as consequence reducing carbon emissions in the local environment. Most of the previous researchers and applications have approached the palm midrib material from an engineering point of view, with a particular focus in the material technical characteristics such as its strength and flexibility [1]. Furthermore, end users may not trust in products made from midrib material as it is not yet familiar or widely appearing. Therefore, there is a lack of existing designed products as references focusing on intangible characteristics such as the emotional and tactile attributes of the material. Positively, we assume that the sensorial properties of the midrib material can capture the imagination of engineers and designers alike to develop trendy products that make a strong impact on purchaser decision.

Essentially, palm midrib is considered a waste which is burned similar to rice ash. Updated manufacturing techniques transform midribs into compressible strips to produce block boards

By-Products of Palm Trees and Their Applications   Materials Research Forum LLC
Materials Research Proceedings **11** (2019) 315-324   doi: https://doi.org/10.21741/9781644900178-27

and other marketable furniture items. El-Mously, a well-known researcher in this field, is working on rediscovering different uses of Palm Trees' secondary products as renewable resources to satisfy national and international demands for specific applications [2]. He confirmed in different publications that midrib material's modulus of rupture (MOR) and tensile strength is similar to pine wood. The midrib material's outer layer offers a specific tensile strength of 196 $(N/mm^2)/(g/c\ m^3)$ to 142 $(N/m\ m^2)/(g/c\ m^3)$ of the European Reed Pine, making it an alternative to imported timber [3]. Another remarkable effort made by Egyptian engineers and craftsmen was a series of handmade furniture pieces developed from palm fronds boards created specifically to present innovative structural properties. They successfully utilized the results of the engineering research to extend the potential of this raw material and explore a new dimension in artisan capability. For example, furniture pieces shown in Fig. 1, developed by start-ups such as Jereed and Jozour [4] are interesting applications not just in terms of the form structure but also regarding the design process in which design and material properties can be compromised together [5].

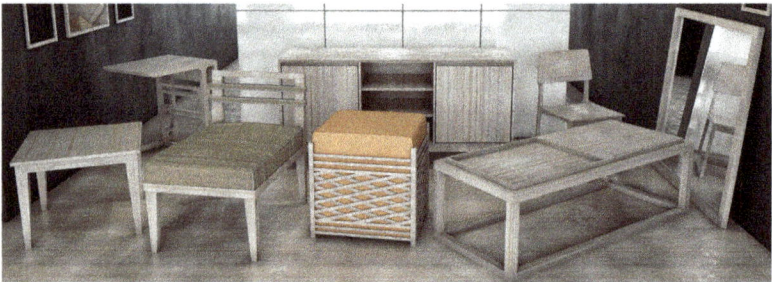

Fig. 1: Jereed Furniture Products made from Palm Midribs [4].

This is an innovative and integrative point of view, since generally designers perceive materials and generate judgments based on sensorial properties resulting from appreciation of the product form. On the other hand, as engineers, mechanical and environmental resistance properties are the main focus. Consequently, there is a huge informational and perceptual gap between the two fields explaining why tensions and clashes between the two disciplines evolve.

Therefore, this paper aims at studying the interpretations of designers and engineers towards the palm tree midrib in order to develop a strategic platform that enables better communication and understanding between both academic and professional views. By doing so, improved products and designs using the newly developed material can be introduced to the Egyptian and international market successfully.

Multi-dimensions of possibilities offered by palm midribs

Palm midribs have numerous possibilities of applications taking in account their sustainable, physical, mechanical, aesthetic and symbolic properties. Previous tests carried out on the midrib show that it has pronounced physical and mechanical properties. The midrib enhanced strength is comparable to Beech and Spruce wood. According to a research conducted at Munich University, it was found that the exterior layer of the midrib has a tensile strength 4 times superior in comparison to steel. The extraction process of the material involves trimming, cutting then compressing it using a storage room to control oxygen, carbon dioxide, and nitrogen levels.

By-Products of Palm Trees and Their Applications   Materials Research Forum LLC
Materials Research Proceedings 11 (2019) 315-324   doi: https://doi.org/10.21741/9781644900178-27

Therefore, the materials increased qualities and uses are due to the midrib's radical transformation from a living material to hardwood [5].

Date palm midribs can be employed to produce wood-cement panels (WCPs). The results showed that the WCPs from C. erectus maintained higher stability in water than those produced from P. dactylifera. The mechanical properties of the WCPs made from the date palm midribs exceeded the minimum requirements; the dimensional stability characteristics of the panels were also lower than the maximum limits of particleboard standards as presented in Table 1 [6].

Table 1. Values of the mechanicals and dimensional properties of WCPs from the date palm midribs and buttonwood.

Wood species	MC	Mechanical properties				Dimensional stability (%)		
		Density⁺	MOR	IB	JHN	Water	Thickness	Linear
	(%)	(kg.m⁻³)	(MPa)	(MPa)	(N)	uptake	swelling	expansion
P. dactylifera	8.45^A	1250^A	11.20^A	1.76^A	5795^A	23.03^A	1..34^A	0.50^A
C. erectus	8.65^A	1255^A	11.75^A	1.74^A	5371^B	19.34^B	0.57^B	0.34^B
Requirement	-	1200	9.00	0.45	2222	-	Max 8.0	-

Characteristics such as high strength and elasticity can inspire product designers to create a wide range of new applications. Another advantage of using palm midribs is in the reduction of carbon emissions by the avoiding of its burning. This material is extracted from the palm tree fronds, in which 10 to 26 leaves grow every year. Without using, it constitutes a waste of approximately 965 tons annually in the Middle East [5]. On the other hand, because the frond renews itself it is a consistent local material resource, reducing the cost to import wood and reducing carbon emissions. It can also save cutting down trees and, in this way, reduce deforestations. Therefore, full utilization of palm midribs supports local manufacturers substituting imported woods and benefits the ecosystems by relieving pressure on forests [2]. Thus, employing palm midribs is possible to create wealth and awareness to the Egyptian market, diminishing costs related to wood importation and the repercussions of exhausting non-renewable resources on the environment.

The date palm midribs were successfully used in different furniture applications such as parquet boards, doors, and "Mashrabia" handicrafts, a design similar to architectural window elements commonly used in Arabic residences for privacy [3], [5]. These applications indicate that characteristics palm midribs present stability and durability to be employed as a skeleton for structures as well as laminated materials. Customers are increasingly looking for sensorial properties and emotional experiences while interacting with products [7]. The sensorial properties of the materials should be applied to create meaningful perceptions and associations that attract the user. The surface of palm midribs gives a unique feeling and impression emphasizing tactile qualities. Special aesthetical advantages of the palm midribs such as natural warmness, calming, kindness and urbanity are crucial for designers to create products that evoke powerful emotional experiences.

Thus, this paper focuses on the tangible and intangible characteristics of palm midribs in order to assist in representing a new market value. These unique tangible material characteristics are based on sensory aspects and can be categorized into visual and tactile attributes. The visual attributes of the material are according to the placement and orientation of the midrib strands, either horizontally or vertically. The horizontal placement creates the image of different shades of wood lines across the product surface. The vertical formation creates a mosaic looking effect that appears similar to marble or stone. The vertical placement shows the square base of the strands and highlights the different wood shades. The tactile attributes are the rough texture of the material. If the strands are placed horizontally without treatment the material is raw with a slightly rough texture. Vertically, the square bases of the strands are normally sanded giving the material a smoother surface.

Experimental methods: Material evaluation from different perspectives
New ideas evolve out of the tension between different disciplines. In the product development process, designers and engineers face a perceptual gap that is driven from their different educational backgrounds [8]. It is important for both professions to adopt a more explorative approach to each other that may help in providing the platform for unconventional innovation. The interpretation of a material like palm midribs from different aspects can give deep insights inherent within the user experience. The experimental methods carried out in this research have a qualitative nature and were developed to evaluate and gather information on the following:

- Process of working in a multidisciplinary team to identify the issues/ conflicts;
- Participants' perceptions and opinions on the materials;
- Similarities and differences between the professional perspectives;
- Complications and difficulties working together.

The method of research was a semi-structured interview conducted on a sample of experts according to their backgrounds, knowledge, and experiences in working with the midrib material. Participants were both males and females, ranging from 24-34 years of age, originating from Egypt. The total sample consisted of 13 participants including:

- 7 designers concerned with palm midrib material and working in the furniture design field.
- Co-founder of Jereed, a start-up that develops products using the midrib material. The participant was from an engineering and machinery background.
- CEO and engineer from Jozour, a start-up that also develops home accessories using midrib material.
- Business developer at Jozour start-up.
- 2 engineers concerned with material sciences from the German University in Cairo.

The qualitative research methods conducted were:
1. Semantic Differential Scale
   The Semantic Differential Scale was developed by Osgood in 1957 [9] to test a users' perception towards a product. This method helps to get an idea of how users perceive an object to further develop or modify it. It employs a set of bipolar adjectives to describe the product or, in this case, the material. Participants are then asked to select on a Likert scale the relevant term that shows how they perceive that object. For example, whether they perceive a product to be safe or dangerous, attractive or ugly, etc. [10]. The keywords used in a scale are derived from the human senses including visual, tactile, haptic, auditory, gustatory, and kinesthetic so that a clear idea of their perceptions can be analyzed. This method was used in this study to identify the similarities and differences between the perceptions of designers and engineers. Fig. 2 shows the four midrib board samples materials produced by different techniques given to the participants. They were asked to rate the sample according to several sensory keywords. The DPLM wood boarding samples were:
A. Midribs arranged horizontally on longitudinal and tangential surface

B. Midribs arranged vertically on vertical and cross section view
C. DPLM board with veneer coating
D. DPLM board as sandwich panel

Results were plotted in a map, as shown in Fig. 3, and analyzed according to the keywords in common between designers and engineers as well as the differences.

Fig. 2: Four samples of different midrib board techniques shown to the participants. From left to right, the samples were DPLM on longitudinal and tangential surface (midribs arranged horizontally), DPLM on vertical and cross section view (midribs arranged vertically), DPLM with veneer coating, and DPLM board as a sandwich panel [1].

2. Focus-group session

A focus group was conducted to address the process of working together and conflicts that would rise between each profession. The sample was focused because it consisted of participants who all worked on developing products using the palm tree midrib material. The group consisted of four participants; two designers and two engineers. Their background formation was their level of experience working with the midrib material. The designers were furniture designers who previously developed midrib products and the engineers were co-founders of Jozour start-up. The participants were asked how they perceived the material, the communication process, and their knowledge about the market through a semi-structured interview. They were asked what was successful with the process of working with another discipline and what they would do to develop it.

Findings: The perceptual gap between two disciplines

According to the Semantic Differential Scale, a clear gap was apparent between the engineers and designers regarding how they perceived the samples in respect to strength, quality, and cost. Fig. 3 shows the mapping of the perceived keywords where it indicates the clear gaps/ skips related to the material and the similarities or agreements are shown when the lines are close to each other. The scale highlighted the words related technically to the material properties were perceived differently. For example, engineers would perceive the material as synthetic, precious, and expensive whereas designers would see it as natural and common. This indicates that due to the expertise and experience of the engineers with the material they can perceive its properties technically in a more realistic sense. By analyzing where the disagreements and agreements are, we can begin to gain an understanding of the similarities and differences between the two disciplines.

The analysis of the results in Fig. 3 showed that there were some differences between the two professions, meanwhile Fig. 4 describes the keywords that designers and engineers individually perceived. The common perceptions between both professions (warm, high quality, friendly, strong, bright, and positive) were highlighted and analyzed as the basis for the platform development. This indicated that participants of both professions perceive the material according to those keywords. It is also important to underline that the term "high quality" was agreed on by the participants in relation to two different samples which were: B. midribs arranged horizontally and D. DPLM board as sandwich panel. This can be due to their experiences with the material or according to their emotions towards the samples.

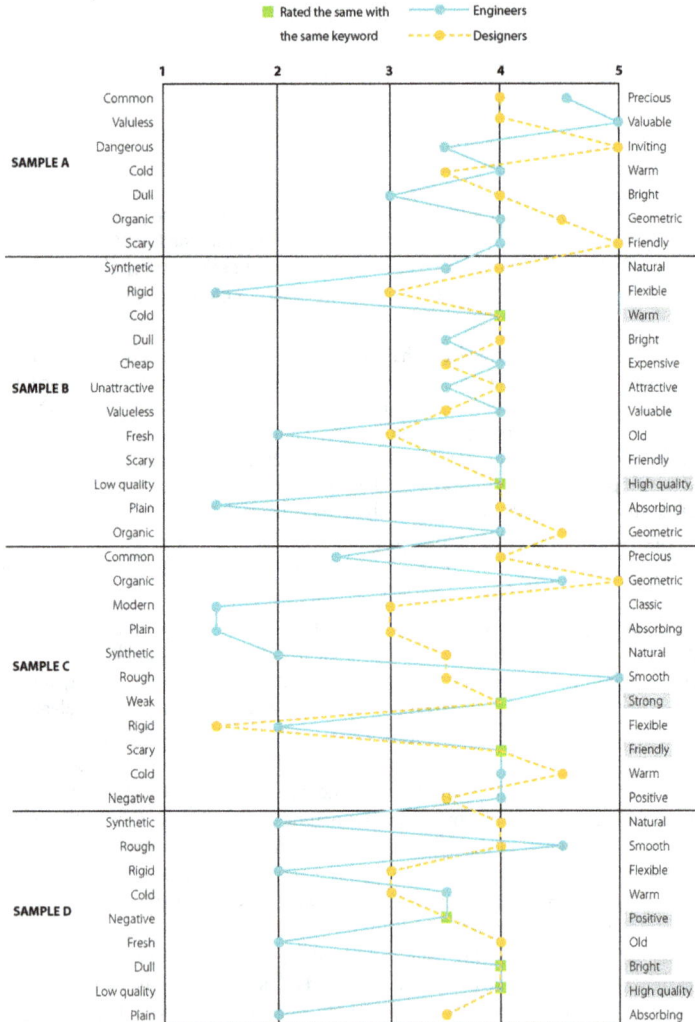

Fig. 3: Semantic Differential Scale analysis of similar keywords employed by engineers and designers. The map shows the words that were in common and the ones that were different.

Materials Research Forum LLC
doi: https://doi.org/10.21741/9781644900178-27

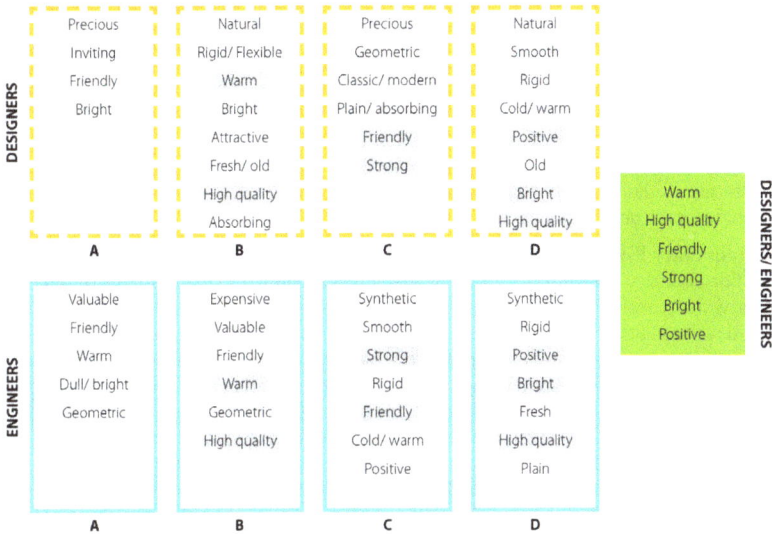

Fig. 4: Keywords perceived by designers and engineers. The highlighted words were the ones found to be in common between participants of the two professions.

When asked in the focus group if they believe to perceive the materials differently a designer stated that although both designers and engineers focus on the function, they do not always perceive the finishing in the same way. This statement was also proven correct using the scale presented in Fig. 3 because there were clear disagreements on the finishing of a material in regard if it was (or not) smooth or modern. Because of the involvement of engineers with material producing they realized the material differently to the designers and the market because they were concerned with sustainability promoting. However, from their business experience, they found that the local market was not interested in this aspect. It is more interested on new functions such as product customizing; adding new colors, illustrations or patterns to useful products.

Designers found the challenges in communication were due to a lack of flexibility with testing or changing the material as well as limited appreciation towards the value of the design. The design is able to transform the product which is an understanding that is difficult to communicate to the manufacturer. Engineers found that highlighting specific terminologies and keywords, which are important to designers, is generally missing in a multidisciplinary project process. They felt that these needs should to be clearly delivered in two or three sessions prior to working on a project so that the process moves forward smoothly without miscommunication.

The design discipline focuses on the emotional elements inherent within a user's experience. Therefore, the technical information about a material such as the midrib should be provided in a form that is matched with people's needs and triggers the designer's sensibility. By developing everyday products for users to get familiar and evoke the market curiosity the midrib material could become accepted. Designers believe that material characteristics and limitations need to be clearly communicated and shown visually by engineers before beginning a project. When marketing and selling the products, its value must be shown in the pricing of it to indicate the quality of the material and the design.

Design understanding for engineers should be enhanced to reinforce feasibility of new palm midrib applications. Engineers should work in developing ideas in an integrative way with the designers early in a project. Based on the experience of the engineers, it was found that the market demand was for a unique, well-designed product. However, in order to create a smoother workflow, designers must reach an understanding regarding the material properties and manufacturing methods. They must also communicate their ideas technically in a logical manner for the engineers to understand their idea interpretation.

Multidisciplinary Platform Description: a perceptual synthesis of midrib properties
The multidisciplinary platform is a proposal developed according to the synthesis of three project elements which consists of: using the emotional and symbolic characteristics of the palm tree midribs, developing the business strategy, and incorporating engineering aspects. In the frame of these three areas, market research is the key factor that aims to understand the needs of customers to discover the market opportunity. Market research is the core of any business strategy that, in the case of products made of palm tree midribs, helps both designers and engineers to diffuse a new meaning to the value of the midrib material from different angles. However, according to **Fig. 5**, the main objective of the suggested platform is to outline the common areas of interest in each field. This implies that, for designers, the properties of the material related to the tangible sensory characteristics including visual tactile, haptic, auditory, gustatory, and kinesthetic perception or the intangible characteristics are mainly focused to evoke a certain emotion that cannot be communicated or described.

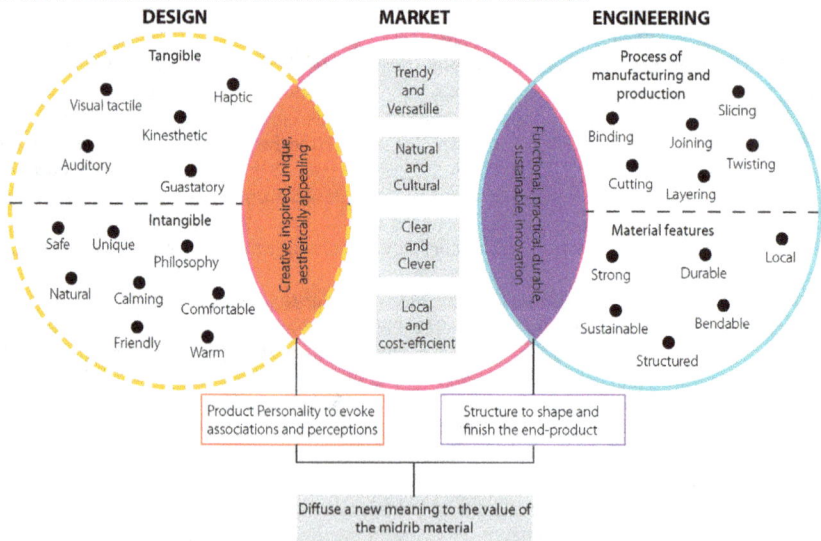

Fig. 5: Multidisciplinary Platform to highlight the main elements of each profession and the four key factors needed by the market.

The co-designing process is an opportunity to develop new experiences through several applications that allow designers and engineers to maximize the value of product creation. In order to track the integration of design and engineering aspects as a business strategy, the multidisciplinary platform emphasizes four inspiring topics to establish a common ground of interest extracted from the different vocabulary and terminology that continually appear among the concerned disciplines using the palm tree midribs: (1) trendy and versatile; (2) natural and cultural; (3) clear and clever; (4) local and cost-efficient (Fig. 5). These four topics cover the impact of the midrib material (tangible and intangible characteristics) on the user's insights that contain language both sides can understand. In other words, the four topics include different types of knowledge and market demands implying a specialized set of properties that should appeal to the senses of users.

Through a specific set of features of midribs, two overlapping roles will represent the way by which the four mentioned topics can be appropriately interpreted. Each role falls to one of the two disciplines:

1. Providing structure to shape and finish the end-product (engineering);
2. Creating product personality that evoke associations and perceptions (design).

This scheme is a perceptual synthesis of properties that can affect more than the direct visual and tactile features of the product. It guides designers and engineers to employ the features of midribs in an unconventional way to evoke feelings of sensory experience related to the user of the product.

Conclusion

According to the research, it was found that both designers and engineers have the same end goal: to produce new products using the midrib material that can be sold to the market. However, both professions focus on different areas of the process according to their disciplines. Therefore, the platform shares the current information that is accessed by both disciplines to aid in the product development process. By developing a strategy that enables a successful multidisciplinary collaboration the value of the midrib material will be enhanced. On the other hand, traditional research methods, mainly engineering testing techniques, used to explore the possibilities of applications for such a material have to be developed in collaboration with designers to gather deep insights into the sensorial characteristics of the midrib material. The intangible aspect of midribs will attract a greater market which will enable an economically stable business approach.

The need and demand for an alternative material to the imported wood is increasing rapidly in Egypt. However, the market is still in the phase of being unsure what to expect from this material. It is the role of the designers and engineers to communicate and work together to bring forward a product to the market with this material that fulfills the market gaps; a trendy, functional, affordable, unique, attractive, and durable locally produced product.

Future recommendations

It is recommended that the study continues testing on a wider scope of users to be implemented onto the market. This will create a greater knowledge of the needs of market to develop a successful multidisciplinary project. The Semantic Differential Scale can be adapted and tested on consumers to get an idea of their perceptions and to analyze if they share the same perceptions as the designers and the engineers. The developed platform needs to be tested in the form of a case study with several designers and engineers to find out more information on the two professions and how to create successful collaboration for the future.

References

[1] CrossLink Technoqlogies. Preliminary Technical Report. Vancouver: U.S. (2016).

[2] H. El-Mously, Innovating green products as a mean to alleviate poverty in Upper Egypt, Ain Shams Eng Journal (2017) 3-7. https://doi.org/10.1016/j.asej.2017.02.001

[3] H. El-Mously, The Industrial use of the Date Palm Residues: An Elequent example of Sustainable Development. 11th International Conference on Date Palms, Al Ain, United Arab Emirates (2001) 866-886.

[4] Jereed Furniture Products [Image] (2015), Retrieved December 3, 2018 from https://www.facebook.com/Jereedeco/photos/a.1488347228138145/1488347191471482/?type=1 &theater.

[5] A. Eldeeb, Recycling Agricultural Waste as a Part of Interior Design and Architecture History in Egypt, The Academic Research Community Publication (2017) 2-6. https://doi.org/10.21625/archive.v1i1.116

[6] R.A. Nasser, Influence of board density and wood/cement ratio on the properties of wood-cement composite panels made from date palm fronds and tree prunings of Buttonwood. Alexandria Sci. Exchange J. 35-2 (2014) 133-145. https://doi.org/10.21608/asejaiqjsae.2014.2588

[7] R. Jensen, The Dream Society. The coming Shift from Information to Imagination, McGraw-Hill Book Company, 1999.

[8] J. Cagan, C. Vogel, Creating Breakthrough Products, Financial Times Prentice Hall, Upper Saddle River, New Jersey, 2002.

[9] C.E. Osgood, C.J. Suci, P.H. Tannenbaum, The Measurement of Meaning. University of Illinois Press, Urbana (1957) 76-124.

[10] S.H. Hsu, M.C. Chuang, C. C. Chang. A semantic differential study of designers' and users' product form perception. International Journal of Industrial Ergonomics. 25 (2000) 375-391. https://doi.org/10.1016/s0169-8141(99)00026-8

By-Products of Palm Trees and Their Applications   Materials Research Forum LLC
Materials Research Proceedings 11 (2019) 325-332   doi: https://doi.org/10.21741/9781644900178-28

The Technical Heritage of Date Palm Leaves Utilization in Traditional Handicrafts and Architecture in Egypt & the Middle East

E. A. Darwish[1,a*], Y. Mansour[1,b], H. Elmously[2,c], A. Abdelrahman[3,d]

[1]Dept. of Architecture, Ain Shams University, Egypt

[2]Dept. of Design and Production, Ain Shams University, Egypt

[3]Dept. of Structural Engineering, Ain Shams University, Egypt

[a]eman.atef@eng.asu.edu.eg, [b]yasser_mansour@eng.asu.edu.eg,
[c]hamed.elmously@gmail.com, [d]amr.abdelrahman@eng.asu.edu.eg

Keywords: date palm leaves, technical heritage, architecture, handicrafts

Abstract. Date Palm Trees enjoy a recognized stature in Egypt since the ancient times. The abundance of Date Palm Trees and their distribution over the Nile valley, Delta, Oases and Sinai in Egypt granted them familiarity with the people that remains until the present. This familiarity is represented in the survival of various traditional techniques in the utilization of Date Palm Trees pruning residues in the fields of handicrafts and construction in rural Egypt. On the top of those pruning residues are the leaves, which rank the highest in the annual quantities. Date Palm Leaves are still widely used in traditional handicrafts and building in the poor rural areas in Egypt due their renewable availability and low cost. This paper aims to analyze the technical heritage behind those traditional utilization fields in order to identify the dominant techniques used. Those techniques, including Bundling, Rope Fastening and friction based assembly, can be introduced as the basis on which the development of those techniques for modern and contemporary uses of date palm leaves should be based in order make use of the surviving skills to sustain the familiarity needed to guarantee the success of the developed uses.

Introduction

Date palm tree acquires great importance historically, economically and socially in Egypt. The pruning residues of Date Palm are utilized in many traditional industries and construction by the cultivators and craftsmen in Egypt; thus playing a huge role in sustaining the rural societies against the immigration to urban cities, as those date palm related industries support over one million families in Egypt [1].

Date Palm Leaves, representing 52.9% of the annual date palm pruning quantities [2], are used in various fields historically. The palm leaves were fundamental in manufacturing baskets, clothing and sandals in ancient Egypt and Nubia. The roofs were constructed by split palm trunks and leaves and the interior walls were covered by palm leaves ornaments [3]. Palm midribs and trunks have used for roofing in a fashion that still survives in Siwa Oasis [4].

Those ancient evidences prove the adaptability of date Palm leaves to our environment [5]. This high adaptability, besides flexibility and low cost, qualified the material to gain popularity and the trust of the rural craftsman in Egypt [3], which opens the door to exploit the potentials of this materials in contemporary uses as a promising fields for small projects. Those new uses ought to originate from that technical heritage in order to help those surviving skills to flourish and make use of the craftsman with that irreplaceable know-how. However, most of the previous researches have not introduced an integrated analysis of that heritage and the detailed processes of the traditional techniques that are still surviving.

Aim and Methodology

Two of the basic elements of sustaining the industrial development of a local material are to benefit from the current craftsman skills in order to attract them to participate in the development, plus saving and expanding the base of the artisans that practice traditional techniques that can be reinforced and developed [6].

Accordingly, the industrial development of date palm leaves should be based on the traditional techniques in order to obtain a solid ground to guarantee its continuance. Therefore, the paper aims to identify the dominant traditional techniques that are most qualified to be the basis of the industrial development of date palm leaves.

This aim is fulfilled by analysis of the traditional techniques of the utilization of date palm leaves in handicrafts and construction. This analysis depends highly on the investigation of those traditional manufacturing processes in order to identify those techniques.

Date Palm Leaves in Traditional Handicrafts

Crates and Bird Cages. Nowadays, most of the date palm leaves, uses in handicrafts, specifically the midribs parts, are specialized in making crates and bird cages [3]. The manufacturing process of a bread crate is as the following as illustrated in Fig. 1:

1. The pruned date palm midribs are shaped into almost rectangular cross sections and cut to the needed specific lengths of the longitudinal and transverse members.
2. The places in the longitudinal members where the holes are to be punctured are marked. Those places are punctured by a wooden hammer.
3. The transverse members are hammered through the holes of the longitudinal members, and 2 short supporters are added on the sides of the middle longitudinal member to increase the stiffness and decrease the deformation under the load carried by the crate.

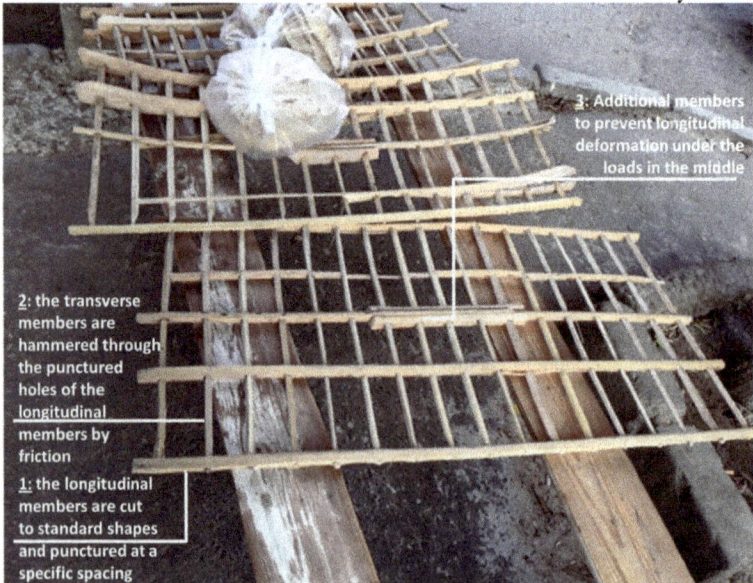

Figure 1 Bread crates made from Date Palm Midribs.

The bird cages made from date palm midribs depend essentially on the same friction-driven fixations adopted in makes the bread crates as discussed earlier. The manufacturing process of a bird cage is as the following [3]:

1. The crate is assembled from the base up where the dowels are fixed vertically in the holes of the punctured strips until a cube is formed.
2. Additional punctured strips are laid horizontally onto the dowels to tie them together until the cube is completely latticed.
3. The top edges of the dowels are hammered to the same level.
4. Most bird cages, as illustrated in Fig. 2 , the corners of the cage are fastened with wires to make increase the overall stiffness of the cage. Another method of fastening the corners is using linen ropes around a traditional corner-block joint as illustrated in Fig. 3.

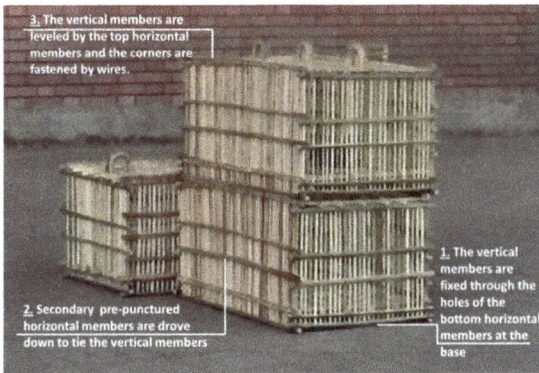

Figure 2 Bird cages made from date palm midribs

Figure 3 Cages made from date palm midribs and linen ropes

Handmade Furniture. In addition to crate making, traditional furniture items such as beds, chairs and tables are being made until the present day using date palm midribs. The manufacturing process of a chair is as the following [3]:

1. The midribs used vary from green to be easily bent to make the chair frame, and dry to be stiff enough to make the latticework. Those midribs are cut according to the needed dimensions with mostly rectangular cross sections.

2. The places where the holes are to be punctured are marked. Those places are punctured by a wooden hammer.

3. The legs of the chair are taken from the broad ends of the midribs. The seat members are cut from the straightest parts of the midribs, then the back and the armrests. Then those parts are assembled in the sequence illustrated in Fig. 2.

4. 4 cm- nails are used to fasten the armrests together, and 10 cm-nails to fix them to the seat.

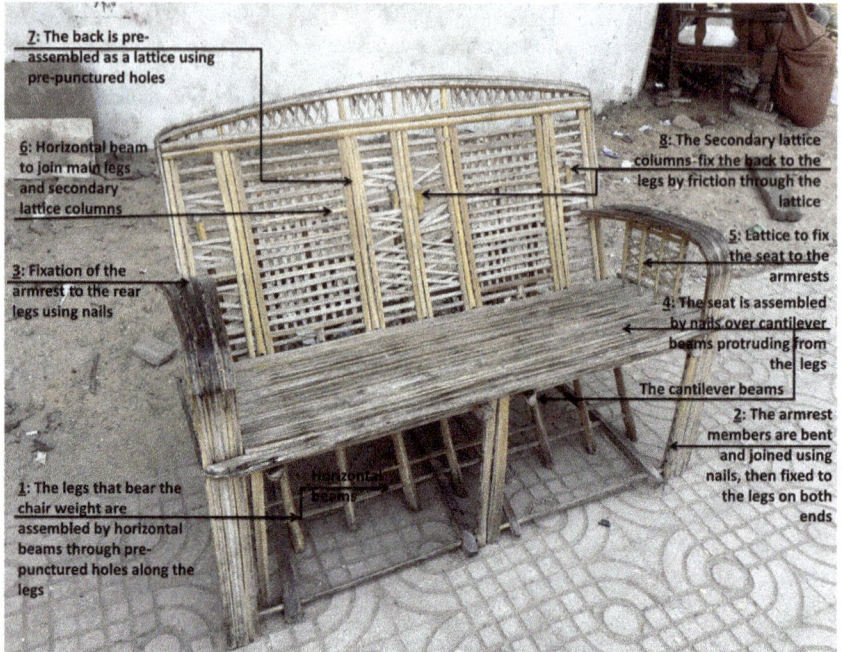

7: The back is pre-assembled as a lattice using pre-punctured holes

6: Horizontal beam to join main legs and secondary lattice columns

3: Fixation of the armrest to the rear legs using nails

1: The legs that bear the chair weight are assembled by horizontal beams through pre-punctured holes along the legs

Horizontal beams

8: The Secondary lattice columns fix the back to the legs by friction through the lattice

5: Lattice to fix the seat to the armrests

4: The seat is assembled by nails over cantilever beams protruding from the legs

The cantilever beams

2: The armrest members are bent and joined using nails, then fixed to the legs on both ends

Figure 4 Chair made of Date palm midribs members

It can be realized that making furniture using date palm midribs depends on the same inherited techniques in making bird cages and crates; reforming the members, drilling and using friction in the connections while depending entirely on the natural strength of the midribs without using external reinforcement.

Leaflet Mats *Hassir* and Ropes. In addition, date palm leaves and midribs fibers are used in making baskets and ropes as illustrated in Fig. 5 & Fig. 6. The manufacturing process of a mat is as the following [7]:

1. The leaves are stripped from the palm branch and laid out under the sun for two days.
2. The leaves are gathered in bunches and soaked in non-saline water for a day to facilitate bending and weaving.
3. The leaves are plaited together according to the desired shape and pattern.

Figure 5 Date palm leaflets plaid

Figure 6 Nets and mats made from date palm leaflets

Date Palm Leaves in Traditional Architecture

Date palm midribs are used in many rural villages in the present day for roofing, fencing and wall sheathing after being woven into mats *Sadda* using sheath ropes that are extracted from date palm also in the pruning process [8].

Wickerwork Mud Layering [5]: However, the true origins of using date palm midribs in architecture go back to the ancient Egyptian rural housing in a more complicated way. Previous studies predicted that the roofs of the small houses of the workers in Ancient Egypt were made of close mats *Sadda* of palm midribs rows over wooden beams. The midribs were covered with a paste of mud that was so thick that it could be rain proof, and then more piles of midribs and straw were put over the mud layer as illustrated in Fig. 7 & Fig. 8.

Figure 7 Piles of date palm leaves over the roof in a house in Upper Egypt

Figure 8 Date palm leaves over wooden poles for roofing in a storage house in Menya, Egypt.

Simple post-beam with palm leaves mats *Sadda:* On the other hand, simple huts and sheds in Egypt and UAE depend more on the skill of the tight weaving of date palm leaves, midribs and reeds for roofs and walls solely without additional mud layers [7]. The main idea of this technique is tying whole date palm leaves and reeds using ropes to a simple post-beam wooden structure system made of tree branches as illustrated in Fig. 9.

By-Products of Palm Trees and Their Applications   Materials Research Forum LLC
Materials Research Proceedings **11** (2019) 325-332   doi: https://doi.org/10.21741/9781644900178-28

Figure 9 Date palm leaves hut in Upper Egypt

Typically, the building process is as the following [7]:
1. The dried date palm leaves are prepared by trimming the petioles and the leaves can be removed if desired.
2. The midribs are then soaked in non-saline water for a night while preparing the primary structural system using palm trunks or tree stems.
3. Making the *Sadda* mats by threading the ropes and the midribs. This process takes about 4 people for the walls and the roof covering.
4. Sometimes, the traditional date palm walls were built to be double layered. A double wall consists of 2 layers of midribs mats *Sadda*, and leaflets mats *Hassir* are added in the middle to provide a space for heat insulation and prevent sand from penetrating the spaces between the midribs in the wall *Sadda* mats.

Simple post-beam with date palm leaves bundles roofs and walls [7]: In hot arid climates such as in United Arab Emirates and in Upper Egypt, the main function of these huts is to provide a cheap shelter from direct sun, dusty winds and winter while sustaining adequate ventilation inside. This simple tying technique is more developed to be denser using weaved date palm midribs bundles as illustrated in Fig. 10 & Fig. 11.

Figure 10 Date palm leaves bundles rural hut   *Figure 11 Dense date palm leaves huts in rural Egypt*

Fencing [9] : The simplest way of using date palm leaves in construction is in building fences. Whole date palm leaves are laid on the ground and tied together using ropes to make a single sheet that is planted in the wet soil. Then, thin date palm midribs bundles are used as horizontal beams to ensure the sheets are straight and tight as illustrated in Fig. 12.

Figure 12 Date palm leaves fence in rural UAE

Analysis of the technical heritage of date palm midribs

Table 1 Summarized Analysis of the traditional techniques in handicrafts and architecture

Experiment	Basic Concept	Main Techniques Used	Major Notes
Crates and bird cages	Lattice of transverse members through holes of pre-punctured longitudinal members with wires/nails at the corners	Puncturing Latticing Friction-based assembly	Extensively used to present day
Handmade Furniture	Beam-joined pre-punctured columns with which secondary members are nailed to hold the lattices	Puncturing Latticing Nails Fastening Friction-based assembly	Extensively used to present day
Leaflet Mats *Hassir&* Ropes	Shredding leaves and fibers threaded and weaved	Threading Weaving	Extensively used to present day
Wickerwork Mud Layering	Timber beams with midribs mats *Sadda* covered with mud and leaves and straw piles	Rope Fastening Mud paste	Limited use in the present day
Simple post-beam with palm leaves mats *Sadda*	Primary timber structure with rope tied mats for roofs and walls	Rope Fastening Threading Weaving	Extensively used to present day in temporary huts
Simple post-beam with whole leaves bundles	Primary timber structure with rope stacked bundles	Bundling Rope Fastening	Extensively used to present day in poor villages houses and huts
Fencing	Vertical implanted leaves mats *Sadda* with horizontal bundled beams	Threading Weaving Bundling Rope Fastening	Extensively used to present day in fencing around cultivated lands

According to the analysis inTable 1, the most dominant and surviving traditional techniques are bundling, rope fastening, weaving and friction-based assembly.

Conclusion & Recommendation

The exploiting of the role of date palm leaves in the development of the rural communities in Egypt and the Middle East depends necessarily on their traditional techniques that originated from the people and survived in the middle of the contemporary industrialization. That surviving technical heritage is the main key by which date palm leaves uses can be modernized to match the need of the youth, while investing, and saving, the skills and the know-how that can be developed as long they still enjoy a wide base of use in the present.

This paper included analyses of the skills and techniques still used in handicrafts and construction in order to identify the most qualified for further development. The most dominant techniques that are identified from those analyses are bundling, rope fastening, weaving and friction-based assembly. It can be concluded that the techniques that are most qualified for further development are more present in the traditional huts in the poor rural areas.

This means that although most of the previous startups and small projects depending on date palm midribs are more concerned with furniture, the development of the utilization of date palm leaves and midribs in architecture is a promising field for researchers and artisans. The challenge this development needs to face is the need to produce sophisticated architecture that meets the modern lifestyle that the rural youth pursue in Egypt.

References

[1] S.A. Bekheet, S.F. El-Sharabasy, Date Palm Status and Perspective in Egypt, in Date Palm Genetic Resources and Utilization: Volume 1: Africa and the Americas, ed. by J.M. et al. Al-Khayri ([n.p]: Springer Science+Business Media Dordrecht, 1 (2015) 113. https://doi.org/10.1007/978-94-017-9694-1_3

[2] Hassan Hosseinkhani, Markus Euring and Alireza Kharazipour, Utilization of Date palm (Phoenix dactylifera L.) Pruning Residues as Raw Material for MDF Manufacturing, Journal of Materials Science Research, 4 (2015) 46-62. https://doi.org/10.5539/jmsr.v4n1p46

[3] Menha El-Batraoui, The Traditional Crafts of Egypt (Cairo: The American University in Cairo Press), 2016.

[4] R.M. Ahmed, Lessons Learnt from the Vernacular Architecture of Bedouins in Siwa Oasis, Egypt, in The 31st International Symposium on Automation and Robotics in Construction and Mining (London, UK: ISARC), 2014. https://doi.org/10.22260/isarc2014/0123

[5] Omar A. Azzam, The Development of Urban and Rural Housing in Egypt (PhD Thesis, The Swiss Federal Institute of Technology, Faculty of Technical Science), 1960.

[6] Joseph Kennedy, *Building Without Borders: Sustainable Construction for the Global Village* (Ontario, Canada: New Society Publishers, 2004).

[7] Sandra Piesik, *Arish: Palm-Leaf Architecture*, 2nd edn (London, UK: Thames & Hudson, 2012).

[8] W. H. Barreveld, 'FAO AGRICULTURAL SERVICES BULLETIN No. 101: DATE PALM PRODUCTS ', in *Food and Agriculture Organization of the United Nations* <http://www.fao.org/docrep/t0681e/t0681e00.htm#con> [accessed 25 April 2018]

[9] Ayah Eldeeb, 'Recycling Agricultural Waste as a Part of Interior Design and Architectural History in Egypt', in *Cities' Identity Through Architecture and Arts* (Helwan: IEREK, 2017), I.

By-Products of Palm Trees and Their Applications   Materials Research Forum LLC
Materials Research Proceedings **11** (2019) 333-342   doi: https://doi.org/10.21741/9781644900178-29

Using Printed Palm Leaflets in Modern Crafts according the International Fashion Trends

Heba Mohamed Okasha Abu Elkamal Mohamed Elsayegh

Faculty of Applied Arts, Benha University, Egypt

heba.okasha2008@gmail.com, heba.okasha@fapa.bu.edu.eg

Keywords: printed leaflets, modern crafts, fashion trends, dyed palm leaflets, textile dyes, textile printing

Abstract. Egypt is one of the most important and oldest countries in the world that is known for producing dates, because date palms are available in all the cultivated areas of the country. Taking in account the date palm by-products (such as palm leaflets), this paper focuses and outlines the utilization of palm leaflets in making modern crafts and accessories in a creative way ("out of box") and at the same time following the international fashion trends. This utilization can result in benefits like achieving added economic value in two ways: creating trendy modern fashionable crafts and achieving the concept of "zero waste". Palm leaflets can be mixed and matched with fabrics and leather waste according the design idea. By using techniques of textile printing and dyeing to create more attractive designs according the trend, which can be employed in the palm leaflets and fabrics, a variety of innovated designs are created and can be applied in modern crafts according the international fashion trends for making accessories according the season. Also, dyed palm leaflets are mixed and matched with printed fabrics and leather wastes generating very unique up-cycled products in modern crafts and fashion accessories. At the end, a boxy bag prototype is implemented according the proposed outlines as a final product made from woven leaflets mixed with waste of fabrics and leather up-cycling. The main purpose of this paper is to highlight making fashionable accessories products according to the international fashion trends using mixed media as up-cycling product from (printed & dyed palm leaflets according the color of the season), waste of fabrics and waste of leather.

Introduction

Cultivation of date palms (*Phoenix dactylifera*) in Egypt goes back to thousands of years. Agricultural operations on date palm, like pollination, are known at least since 2,500 BC. In Egypt, date palm is cultivated and grown everywhere. Date palm grows in warm weather countries and has a tall trunk with a mass of long pointed leaves at the top [1]. Egyptian people use the parts of date palms in many things, like using palm leaflets in plaited cradle [2]. Nowadays date palm plantations are spread out all over country; wherever water is available. Egypt has more than 15 million date palm trees and is considered the world biggest producer of dates. It has a long heritage of utilization of date palm by product since ancient times [3]. The palm leaflets have been collected from the palm trees and dried by a traditional way, then woven as palm leaves folding (Fig. 1), so after that they can be used in many things, such as creating and making new ideas in modern crafts (such as, bags and accessories), furniture, tables sets and many other creative productions. These creative industries may be made by traditional or modern way with a unique and different concept according the international fashion trends for making accessories according the season.

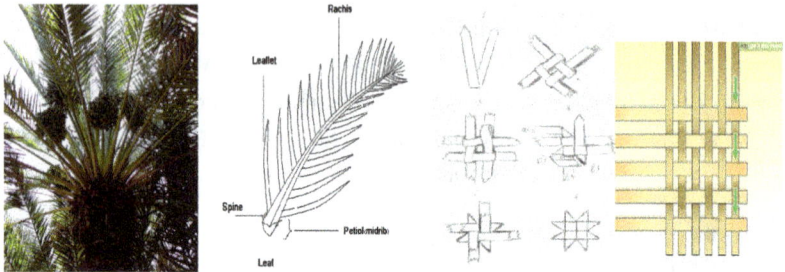

Fig.1: Egyptian date palm tree - parts of the leaf- ways of weaving leaflets [4][5].

Purpose of the research

- Making unique modern crafts (bags and accessories) from printed and dyeing date palm leaflets according to the international fashion trends for the season spring/summer (S/S) 2018, So printing and dyeing the palm leaflets is just for making catchy fashionable colors .
- Applying textile printing, dying methods and techniques as only helping method in coloring and decorating the date palm leaflets, which could be mixed with fabrics and leather waste, for creating modern crafts (bags and accessories) by implementing the concepts of "Zero Waste" and "up-cycling" using these materials.
- The research deals with only the innovation and creativity in using different materials in designing bags according the season trends focusing mainly on Palm leaflets.

Methods and (Methodology) of the research

The research used a mixture of experimental and analytical methods:

Experimental methods

- The researcher used experimental method by implementing textile dyeing and printing (methods and techniques) in many kinds of palm leaflets weaving according to the design of the type of bag or accessory, and the purpose of this step is to produce many colors of palm leaflets that meet the international fashion trends of the season.
- The researcher made some suggested sketches ideas for bags according to the international fashion trends S/S 2018.
- The researcher implemented <u>one only prototype</u> from the ideas, using printed palm leaflets, waste of fabrics and leather to make an "up-cycled" product according the international fashion trends in bags and accessories .

Analytical method

- The researcher used the analytical method in studying the results of using traditional manual dyeing and printing methods in coloring palm leaflets to meet the colors of the season (international season colors S/S 2018).
- The researcher chose the best result and the most suitable printed palm leaflets for the accessory piece (Suggested implemented bag) in both sides (design, color trends and performance requirements).

Literature review

The international fashion trends in bags and accessories

The international fashion trends are very relevant for fashion industry. International fashion trends mean what is hip or popular at a certain point in time. While a trend usually refers to a certain style in fashion or entertainment [6]. The international fashion trends lead the market to meet the client requirements and needs. Bags and accessories are very important elements for fashion industry. Palm bags are an important essential for Summer Fashion Trends 2018 according to many of the international fashion trends reports and many other sources in the fashion industry. Many brands have shown their new collections summer 2018 and lay emphasis on the bags and accessories made from hand-woven palm as environmentally friendly products. This can diminish the wastes derivate from palm trees and also follow the international fashion trends, sustainability and environment requirements (Fig. 2). According to 20 fashion trends for spring/summer 2018, the usage of palm leaf in handcrafts and manufacturing is essential in this season. Also, according fashion trends 2019, palm leaf is also an important and essential fashion trend and very good inspiration source for textile printing, which can be used in bags and accessories beside other trends like tie-dye [7–13] (Fig. 3).

Fig. 2: Bags and accessories made from palm leaflets according the international fashion trends (international brands S/S 2018) [8] [12] [15].

Fig. 3: Bags and accessories made from palm leaflets as an important source of inspiration for modern handcrafts and accessories designers through many seasons [13].

The international fashion trends in bags materials as modern crafts

The international fashion trends in bags and accessories deal with everything related with colors, silhouettes and collection materials, which were presented at some important trade fairs in Milano (Italy), offering a preview of what designers have got as inspirations for their spring/summer 2018 footwear and leather accessory designs. And there was plenty of inspiration to be found. Leather was still the main attraction with a multitude of new patterns and finishes. But the rise of non-leathers could not be ignored. New meshes and neoprene versions catapulted materials into the future, whereas natural straws, raffia and hemp promoted eco-friendliness and sustainability [14] [15].

So, using natural materials like palm leaflets is so trendy in fashion industry and can add value to the product by implementing the concepts of sustainability and eco-friendliness. The international fashion trends and the big brands realized very well the raise of consumer awareness of these concepts and how implementing them, can help increase consumer loyalty to the brand, especially sustainability and eco-friendliness, which become a worldwide demand and necessity.

Experimental methods

Using textile dyestuff in dyeing palm

The researcher implemented some methods and techniques of textile dyeing and printing in coloring the weaving palm leaflets, to make different colors and effects on them. These obtained colors and designs differ from each other according to the idea and the implementation of the designs. Also, the researcher made these designs according the international fashion trends in patterns and colors of bags and accessories.

The researcher used reactive dyes in coloring raw palm leaflets, fiber reactive (Procion type Dyes). The dye is powder; the employment of dyes as follows:

The reactive dyes were in powder form and then dissolved in water. The liquor ratio was 15 g dyestuff reactive dye: 1 liter of water.

Step 1: As known for reactive dyes, it should be started in a neutral medium when the dye does not react either with the fiber or with the water to prepare the dyestuff solution.

Step 2: After washing the woven palm leaflets with warm water (Fig. 4), they were immersed into the dyestuff bath for (25-30) min.

Step 3: Common salt (NaCl) was added to the reactive dyes, dissolving it in the bath, and then the woven palm leaflets were left in the bath extra (25-30) min.

Step 4: Finally, the fixation of the dye was done by alkali addition (sodium carbonate or soda ash -Na_2CO_3). As the dye is already exhausted into the palm leaflets, it is not be available for reacting with water. As known the hydrolyzed dye, due to affinity forces, is absorbed by the palm and retained in it [16] [17].

Result: by Implementing this way of dyeing, we can get light coloring and many color tones depending on the following factors:

- The kind of weaving palm leaflets (like warp and weft weaving and according this weaving is narrow or wide or open little bit, so we can obtain more results of light tones and shading), this is can be changeable according the required designs of accessories and bags .

- The rate of dye exhaustion into the palm, by using fiber reactive (Procion type Dyes) because this kind of dye can be very effective in dyeing. This way achieves very nice tones of colors (Fig. 5).

Fig. 4: Egyptian data palm leaflets - Raw Woven leaflets- before dyeing and printing.

Fig. 5: Steps of dyeing and printing the Raw Woven leaflets - after dyeing and printing in the dye bath in home

The researcher didn't use the natural dyes instead of reactive dyes as synthetic dyes because of the following reasons:
- It is more expensive than the reactive dyes as synthetic dyes.
- Although natural dye sources are renewable, but sustainability can be big issue for natural dyes because producing them require vast areas of land and this is can't be achieved easily
- The availability of raw materials of the natural dyes can vary from season to season, place, and species, whereas reactive dyes as synthetic dyes can be produced in all year round.
    - Sometimes Color pay-off from natural dyes and going to fade quickly.

Using textile printing methods and techniques in coloring palm leaflets
There are a lot of textile printing methods and techniques that can be used it in coloring and decorating palm leaflets like: stencil, direct painting and drawing, silk screen and block printing with some motifs and patterns. In this research, direct drawing and stencil were used to make a sample (prototype). Also, in this field of fashion industry, a lot of previous printing methods and techniques can be used to create and implement many designs of bags and accessories according

the international fashion trends (color theme, materials and silhouettes, etc.). The implementation of these techniques was as follows:

- Direct painting and drawing: we can draw and indicate the outline of the motifs and patterns that we want to apply in the weaving palm leaflets. Then colors of parts or textures are according the design coloring theme and plan (Fig. 6 and Fig. 7).

- Stencil: stenciling produces an image or pattern by applying pigment to a surface over an intermediate object with designed gaps in it, in which is created the pattern or image, only allowing the pigment to reach some parts of the surface. The stencil is both the resulting image or pattern and the intermediate object [18]. Stencil can be applied at any surface, not only in the palm leaflets, like wood, furniture, wall, garment and textile, etc. It is a very effective way to make fast and unique results in decorating and coloring surfaces (Fig 8).

Fig. 6: Experimental results of dyeing and printing - the raw woven leaflets- after dyeing and printing (direct painting).

Fig. 7: Experimental results of dyeing and printing the raw woven leaflets- after dyeing and printing (direct painting).

Fig. 8: Stencil technique and the implementation of the pigment and how it can be applied in many surfaces to create unique designs [18].

Up-cycling trend worldwide:
Up-cycling is known also as creative reuse. Up-cycling is defined as the process of transforming by-products, waste materials, useless, or unwanted products into new materials or products of better quality or for better environmental value, and the opposite of the up-cycling is down cycling [19]. Up-cycling nowadays is a growing trend in every industry and since the necessity to make creative reuse become urgent for unwanted products or waste materials. Up-cycling is an innovated method of recycling waste into products of higher quality. So, up-cycling is a design solution to an environmental problem.

Topics to up-cycle design are indicated as following [20]:
- Always know what is on offer before you start designing.
- Follow your own waste stream.
- No scrap is too small.
- Ugly can be beautiful too.
- Finishing is the key to good design.

Up-cycling ideas in the research:
This research suggested and created new ideas of modern crafts, like bags designs which are made from printed and dyed woven palm leaflets as basic and main material in implementation the design, because Egypt is the first country in producing the data palm worldwide, and this has been achieved according to the previous topic ("always know what is on offer before you start design"), using these palm leaflets in achieving sustainability of the products, as palm leaflets is available raw materials and also coloring this palm leaflets by dying and printing create more added value to the palm leaflets to use in more fashionable crafts and products like fashion accessories (Bags – belts …………..) .

-Also, according to "follow your own waste stream using waste of materials", the researcher used waste of fabrics and leather to reuse and up-cycling in order to create new and innovated products.

In fact, "no scrap is too small", because very small waste of fabrics and leather were employed to make the first prototype (sample).

- By mixing and matching colors and waste of materials in one design, so you can create beauty and innovation from ugly ("ugly can be beautiful and fashionable too").

- According to "finishing is the key to good design", the sample of the bag (first prototype) was made with high finishing quality and accurate details.

Results and discussions

Making the first sample (prototype):

After all the previous experiments in coloring and printing the palm leaflets, the researcher will turn to the main core of the paper which is using this dyed and printed palm leaflets in making modern crafts according the international fashion trends. The idea of the first sample (prototype) depended on using the previous result of printed palm leaflets (Fig. 5) and then mixing and matching them with other materials to make and implement the idea, as follows:

- The researcher also dyed the waste of the Egyptian heavy fabrics with green colors by using also the reactive dyes, to match with the printed palm leaflets color as one of the color trend of the year S/S 2018. The heavy Egyptian fabrics have been dyed using tie-dye technique to make texture on fabrics matching with printed woven palm leaflets.
- Mixing and matching with the three materials: printed palm leaflets, dyed waste fabrics and the waste of leather (Fig. 8).
- Making fast design sketches of bags to choose the most suitable design to be implemented by using the three materials (up-cycling), and also following the international fashion trends summer 2018 (Fig. 9).
- Choosing the most suitable design to be implemented by using the three materials (Fig. 10).

Fig. 8: Steps of making (mix and match) dyed woven leaflets with waste of fabrics and leather up-cycling

Fig. 9: Some suggested sketches of bags suitable to be implemented from printed palm leaflets and (waste of fabrics and leather) up-cycling according the international fashion trends S/S 2018.

Fig. 10: Sample of the implemented bag (prototype) as a final product which was made from woven leaflets mixed with waste of fabrics and leather up-cycling

Product (Prototype) Specifications

Product name	Up-cycling bag as shown in Fig.10
Materials & techniques	Printing palm leaflets (side), waste of fabrics dyed with (tie-dye) method using the same procedure of palm leaflets dyeing in another side). Using waste of leather in the (two sides), as shown in Fig.10
Size	20 cm (L) * 17 cm (H) * 5 cm (W) dimensions of the bag
Description	The idea of making this boxy bag depended on palm leaflets as essential and main material and this palm is dyed or printed, waste of dyed heavy fabrics and leather and the inner fabrics is natural cotton.
Edge material and color	Waste of mustard leather in two sides
Target client	The targeted client age is (20-45) years, loves eco-friendless and sustainability, craze for changing bags, fashion materials and afternoon boxy bag.

Different between this study and the previous attempts in this field (brief critical view):
This study focuses on making fashionable handicrafts from dyed and printed palm leaflets, according the international fashion trends S/S 2018, also using this dyed palm leaflets in designing and making up-cycling crafts like the prototype (bag) which has been implemented by using the printed and dyed palm leaflets and waste of (leather and dyed fabrics) and up-cycling also is very trendy worldwide.

Meanwhile, the previous studies in this field related and focused more in the technical side of using dyes in palm leaflets, yarns and fabrics. Also measuring and enhancing color fastness for all of them.

Otherwise there are some previous studies that deal with coloring and dyeing the palm leaflets without focusing and taking attention to the international fashion trends of the season (colors, designs and shapes of bags.......)

References

[1] Information on: https://dictionary.cambridge.org/dictionary/english/palm.

[2] Loutfy I. El-Juhany- Degradation of Date Palm Trees and Date Production in Arab Countries: Causes and Potential Rehabilitation-Prince Sultan Research Centre for Environment, Water and Desert, King Saud University Australian Journal of Basic and Applied Sciences, ISSN 1991-8178 © 2010, INSInet Publication 4-8 (2010) 3998-4010.

[3] Information on: https://www.bypalma.com/.

[4] Information on: https://www.pinterest.com/pin/308778118171108000/?lp=true.

[5] Information on: http://b-c-ing-u.com/2017/07/25/secrets-sahara-dates-part-2-date-palm-fronds-industry/fig-1-structural-parts-of-date-palm-tree/.

[6] Information on: https://www.vocabulary.com/dictionary/trend.

[7] Information on: https://www.elitedaily.com/p/the-summer-2018-beach-bag-trend-you-need-right-now-8945190.

[8] Information on:
https://i.pinimg.com/originals/29/4f/a2/294fa2a82e349e4cbd7c5a1935f56827.jpg.

[9] Information on:
https://i.pinimg.com/originals/34/3b/a6/343ba6fafe6ddf992cf9e3d60486eda1.jpg.

[10] Information on:

https://www.marieclaire.co.uk/fashion/summer-fashion-trends-2018-553795.

[11] Information on:

https://www.marieclaire.co.uk/fashion/summer-fashion-trends-2018-553795.

[12] Information on:
https://www.pinterest.com/pin/AQbA3m2uwy6ogS2Y868ztHmCGyV9yFztjSqUzY9eBqH7LVc
YYvU-pTs/.

[13] Information on:
https://i.pinimg.com/originals/34/3b/a6/343ba6fafe6ddf992cf9e3d60486eda1.jpg

[14] Information on: https://www.colourandtrends.com/.

[15] Information on:

https://www.colourandtrends.com/blog/2017/3/15/lineapelle-is-enthusiastic-about-springsummer-2018.

[16] Information on:
https://www.researchgate.net/publication/268881605_The_chemistry_of_reactive_dyes_and_thei
r_application_processes.

[17] Information on:

http://textilelearner.blogspot.com/2012/01/dyeing-of-cotton-fabric-with-reactive.html.

[18] Information on: https://en.wikipedia.org/wiki/Stencil.

[19] Information on: https://en.wikipedia.org/wiki/Upcycling.

[20] Information on: www.ecochicdesignaward.com. pdf.

Keyword Index

About the Editors

Dr. Hamed El-Mously, Ain Shams University, Egypt
The Legendary Professor Dr. Hamed El-Mously, is an Emeritus Professor at the Faculty of Engineering – Ain Shams University, Egypt. Dr. El-Mously is recognized as the founding father of Date Palm by-products research and development, and one of the warriors of sustainable development of local communities. He received numerous prestigious awards for his work, such as Khalifa International Award for Date Palm and Agriculture Innovation 2013. El-Mously has been working on projects aiming at developing local communities in all villages of Egypt, by applying developing projects using their local resources since 1995. Projects include production of blockboard, parquet, and arabesque from palm midribs, production of non-traditional animal feed from agricultural residues, production of organic fertilizers from pruning-products of date palms, doum palms, and mango trees, and production of fig jam. El-Mously contributed in the foundation of several research and societies:

- Small Industries and Local Technology Development Center.
- The Egyptian Society for Endogenous Development of Local Communities.
- Network of Fiber-Plastic Composites, Tree-Free Wood Innovation.
- Network, Foundation for Renewable materials Research, Technology and Applications.
- He is a member in several strategic consulting and scientific committees', Ain Shams
- University. He obtained the Ph.D. from Metal Cutting Machine Tools Institute, Moscow
- The International Palm By-Products Association (under registration)

Dr. Mohamad Midani, German University in Cairo, Egypt
Dr. Mohamad Midani is an Assistant Professor in the Materials Engineering Department at the German University in Cairo, he is also the CEO of inTEXive Technical Consulting, and he serves on the advisory board of Middle East Co. for Textiles. His research mostly focuses on natural fibers and natural fiber composites, and he is the inventor of the proprietary long micro-fibrillated date palm midrib fibers. He teaches courses on Fiber Reinforced Polymer Composites as well as New Product development and Innovation Management. In 2016, Dr. Midani received a recognition for excellence in teaching from NC State University. Dr. Midani has his BSc in Mechanical Engineering from Ain Shams University, MT in Textiles Technology and Management, and Ph.D. in Fiber and Polymer Science from the College of Textiles NC State University. He is a member in the Product Development and Management Association (PDMA), as well as the American Society of Mechanical Engineers (ASME).

Mohamed Wagih, Ain Shams University, Egypt
Mohamed Wagih is a Teaching Assistant at the Faculty of Engineering: Department of Design and Production Engineering, Ain shams University since 2013. He was awarded his Master of Science degree in the field of date palm midrib processing. He worked as a Research Assistant at the Faculty of Engineering, Ain Shams University and the American University in Cairo. He participated in several projects related to light harvesting, date palm residues processing, and acoustics and vibrations measurements. He has a good experience in the field of teaching, and design and manufacturing of machines. He published one patent and four papers from these projects and his M.Sc.thesis.

www.ingramcontent.com/pod-product-compliance
Lightning Source LLC
Chambersburg PA
CBHW061203220326
41597CB00015BA/1296